Confined Space

Principles and Practice

SECOND EDITION

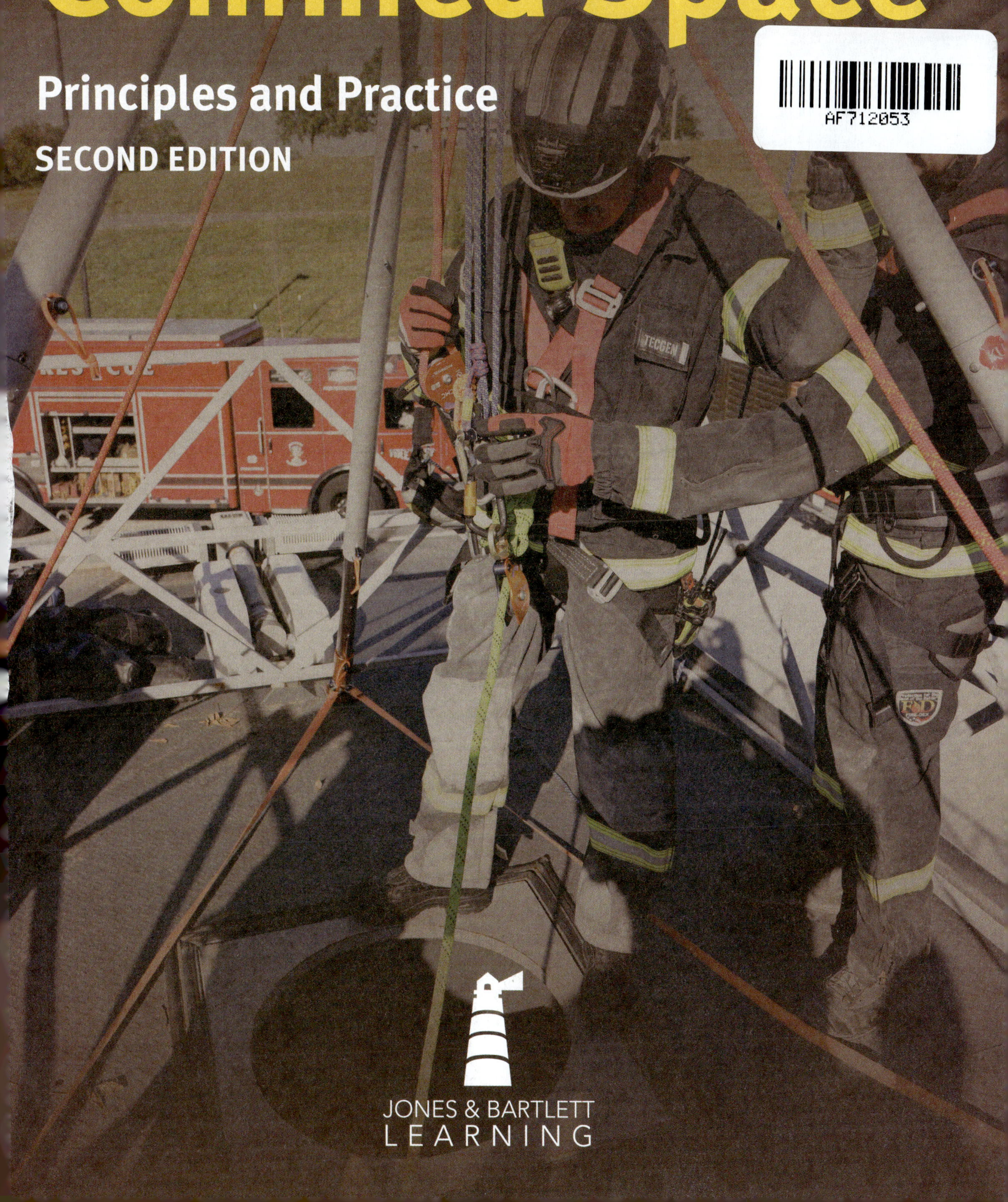

JONES & BARTLETT
LEARNING

World Headquarters
Jones & Bartlett Learning
25 Mall Road
Burlington, MA 01803
978-443-5000
info@jblearning.com
www.jblearning.com
www.psglearning.com

Jones & Bartlett Learning books and products are available through most bookstores and online booksellers. To contact Jones & Bartlett Learning directly, call 800-832-0034, fax 978-443-8000, or visit our website, www.jblearning.com.

Substantial discounts on bulk quantities of Jones & Bartlett Learning publications are available to corporations, professional associations, and other qualified organizations. For details and specific discount information, contact the special sales department at Jones & Bartlett Learning via the above contact information or send an email to specialsales@jblearning.com.

24010-8

Production Credits
Vice President, Innovative Learning and Assessment Solutions: Ada Woo
Senior Director, Content Production and Delivery: Christine Emerton
Director, Product Management: Cathy Esperti
Product Manager: Janet Maker
Manager, Content Development: Tiffany Sliter
Content Coordinator: Michaela MacQuarrie
Content Manager: Elena Sorrentino
Manager, Intellectual Properties and Content Production: Kristen Rogers
Content Production Manager: Belinda Thresher
Senior Intellectual Property Specialist: Angela Dooley
Director, Marketing: Andrea DeFronzo
Senior Product Marketing Manager: Elaine Riordan
Director, Sales: Brian Hendrickson
Director, Product Fulfillment: Aaron McKinzie
Purchasing Manager: Wendy Kilborn
Composition: S4Carlisle Publishing Services
Project Management: S4Carlisle Publishing Services
Text and Cover Design: S4Carlisle Publishing Services
Intellectual Property Specialist: Faith Brosnan
Intellectual Property Specialist: Maria Leon Maimone
Cover Image (Title Page, Section Opener, Chapter Opener): © Jones & Bartlett Learning
Printing and Binding: Lakeside Book Company

Library of Congress Cataloging-in-Publication Data
Library of Congress Cataloging-in-Publication Data unavailable at time of printing.
LCCN: 2024052413

6048

Printed in the United States of America
29 28 27 26 25 10 9 8 7 6 5 4 3 2 1

Brief Contents

Contents

Skill Drills

CHAPTER **10**

CHAPTER **11**

CHAPTER **12**

Reviewers and Contributors

The Jones & Bartlett Learning Public Safety Group would like to thank all the contributors, reviewers, and advisors involved in the creation of *Confined Space: Principles and Practice, Second Edition.*

Contributor

Dr. Ken Fowler, DEd, CFO

Assistant Director of Operations and Accreditation
Kansas Fire & Rescue Training Institute (KFRTI)
Lawrence, Kansas

Reviewers

Keene J. Black

Instructor
Virginia Department of Fire Programs
Glen Allen, Virginia

Fire Department Captain (retired)
Virginia Beach Fire Department
Virginia Beach, Virginia

Kip J. Bordeau

Confined-Space Rescue Lead Instructor
Illinois Fire Service Institute
Champaign, Illinois

Sean Broyles

Captain/Paramedic (retired)
Kingsport Fire Department
Kingsport, Tennessee

Co-Owner and President
TechResQ Training Specialists, Inc.
Kingsport, Tennessee

Anthony Chen

Rescue Technician–Confined Space, Trench, Ropes, Structural Collapse, Swiftwater
Chambersburg Fire Department
Chambersburg, Pennsylvania

DeAngelo M. Cooke

Confined Space Rescue Program Manager
Illinois Fire Service Institute
Champaign, Illinois

Katie C. Cressionie

Certified Fire and Emergency Services Instructor III, Certified Rescue Technician, Technical Rescue Instructor
Louisiana Fire and Emergency Training Academy (FETA)
Baton Rouge, Louisiana
Plaquemine Fire Department
Plaquemine, Louisiana

Spence J. Cressionie

Instructor II, International Fire Services Accreditation Congress (IFSAC); Confined-Space Rescue I/II, IFSAC
Vacherie-Gheens Volunteer Fire Company, Inc.
Gheens, Louisiana

Senior Electrical Engineer
Bollinger Shipyards, LLC
Lockport, Louisiana

Douglas Fuller

AAS Fire Science
Pennsylvania State Fire Academy
Lewistown, Pennsylvania

Benjamin Green

Fire & Rescue Program Manager
Kansas Fire & Rescue Training Institute (KFRTI)
Lawrence, Kansas

Russ Kane, MSc, CFO, CWMS

Fire Chief
Galeton Fire Protection District
Galeton, Colorado

Colin LeGrow

Owner/Operator
Technical Rope & Rescue, Inc.
Bauline, Newfoundland and Labrador, Canada

Jason R. Probst

Virginia Beach Fire Department
Virginia Beach, Virginia
Virginia Department of Fire Programs
Glen Allen, Virginia

Christopher Riale
Rescue Program Manager
Delaware State Fire School
Dover, Delaware

Chad Riddleberger
Division Chief, Special Operations
Virginia Department of Fire Programs
Glen Allen, Virginia

Michael A. Smith, MPA, CFEI
Head, Fire Investigation Unit
Lynn Fire Department
Lynn, Massachusetts

Adjunct Professor
Anna Maria College, Paxton, Massachusetts; North Shore Community College, Danvers, Massachusetts; Bunker Hill Community College
Boston, Massachusetts

Thomas Townsend
Chief of Technical Rescue
Louisiana Fire and Emergency Training Academy (FETA)
Baton Rouge, Louisiana

Curvin L. Wolfgang, Jr., MA, EFO, CFO
Battalion Chief
York Area United Fire and Rescue
York, Pennsylvania

Photographer

Nick Verbenec
Owner/Operator
Nick Verbenec Photography
Basehor, Kansas

Fire Apparatus Operator/Driver
City of Leavenworth, Kansas

Acknowledgments

The Jones & Bartlett Learning Public Safety Group would like to recognize Dr. Ken Fowler for his significant contributions to this book. Ken is a consummate fire professional whose commitment to public service is surpassed only by his commitment to educate and share his knowledge with others.

Many thanks to the Kansas City Kansas Fire Department (KCKFD) for hosting a photo shoot that resulted in many of the images in this edition, with special thanks to Chief Dennis L. Rubin for supporting our use of the KCKFD facilities and personnel. We also thank the Kansas Fire & Rescue Training Institute for allowing the use of its equipment at the photo shoot, with special thanks to Ben Green and Clint Hornberger.

Special thanks to Chad Riddleberger, chief of special operations at the Virginia Department of Fire Programs, for appointing and managing a working group to review content, which included Keene J. Black and Jason R. Probst.

Additional thanks to Perrin Davis and Rachel Claff at KTD+ for their invaluable contributions to the development of the manuscript.

Foreword

Confined-space rescue operations represent some of the most challenging and hazardous tasks in the realm of emergency response. These operations demand a precise combination of technical expertise, situational awareness, and teamwork. Confined spaces, by their very nature, are fraught with potential dangers, such as limited entry and exit points, poor ventilation, hazardous atmospheres, and unstable environments. The complexity of these conditions underscores the critical importance of having a well-qualified and certified rescue team ready to respond.

Every confined-space rescue incident is unique, with its own set of challenges and risks. Whether it's a worker trapped in an underground vault, a maintenance technician immobilized in an industrial tank, or an unfortunate fall into a utility shaft, the stakes are incredibly high. A successful outcome hinges not only on the quick response of rescuers but also on their advanced training, knowledge of confined-space dynamics, and adherence to safety protocols.

This book provides a comprehensive overview of confined-space rescue operations with a focus on *NFPA 1006: Standard for Technical Rescue Personnel Professional Qualifications*. Although this standard serves as a foundational guideline for achieving operational objectives, it is important to acknowledge that other techniques and methods exist. Because of the scope of this book, not all possible approaches could be included; instead, the content is designed to serve as a reference to help rescue teams meet the objectives outlined in NFPA 1006.

Qualified rescue team members bring more than just skills to the scene; they bring confidence, precision, and preparedness. Certification ensures that team members are proficient in key areas, including hazard identification, atmospheric monitoring, rope and rigging techniques, and patient extrication. Moreover, these rescuers are trained to operate within the framework of safety standards, such as those established by the Occupational Safety and Health Administration (OSHA) and the National Fire Protection Association (NFPA).

The importance of preparation cannot be overstated. Regular training, simulated exercises, and meticulous preplanning are essential to maintain readiness. When an emergency arises, these efforts translate into rapid, effective, and safe rescue operations that can mean the difference between life and death.

This book serves as a testament to the indispensable role of a highly skilled confined-space rescue team. It is a reminder that investing in proper training, equipment, and certification is not just a regulatory necessity but a moral obligation to safeguard lives. As you explore the intricacies of confined-space rescue, remember that excellence in preparation is the foundation of success in any rescue mission.

Dr. Ken Fowler, DEd, CFO
Assistant Director of Operations and Accreditation
Kansas Fire & Rescue Training Institute (KFRTI)

SECTION

1

Awareness

You Are the Rescuer

Upon your arrival at the emergency scene, a dairy farm, a frantic worker explains the situation: the dairy farmer had been attempting to drain farm biological waste from a small waste pit to a larger one, but the pipe became clogged. The clog had been a long-standing problem, and the farmer had been forced to do the clean-out process on a weekly basis for some time.

On this fateful day, the farmer entered the smaller pit, which contained about 12 inches (30.5 cm) of biological waste, through a four-foot-wide (1.2-m-wide) opening and began attempting to unclog the drainpipe. Shortly after he entered the pit, he collapsed. A worker who witnessed the farmer's collapse immediately entered the pit to assist, and a second worker notified the farmer's wife to call for assistance. The farmer's wife, along with his two young daughters, ran to the pit and entered it to help. Now, your crew has arrived on the scene to find the farmer, his wife, his daughters, and the first farmworker still in the pit—and all of them unresponsive.

This confined space was not new to the farm, and the farmer had been in there on a weekly basis for quite some time. What happened differently this time that led to the deaths of these people? Some of the workers speculated that a recent change in the animal feed may have been to blame: The new feed contained brewer's yeast, and a recent rainstorm had washed some of the feed into the pit. This may have changed the conditions in the space because the interaction of the brewer's yeast and the biological waste could have spurred a fermentation process and generated carbon dioxide. Others ventured that methane gas from the biological waste may have caused the deaths by displacing oxygen in the pit.

No matter what may have made this incident lethal, five people are now dead.

You Are the Rescuer Questions

1. One of the dead, the farmer, was the original victim; the other four were would-be rescuers. What would you, as the rescuer, do differently than the ill-fated four who were first on the scene?
2. Would you be able to spontaneously recognize the potential hazards of this confined space?

CHAPTER

1

Awareness

Understanding Confined Spaces

KNOWLEDGE OBJECTIVES

After studying this chapter, you should be able to:

- Define the meaning of the term *confined space*. (**NFPA 1006: 7.1.1**, pp. 4–5)
- Define the terms *awareness*, *operations*, and *technician* in the context of confined-space rescue, and identify the roles appropriate to these terms. (**NFPA 1006: 7.1.2**, pp. 8–9)
- Recognize immediate hazards. (**NFPA 1006: 7.1.1**, pp. 12–17)
- Describe the process of access control to a scene. (**NFPA 1006: 7.1.1**, pp. 17–20)
- Describe local policies relating to confined-space access areas. (**NFPA 1006: 7.1.2**, p. 21)
- Describe the process for selecting appropriate personal protective equipment (PPE). (**NFPA 1006: 7.1.2**, pp. 20–21)
- List different types of technical references. (**NFPA 1006: 7.1.1**, pp. 21–24)
- Explain how awareness level personnel support operations or technician level personnel. (**NFPA 1006: 7.1.7**, pp. 8–9)

SKILL OBJECTIVES

- Predict the potential hazards associated with a confined-space incident. (**NFPA 1006: 7.1.1**, pp. 12–17)
- Utilize reference materials to ensure that all needed information is obtained and communicated. (**NFPA 1006: 7.1.5**, p. 24)

Introduction

In the 1980s, the **Occupational Safety and Health Administration (OSHA)**, a regulatory agency of the U.S. government, developed a set of regulations for safe operation in **confined spaces**. These are spaces that are large enough for a person to enter to perform a specific job, have limited means of access and a restricted entrance or exit, and are not designed for people to spend significant time within or for continuous occupancy. In January 1993, OSHA set forth, in the ***Code of Federal Regulations* (C.F.R.)** (a codification of general and permanent rules published in the *Federal Register* by departments and agencies of the U.S. government), its rule 29 C.F.R. 1910.146, entitled "Permit-Required Confined Spaces." With the development and implementation of this standard and decades of education and enforcement, you might think that deaths in confined spaces would have become isolated incidents. Unfortunately, they have not.

Just prior to the development of OSHA's confined-space regulation, the **National Institute for Occupational Safety and Health (NIOSH)**—the agency within the U.S. Department of Health and Human Services that identifies work-related diseases and injuries and the potential hazards of new work-related technologies and practices—studied eight cases involving deaths that occurred in confined spaces. In these events, 16 fatalities and 53 injuries occurred; 10 of those fatalities were people attempting a **rescue** (the process of safely extricating a living person from a confined space). Most of the time, these would-be rescuers were untrained people who were either part of the confined-space work party or new arrivals on the scene after the original victims were found in the space. As well intentioned as their efforts were, these would-be rescuers' lack of awareness of the hazards of confined spaces cost them their lives. As rescue personnel, how many of us might find ourselves in that same position?

This text has been written and reviewed by highly experienced, field-oriented rescuers and drafted in accordance with requirements for **technical rescue**—the application of special knowledge, skills, and equipment to resolve unique and/or complex rescue situations—established by the National Fire Protection Association (NFPA).

Rescuers must be able to identify confined spaces and the potential hazards present in a confined-space incident and must protect themselves and others from these hazards. Deaths in confined spaces are terrible tragedies compounded by the fact that they are preventable. The incident recounted in "You Are the Rescuer" is based on an actual incident that involved multiple deaths, including would-be rescuers. Could a similar scenario happen again? Read on to learn more.

Defining *Confined Spaces*

What exactly is a confined space? For a space to be considered a confined space, it must have all three of the following characteristics: It must be large enough for a person to enter to perform a specific job, have limited means of access and a restricted entrance or exit, and not be designed for people to spend significant time within or continuous occupancy.

According to OSHA, confined spaces include, but are not limited to, tanks (**FIGURE 1-1A**), vessels, silos (**FIGURE 1-1B**), storage bins, attics, crawl spaces (**FIGURE 1-1C**), hoppers, vaults, pits, manholes (**FIGURE 1-1D**), tunnels, equipment housings, ductwork, pipelines, and similar spaces not designed for continuous occupancy (CDC 1986).

Another important term to know is **permit-required confined space**, or *permit space*. This term is used by OSHA to describe a confined space that has the three basic characteristics mentioned earlier, plus has one or more of the following characteristics:

- Contains or has the potential to contain a hazardous atmosphere
- Contains material that has the potential to engulf an entrant
- Has walls that converge inward or floors that slope downward and taper into a smaller area that could trap or asphyxiate an entrant
- Contains any other recognized safety or health hazard, such as unguarded machinery, exposed live wires, or heat stress

In **Appendix B: Sample Confined-Space Entry Permit**, you can review an example of an entry permit form.

So why do confined spaces have so much potential for danger? How can rescuers safely operate in a confined space? OSHA's criteria clearly define the inherent problems confined spaces can introduce.

If a confined space is large enough to enter and work inside, people can enter the space and be directly exposed to any hazards it may contain. The entry itself initiates the process that exposes the worker to the hazards. Did the biological waste pit mentioned in "You Are the Rescuer" fit this description? Did all three criteria exist within that space?

Once workers enter a confined space and recognize or come into contact with hazards, they must either leave the space on their own or be physically removed

A.

B.

C.

D.

FIGURE 1-1 Examples of confined spaces include **(A)** a tank, **(B)** a silo, **(C)** a crawl space, and **(D)** a manhole.

by trained rescuers. If the hazard is not obvious, such as an **atmospheric hazard** (a hazard present in the air of a confined space, such as a toxic gas, flammable gas, or oxygen deficiency), workers entering the space could be overcome by exposure to the contaminated atmosphere and unable to escape.

TIP

The sense of urgency many of us feel during rescue work may blind us to the true nature of hazards and can lead us to become victims.

Confined spaces may also have limited openings—some as small as 18 inches (48 cm) in diameter, or even smaller. They may be elevated or otherwise make entry and exit difficult. Anyone entering the space, including rescuers, must enter through those openings. If hazards in the space require the use of a **self-contained breathing apparatus (SCBA)** (**FIGURE 1-2**) or other respiratory protection to safely enter and work there, the space becomes that much more difficult to enter through those limited or restricted openings. Technical rescue teams often use specialized lower-profile respiratory protection, such as a **supplied-air respirator (SAR)**.

For that reason, workers and rescuers may decide to enter the space without the proper respiratory protection in place, with the intent to put on the equipment after they enter. Entering a confined space without an SCBA exposes a rescuer to whatever atmospheric hazards exist; a rescuer may easily fall victim to the hazards before putting on the SCBA. Rescues that are not preplanned or that have poorly developed action plans can lead to these types of tragic results. In a rescue, the urgency of the moment can blind us to the true nature of the hazards and can lead us to become victims as well.

FIGURE 1-2 An SCBA unit.
Courtesy of 3M.

SAFETY TIP

Never remove your SCBA to enter a confined space. Even if you plan to keep your face piece in place and continue breathing from the unit while carrying it in your hands, it is easily possible to drop the cylinder and backpack, resulting in the face piece detaching from your face.

A confined space is not intended for continuous occupancy by humans. Conditions inside the space can change rapidly. Confined spaces that are known to contain hazardous materials or chemical processes are expected to contain hazards, but what about simpler spaces, such as an underground storm sewer? Road runoff, decomposing organic material, and exhaust gases from vehicles can introduce hazards to a storm sewer without anyone being aware of their presence.

Ultimately, the problems with confined-space entry for normal work practices, such as maintenance, and rescue operations, stem from a lack of preparation before entry. There may be unknown issues and problems, and they may be complicated. A simple medical emergency in a confined space can turn into a complex rescue. For example, sudden-onset cardiac arrest is a leading workplace killer.

Having an awareness of potential hazards and a plan to work within limits set by the hazards present can help make the operation a safe one. Preparation prior to entry is a requirement of any confined-space entry. Just because someone has entered a confined space 100 times previously without encountering a problem doesn't guarantee the next time will be safe.

SCBA, SABA, and SAR

An SCBA is a portable device that provides breathable air in hazardous environments. It consists of a high-pressure air cylinder, a pressure regulator, a harness, and a face piece or mask, all integrated into a portable unit. Its advantages include its portability and the independence of its wearable, self-contained air supply (the cylinder). Its disadvantages are the limited air supply it contains (typically up to 60 minutes) and its weight and bulk. It is typically used in firefighting, confined-space entries, hazardous material handling, and rescue operations in **immediately dangerous to life or health (IDLH)** environments.

A **supplied-air breathing apparatus (SABA)** provides breathable air from a stationary source via an air supply hose. It is used in environments where the air is not immediately hazardous but may be contaminated or oxygen deficient. SABA was commonly used for rescue in the past, but it is considered outmoded today. It is still used in some industrial settings. Its advantages are its unlimited supply of air (as long as the air source remains connected and active) and its relatively light weight. Its disadvantages are its limitations on movement (the user's movement is restricted by the length of the supply hose) and the supply hose's capacity for failure, entanglement, or trip hazard. It is typically used for long-duration work in contaminated but non-IDLH environments, maintenance and repair operations in hazardous areas, and operations in confined spaces with known air quality issues.

A SAR is similar to a SABA in that it provides breathable air from a remote source via an air supply hose. However, SARs can also include variations that have specific applications and configurations, such as continuous-flow and pressure-demand systems. Like a SABA, its advantages are its unlimited supply of air (as long as the air source remains connected and active) and its relatively light weight. It is more versatile because it can be configured for different applications and levels of protection. It shares the same disadvantages as a SABA as well; an additional disadvantage is its complexity, which requires more sophisticated setup procedures and management. Technical rescue teams often use SARs for confined-space events, and other applications include industrial painting and coating

TABLE 1-1 Differences Among SCBAs, SABAs, and SARs

Feature	Self-contained breathing apparatus (SCBA)	Supplied-air breathing apparatus (SABA)	Supplied-air respirator (SAR)
Air supply	Self-contained cylinder	Remote air supply via hose	Remote air supply via hose
Mobility	High, fully portable	Limited by hose length	Limited by hose length
Duration of air supply	Limited by cylinder size (up to 60 minutes)	Unlimited, as long as air source is active	Unlimited, as long as air source is active
Weight	Heavy and bulky	Lightweight	Lightweight
Airflow	Demand-regulated	Demand-regulated or continuous flow	Demand-regulated or continuous flow
Emergency escape option	Not typically equipped with escape bottle	Often includes an escape bottle	Often includes an escape bottle
Typical environments	Immediately dangerous to life or health (IDLH) environments	Contaminated but non-IDLH environments	Industrial environments with hazardous substances
Used for rescue	Yes	Industrial use	Yes

operations; sandblasting; environments containing hazardous dust, fumes, or gases; and other situations requiring prolonged use of respiratory protection. See **TABLE 1-1** for a summary of the differences among SCBAs, SABAs, and SARs.

Permit-Required and Non-Permit-Required Confined Spaces

So far, we have discussed confined spaces as a general concept. By definition, confined spaces must be large enough and configured to allow a person to enter and work in them, have limited or restricted means for entry or exit, and not be intended for continuous occupancy, but not all confined spaces contain hazards.

OSHA classifies confined spaces into two categories—permit-required confined spaces and **non-permit-required confined spaces**—to differentiate between those that are expected to contain hazards and those that are not. Non-permit-required confined spaces do not contain—or in the case of atmospheric hazards, do not have the potential to contain—any hazard capable of causing death or serious physical harm.

As a rescuer, you must be aware of the difference between permit-required and non-permit-required confined spaces. A permit-required confined space requires different precautions to protect workers and rescuers in the space than a non-permit-required confined space does. The requirements to protect people in a permit-required space may include ventilation, respiratory protective equipment, monitoring equipment, retrieval equipment, and **personal protective equipment (PPE)**—equipment that shields or isolates a person from the chemical, physical, or thermal hazards that can be encountered at a specific rescue incident (covered in detail later in this chapter; see Personal Protective Equipment for Confined-Space Rescues).

SAFETY TIP

Rescuers must be aware of the difference between permit-required and non-permit-required confined spaces. Permit-required spaces contain or have the potential to contain serious safety and health hazards, including hazardous atmospheres, engulfment, or an internal configuration that could trap or asphyxiate a person.

If you are called to the scene of an incident in a non-permit-required confined space, you may not find this equipment present. In such a situation, you may be tempted to consider that you are not about to embark upon a confined-space rescue—particularly if you have already been briefed that it has been classified as a non-permit-required confined space.

Never give in to this temptation because an initial impression does not necessarily indicate that a space contains no hazards. Depending on the type of work being performed in the space, new or unknown hazards may have been introduced into it that can easily turn that non-permit-required space into one with hazards so serious that it should be reclassified as a permit-required space.

A further complication is the fact that the OSHA general industry regulation 29 C.F.R. 1910.146 ("Permit-Required Confined Spaces" 1993) does not apply to agriculture, construction, or shipyard employment (for shipyards, reference 29 C.F.R. 1915, Subpart B, "Confined and Enclosed Spaces and Other Dangerous Atmospheres in Shipyard Employment" [1994], which defines confined spaces differently and also adds enclosed spaces). These sites may have confined spaces present, but you may have to identify them as such or identify the hazards unique to those sites ("Confined and Enclosed Spaces and Other Dangerous Atmospheres in Shipyard Employment" [1994] uses a limit of 22 percent oxygen by volume to define oxygen-enriched atmospheres because of work involving electric or gas welding, cutting, brazing, or similar flame- or spark-producing operations, which is known as **hot work**).

Changing or unknown conditions could create an unsafe environment that was initially deemed safe as the rescue proceeds. For example, disturbing materials in a space could create a toxic environment. As a rescuer, you should treat *all* confined spaces as if they are designated as permit-required spaces.

TIP

It is in your best interest to treat all confined spaces as permit-required spaces.

Defining the Terms *Awareness, Operations,* and *Technician*

This text is organized into three separate sections—awareness, operations, and technician. The terms *awareness, operations,* and *technician* are significant because they differentiate between training levels for responders in both the emergency services world and within private industry.

NFPA 1006: Standard for Technical Rescue Personnel Professional Qualifications, Chapter 7, 2021 Edition (NFPA 2021), provides a comprehensive understanding of the requirements to meet each level of training, including how to recognize hazards, access victims, and stabilize victims, along with proper victim retrieval techniques. The NFPA 1006 standard defines the qualifications for technical rescue personnel, including those involved in confined-space rescue (NFPA 2021). **Confined-space rescue technicians** are skilled individuals trained to operate at both the operations and technician levels within confined spaces.

The **awareness level responder** understands the basics of confined-space safety and knows how to safeguard personnel from fire, explosion, unqualified rescue attempts, and other health hazards that are uniquely associated with confined spaces. They have the training, qualifications, and competencies required for personnel responsible for confined-space hazard identification, hazard evaluation, and hazard control. They communicate with victims, perform nonentry rescue, recognize the need for technical rescue resources, and support an operation or technician level incident.

The **operations level responder** gains foundational knowledge and skills related to confined-space rescue. They learn about rescuer safety, team evaluation, and preplanning. Additionally, they become proficient in **atmospheric monitoring**, a method of evaluating the ambient atmosphere of a space, and the proper use of PPE. They must be capable of assembling a portable anchor system for the application of a high-point attachment, as well as following **lockout/tagout** procedures, which means that they must be able to eliminate and secure sources of potential physical or chemical hazards by isolating them from a confined space and/or controlling flowable materials in the space. They must also be able to prepare for entry into a vertically oriented confined space and enter horizontally and vertically oriented confined spaces. The operations level also covers techniques for patient packaging and **recovery** (the process of retrieving the body of a deceased person from a confined space).

The **technician level responder** has skills that build on the operations level, as well as more advanced training. Technicians must be capable of preplanning a confined-space incident; they must apply and use SARs and SABAs as a **rescue entrant**—an individual qualified to enter a confined space and assist in a rescue attempt. They must be well versed in using **tripods** (portable anchor system devices typically meant to support people as they are raised or lowered into and out of a space) and other artificial high anchor points, as well as lockout/tagout procedures. They become skilled in using rope systems specifically designed for confined-space entry and rescue scenarios and using communications systems. Overall, the technician gains confidence through hands-on training and exposure to real-life scenarios in such a potentially dangerous environment.

In summary, an awareness level responder is responsible for recognition, scene control, and **size-up** (the ongoing evaluation of problems confronted in a confined-space rescue operation). An operations level responder is responsible for initiating entry or rescue into areas that are immediately visible. A technician level responder is responsible for entry or rescue into areas that are not immediately visible.

The Confined-Space Entry Program

A **confined-space entry program** is an employer's overall program for controlling and protecting employees from permit space hazards and for regulating employee entry into a permit space. The first step in formulating a program is to evaluate the workplace to determine if any permit-required confined spaces exist. According to 29 C.F.R. 1910.146(c)(2), "if a workplace contains permit-required spaces, the employer shall inform exposed employees, by posting danger signs or by any other equally effective means, of the existence and location of and the danger posed by the permit spaces" ("Permit-Required Confined Spaces" 1993). These signs may say something like, "Danger: Permit-Required Confined Space: Do Not Enter," or use other language that effectively identifies it as a permit-required confined space (see **FIGURE 1-3**).

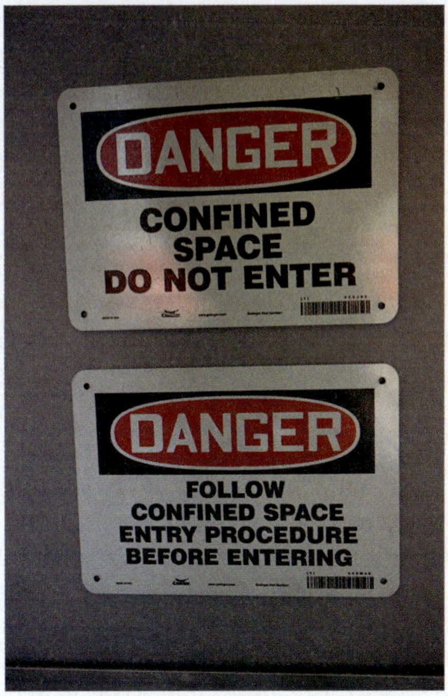

FIGURE 1-3 As part of a confined-space entry program, identification of the confined space is required.

But what if the evaluation reveals only non-permit-required confined spaces? You may or may not find signs posted in such areas. Should you interpret that to mean that those non-permit-required spaces do not pose a hazard? No. As a rescuer, you should treat all confined spaces as potentially hazardous until a thorough evaluation of the conditions shows that no hazards exist.

In some cases, an employer may have decided that its own employees will not enter confined spaces, leaving work requiring confined-space entry to contractors. If this is the case, and you have been called to make a rescue at this site, you have to wonder: How did the victim get into the confined space? Was this entry in violation of the company policy, or is the victim an outside contractor working at the site? Regardless of why you have been called to the incident, as part of the rescue team, you must take control of the situation: Whom do you talk to, how do you evaluate the incident, and where do you start?

To simplify the process of creating rescue procedures for confined spaces, you should begin by looking at the relevant OSHA requirements. A key place to start is with the roles and responsibilities that people are given during a permit-required confined-space entry. Although the roles were originally intended for workers entering the space, these same roles can be adapted for rescue purposes and should be used to identify competencies and provide for the safety of rescue team members. There is no need to reinvent the wheel, and to simplify the process, the roles and responsibilities identified in this chapter will only be those required by OSHA for workers who will enter and work in confined spaces. As rescuers, your team's roles and responsibilities will closely parallel those of the workers, but as you work your way through this text, the differences will become clearer.

Roles of a Confined-Space Entry Program

Supervisor

According to the requirements of a confined-space entry program, someone must be in charge of the program during permit-required entries. OSHA identifies this person as the **confined-space supervisor**, who has a variety of tasks and obligations to fulfill. Supervisors are required to know the hazards that may be faced during entry, including the modes, signs (or symptoms), and consequences of the exposure. They are also responsible for informing the **attendant**—the individual stationed outside the confined space who monitors the people within and the conditions in the area around the space—and entrant(s) of these hazards (**FIGURE 1-4**).

Confined-space entry supervisors have a tremendous amount of responsibility, and those responsibilities start long before trained employees or rescue technicians enter a confined-space environment. First and foremost, the confined-space supervisor must have training and knowledge in, and understanding of, all hazards that could be present within a confined-space environment. Because it is paramount for the supervisor to read and understand a confined-space entry permit, the supervisor must also have a full understanding of the required hazard controls and hazard exposures, signs, and symptoms, as well as the ability to understand and verify those conditions prior to allowing any entry into the space.

A supervisor's other areas of required knowledge must include the **communications plan** detailing how communication will flow between rescuers, attendants, and all other parties involved; any and all rescue plans devised in case something goes wrong; the availability of resources that are or would be available; how to verify that entry conditions are within acceptable parameters; and knowledge of whether to continue ongoing verification of conditions and what to do if changes occur. The confined-space supervisor has the final word on entry into the space and bears the responsibility of entry termination, should a safety issue become present.

FIGURE 1-4 The confined-space supervisor is responsible for making sure that the entry permit is properly filled out.

Supervisors must also be familiar with and trained on permit systems, which are defined as the space owner's or operator's written permission to enter a permit-required confined space. The permit system must include the following:

- Identification of the space: what it is and where it is
- The purpose of entry
- Dates and times of entry and the duration of entry allowed
- Names of authorized entry personnel, the supervisor of the confined space, and the confined-space attendant
- All known hazards within the space
- Atmospheric testing requirements and the results of pre-entry testing
- The acceptable entry conditions of the space
- Any available rescue services (onsite, offsite, etc.)
- The communications plan
- Equipment and PPE requirements
- Any additional permits that are required (such as hot work)

An entry supervisor should be the most well-informed person on the scene of a confined-space accident, but a confined-space entry program must already be in place if a supervisor is to be so well informed. The intent of such a program is to prevent accidents, but if an accident does occur, then the question is: Who or what was the cause? Was a hazard introduced into the space that caught workers off guard, or was there no confined-space entry program in the first place? Was the permit completed correctly? If no program exists, that means no entry supervisor, no trained entrants, and no trained attendant will be present. Without trained people at the scene, how can rescuers gather the accurate information they require to evaluate what has happened and then develop an action plan?

Without accurate information, you will have to begin at the most basic level to protect yourself and other emergency responders. Your most essential goals must be to prevent accidents during the rescue, minimize the impact of potential accidents, and efficiently complete your rescue operations. The information you need to make decisions during a rescue can only be as good as the source of that information.

Attendant

What do attendants do, and how does their work add to the safety of the rescue team? The OSHA regulation 1910.146 ("Permit-Required Confined Spaces" 1993) defines the duties of and the requirements for the role of attendant, as shown in **FIGURE 1-5**. Attendants may not enter the space for any reason. They may not bring tools or equipment into it, perform rescue by entering the confined space, leave the area, or enter the space for any other reason unless they are relieved by another trained attendant.

A confined-space attendant can be any worker who has the required training within all confined-space hazards and who has the responsibility to monitor and assist all workers who are entering a confined space. Types of assistance include air monitoring, maintaining a clear route of communication between those inside the confined space and those outside the confined space, and providing required PPE and tools that might require monitoring of the air supply. Other training requirements include understanding communication protocols, entry and rescue procedures, permit requirements, and all hazards associated with a confined space.

Attendants are also responsible for monitoring the people within the space and providing **standby assistance** per the American National Standards Institute (ANSI) standard Z117.1-2003 (American Society of Safety Engineers 2003)—that is, ensuring the safety of the people within the space and maintaining communication with the workers in the space. Attendants can call for help if an accident occurs and can order the space to be evacuated if conditions within or outside of the space pose a hazard to any person.

The critical nature of the role of attendants is that they are the outside link for workers in the confined space. Attendants not only monitor what is going on inside the space (atmospheric hazards, worker behavior, physical condition, and so on), but they also watch for hazards outside. Without an attendant, how would the workers in the space know that a fire alarm was sounding in the building where the confined space was located? Without an attendant, who would be responsible for ensuring that the area around the opening to the space is clear of tools and equipment, as shown in **FIGURE 1-6**? These tools could easily be kicked or knocked into the space and onto any workers there.

If you are called to a confined-space emergency, the attendant may be the single most valuable source of information you have. This person witnessed the incident and may also have made initial rescue attempts. However, the attendant is required to remain alert to the situation and may feel responsible for the accident; as a result, they may provide less-than-complete information when you interview them.

Attendants can legitimately attempt a rescue from outside of the confined space. In that case, the employer must develop nonentry rescue procedures to be followed by attendants, provide appropriate rescue equipment (e.g., tripods, winches), and properly train

FIGURE 1-5 Attendants communicating with the entrants working inside a confined space.

FIGURE 1-6 Tools and equipment lying around the outside of a confined space present a danger because they can fall or be kicked into the space.

Courtesy of Illinois Fire Service Institute.

attendants to follow the procedures and use the equipment. This process is called **attendant-based rescue**.

As part of the procedures, roles, and responsibilities you develop for your confined-space rescue team, you should also identify the functional position of the attendant as part of your team for any confined-space rescue. The attendant is identified as a **rescue attendant** and should carry out all the attendant duties for the rescue team.

In addition to the rescue attendant and rescue entrant, someone must supervise the rescue operation. This functional role might easily be performed by the incident commander or the operations section chief, but the individuals selected must be qualified to take on these supervisory roles. The qualifications for the individual assigned the role of rescue supervisor are similar to those defined by OSHA for a confined-space supervisor, but a rescue supervisor must also fully understand the roles and responsibilities of the rescue attendant and rescue entrants, the limitations of the rescue team's procedures and equipment, the OSHA confined-space regulations, and their position within the incident command system. Incident command is covered in detail in Chapter 2.

Authorized Entrants

When entrants are properly trained to be aware of the hazards, controls, roles, responsibilities, needed equipment, and other requirements for safe entry to a confined space, they are considered **authorized entrants**. Essentially, authorized entrants must be informed of the hazards that exist or that can occur in the confined space, must know how to use required PPE, must recognize any warning signs or symptoms of a dangerous condition, must alert the attendant as to when a dangerous condition is detected, and must exit from the space as quickly as possible when necessary (**FIGURE 1-7**).

FIGURE 1-7 Any person who makes entry into a confined space—including a rescuer—is an entrant and must be trained as an entrant.

At least one of the entrants at a confined-space accident is the victim (or victims). This entrant may not be able to leave the space without the assistance of an outside rescuer. Authorized entrants are expected to be trained to enter such a space safely, so if authorized entrants are among the victims, the situation should be investigated immediately to determine what went wrong. Questions to consider include the following:

- Were they, in fact, authorized to enter the space?
- Did a piece of equipment fail and change the conditions in the space?
- Was there a change in the materials to be used or the work being done in the space?

Remember, not everyone in a confined space is necessarily an authorized entrant. It's critical to determine whether victims were authorized entrants or instead were unqualified people who were unaware of the hazards they faced.

Just as you must develop procedures, roles, and responsibilities for a rescue attendant as a part of a confined-space rescue team, you must do the same for the functional position of an authorized rescue entrant, whose duties parallel those of authorized entrants in the OSHA confined-space regulations, except that they are entering the space for the sole purpose of rescuing victims.

Recognizing Hazards

It is plain to see why and how atmospheric hazards alone can create life-threatening conditions within a confined space, and it is just as easy to recognize how an unprepared rescuer can become a victim. Firefighters would never start fire operations without first sizing up the situation, and emergency medical technicians (EMTs) would never begin treatment on patients without assessing them first. As a rescuer arriving at a confined-space accident, why would you do any less?

If you saw a person drowning, what would you do to attempt a rescue? Would you immediately jump into the water and attempt to swim out to the victim, or would you call for help and then locate equipment to assist in the rescue, such as a boat or a rope? In a confined-space incident, the victim could be drowning in an atmosphere that is toxic, oxygen deficient, or explosive. What resources will you need to complete a safe and successful rescue?

Atmospheric Hazards

How many times have you been driving down a road or walking along a street and suddenly encountered an odd smell in the area? It could be food odors emitting from a nearby restaurant, the scent of fresh-cut grass, or the stench of a trash dump. Likely, you didn't also encounter a vapor cloud or other visual cue, so you didn't become aware of the odor until you inhaled it.

The atmosphere of a confined space could be invisible as well. You generally can't tell by looking into a confined space that its atmosphere is oxygen deficient or contaminated, and you certainly don't want to find out by inhaling it. Atmospheric hazards in confined spaces are among the most critical elements affecting confined-space operations. These hazards may include toxic gases, flammable gases, oxygen deficiency, or all three. You cannot simply look into a space to determine if an atmospheric hazard exists. You cannot effectively measure the presence of a hazardous atmosphere by sniffing it, either—that alone can kill you.

At one time, coal miners used canaries to detect hazardous atmospheres in mines they were working in. Because a canary is very small, a relatively small amount of carbon monoxide would kill it; the amount it would take to kill a canary would certainly be less than what would kill a person. A dead canary provided an early warning that carbon monoxide was present, and the miners could thus escape.

Candles were also used in the past as monitoring devices for confined spaces. If a candle was lowered into a confined space and the flame went out, it would be clear that the space had an oxygen-deficient atmosphere. But imagine what would happen instead if that space contained a flammable atmosphere. Fortunately, today's monitoring devices are much more sophisticated, safe, and accurate than either canaries or candles.

OSHA 29 C.F.R. 1910.146 ("Permit-Required Confined Spaces" 1993) defines a hazardous atmosphere as follows:

> an atmosphere that may expose employees to the risk of death, incapacitation, impairment of the ability to self-rescue (that is, escape unaided from a permit space), injury, or acute illness from one or more of the following causes:
>
> 1. Flammable gas, vapor, or mist in excess of 10 percent of its **lower flammable limit (LFL)** [the lower end of the concentration range over which a flammable mixture of gas or vapor in air can be ignited at a given temperature and pressure; also known as the **lower explosive limit (LEL)**].
> 2. Airborne combustible dust at a concentration that meets or exceeds its LFL. **Note:** This concentration may be approximated as a condition in which the dust obscures vision at a distance of five feet (1.52 meters) or less. Airborne combustible dust at a concentration that meets or exceeds its LFL is explosive when it is present in a confined space.
> 3. Atmospheric oxygen concentration below 19.5 percent or above 23.5 percent.
> 4. Atmospheric concentration of any substance . . . that could result in employee exposure in excess of its dose or **permissible exposure limit** [**PEL**; the regulatory limit on the amount or concentration of a substance in the air].
> 5. Any other atmospheric condition that is IDLH. **Note:** For air contaminants for which OSHA has not determined a dose or PEL, other sources of information, such as safety data sheets (SDSs) that comply with the Hazard Communication Standard, published information, and internal documents can provide guidance in establishing acceptable atmospheric conditions.

Flammable atmospheres can be created when gases, liquids, dusts, or vapors are present in a confined space. These materials can be introduced during normal process operations, when work is being performed in the space, or as a result of accidental leaks or spills.

Liquids, Gases, and Vapors

To fully understand the hazards of a flammable atmosphere within a confined space, one must first understand certain basic principles about the fuel that is present. Flammability differs from material to material; for these materials to ignite, a minimum concentration of the materials, an ignition source with enough energy to ignite the material, and sufficient oxygen must all be present for combustion to occur.

> **TIP**
>
> Sufficient fuel, oxygen, and heat must all be present for combustion to occur.

To safely work around fuel vapors (from gases, certain liquids, and solids), you must understand the concepts of **flammable range** or **explosive range**. The flammable range has lower and upper limits. These limits are the minimum (LFL or LEL) and maximum (**upper flammable limit [UFL]**; also known as the **upper explosive limit [UEL]**) concentrations of the fuel in air needed to support combustion. Many flammable materials are also toxics and asphyxiants at much lower levels than their LFLs.

TIP

The NFPA and OSHA use the terms *LFL* and *UFL*, whereas the fire service often uses the terms *LEL* and *UEL* instead. In this text, these terms are used interchangeably.

If the mixture is below the LFL, it is considered too lean to burn. If the mixture is above the UFL, it is considered too rich to burn. In addition, if the temperature of the fuel and air mixture is increased, the flammable range will increase, and if the temperature is decreased, the flammable range will decrease (**FIGURE 1-8**).

In discussing flammable limits, we must also recognize that certain liquids produce enough vapors at normal temperatures and atmospheric pressures that the presence of these liquids within a confined space can create a flammable mixture. Typically, these flammable liquids will have a **flash point**, as shown in **FIGURE 1-9**, at or below the ambient temperature within the space. The flash point of a material is the temperature at which an ignitable mixture is created. Flash points are determined in a laboratory and can vary slightly depending on the method used. A good rule to follow is that when the temperature of a liquid is at or above its flash point, it should be considered at or above its LFL.

Not all materials have the same flammable range. Gasoline has a given flammable range of 1.4 to 7.6 percent in air, whereas anhydrous ammonia has a flammable range of 15.5 to 27 percent in air, and acetylene has a flammable range of 2.5 to 100 percent in air under certain conditions, as shown in **FIGURE 1-10**. As the flammable range of materials varies, so do the potential hazards the materials present. A 2 percent concentration of gasoline in air is within the flammable range, and with an appropriate ignition source, you can expect a fire to occur. A 2 percent concentration of either anhydrous ammonia or acetylene in air is outside the flammable range.

TIP

It is important to realize that as the flammable range of materials varies, so do the potential hazards the materials present.

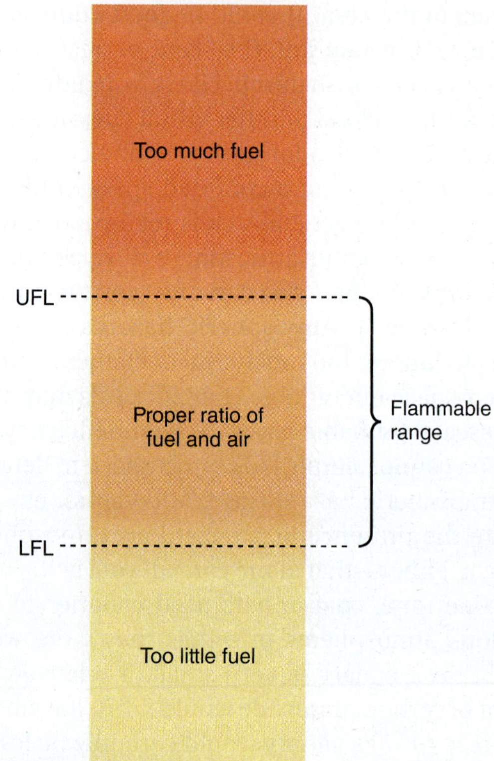

FIGURE 1-8 An illustration of flammable range. The fuel mixture underneath the LFL is too lean (not enough fuel), and the fuel mixture above the UFL is too rich (too much fuel, not enough oxygen).

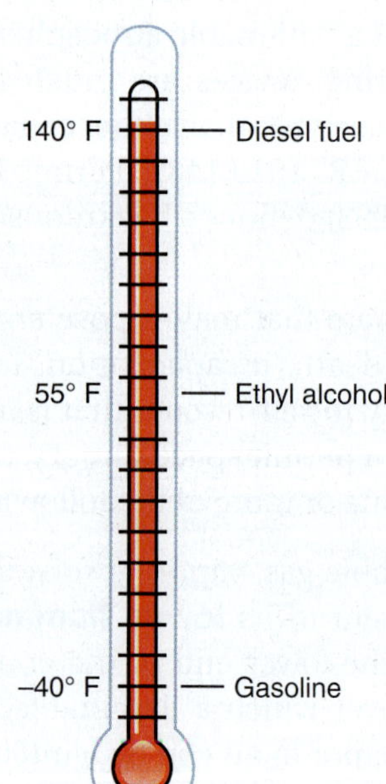

FIGURE 1-9 Different materials have different flash points. Knowing the identity of the material, the flash point, and the temperature of the material will give you an idea of the fire hazard.

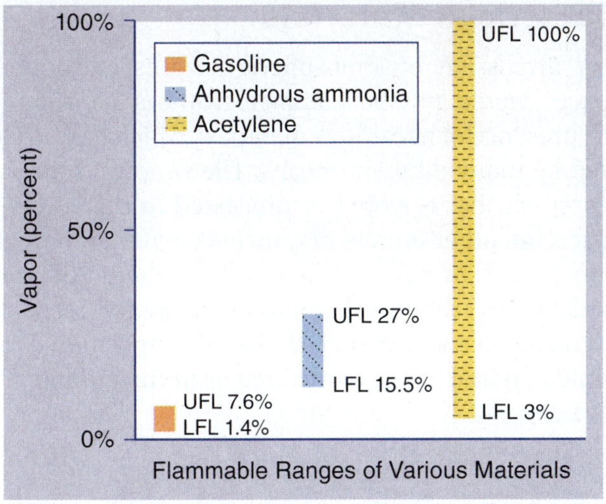

FIGURE 1-10 Flammable ranges of various materials. Some materials have a narrow flammable range, whereas others have a broad range.

FIGURE 1-11 Calibration gas and a docking station for calibration.

Courtesy of MSA Safety.

OSHA uses a limit of 10 percent of the LFL to accommodate the variations in materials. Observing a 10 percent LFL action limit creates a greater margin of safety, in that most metering equipment for measuring combustible gases is carefully assessed, set, or adjusted (or **calibrated**) on a single gas (**FIGURE 1-11**). The calibration gas has a specific known LFL. Staying below 10 percent of the LFL accommodates the notion that other gases may be present with an LFL lower than the LFL of the gas used to calibrate the meter.

Dusts

Flammable atmospheres from solids generally become a problem when the solids are present in the form of dust. Dusts are minute particles of solids; because these particles are so finely divided, they are easily heated by an ignition source. Once heated, these dusts can ignite and burn. When contained in a confined space, the resulting flash fire creates a rapid buildup of pressure and an accompanying pressure wave. This is known as a *dust explosion*.

SAFETY TIP

A flash fire in a confined space constitutes an explosion, which creates a pressure wave because it is contained. Depending on the force of this pressure wave, it can injure and kill people and destroy property.

To categorize the hazard presented by the presence of a combustible dust, OSHA *approximates* the condition as when the dust obscures vision at a distance of five feet (1.52 m) or fewer. Although this is an approximation, the presence of a dust condition of this magnitude should be cause for immediate concern. To help you better understand dusts and their combustibility, some basic information about combustible dusts includes the following:

- The dust particles must be finely divided and be about 420 microns or smaller (about the size of table salt or sugar particles).
- The smaller dust particles are, the easier they are to ignite; typically, this also means that the smaller they are, the larger the explosion will be.
- A dust explosion may not be a single event. When dust ignites, the resulting pressure wave can dislodge other dust in the area, causing a second, a third, or even more explosions.
- As a rule of thumb, if combustible dusts are present and the layer of dust is so deep that you can write your name in it, there is sufficient dust to create an explosion.

SAFETY TIP

The use of mechanical ventilation can disturb settled material, creating a hazardous or even explosive dust condition. For example, corn flour, wood chips, or other hazardous dust in a confined space may have the tendency to move with air currents, thus affecting visibility and possibly creating an explosive environment.

Oxygen

As discussed previously, adequate oxygen must be present for combustion to occur and, more importantly, for the support of human life within the confined space. The atmospheric air we breathe is normally approximately 21 percent oxygen. When oxygen concentration falls below 19.5 percent, the air is considered oxygen deficient and poses a danger to human life, as shown in **FIGURE 1-12**.

Low oxygen concentrations can also affect the flammability of fuels in a space. Even if a potentially explosive atmosphere is present in a space, the fuel–air mixture may lack enough oxygen to ignite. An increase in oxygen can raise the mixture into the explosive range and bring together all the ingredients needed to generate a fire or explosion.

An oxygen-deficient atmosphere is not the only oxygen-related problem in a confined space. Too much oxygen can create an oxygen-enriched atmosphere; when the oxygen concentration rises above 23.5 percent, other hazards come into play. Oxygen-enriched atmospheres rapidly accelerate combustion, and materials that do not easily ignite in normal air conditions or that are not normally considered combustible can be ignited and burn in an oxygen-rich atmosphere. The oxygen levels within a confined space can be affected by what occurred in the space prior to entry, by the type of work being done, or even by sources outside of the space.

SAFETY TIP

Oxygen concentrations below 19.5 percent are considered to be oxygen deficient, pose a danger to human life, and are among the most common causes of confined-space emergencies. Oxygen concentrations above 23.5 percent are considered oxygen enriched and have their own inherent dangers. This range also ensures accurate air monitoring because some readings are referenced by oxygen.

Effects of Reduced O_2

21%	Typical atmosphere
19.5%	Minimum oxygen concentration required for human breathing
17%	Impaired judgment and coordination
12%	Increased breathing rates, accelerated heartbeat, emotional upset, accelerated fatigue
9%	Seizure, coma
6%	Brain death

FIGURE 1-12 Effects of varying levels of reduced oxygen on people.

Hazardous Materials

Any discussion of atmospheric hazards of confined spaces would not be complete without considering the presence of hazardous materials, which can be created by many different means. The simplest source is a product that is stored or processed in the confined space, but other sources may include materials brought into the area by those performing work there. Still other possibilities arise from chemical or biological reactions occurring in a space, such as the decomposition of organic material either in the area or in the ground surrounding it.

SAFETY TIP

Hazardous materials can create toxic atmospheres that can injure or kill victims and rescuers through exposure to the materials (via inhalation, ingestion, or direct skin contact), displacement of oxygen, or a combination of the two.

Physical Hazards

We have spent some time discussing atmospheric hazards, but not all hazards you will face in a confined space will be atmospheric in nature. **Physical hazards** may also be present, and they must also be considered in your assessment. Potential physical hazards include electrical equipment, agitators or paddles, fire-suppression equipment, sharp edges, areas that are difficult to access because of height or other limitations, and the physical state of the material present in the space, as well as other considerations.

Engulfment by materials within a space is a common physical hazard—a liquid or finely divided (flowable) solid substance can quickly surround and trap a person. For example, a victim engulfed by a product, such as sawdust, can aspirate the material and die quickly as the respiratory system becomes overwhelmed. Other materials can flow with gravity, engulf a person, and exert enough force on the body to cause death by strangulation, constriction, or crushing.

Never fail to consider physical hazards as you assess the conditions presented by a particular confined-space emergency. These hazards endanger victims, and they must be controlled so that they do not endanger the rescuers as well. Among the methods to control physical hazards is the lockout/tagout procedure, which is covered in Chapter 8.

Guidelines for pH Monitoring

When monitoring a confined space, OSHA regulation 29 C.F.R. 1910.146(c)5(ii)(c) ("Permit-Required Confined Spaces" 1993) states that the internal atmosphere shall be tested with a direct-reading instrument for oxygen content, flammable gases and vapors, and potential toxic air contaminants, in that order. Emergency responders must always be concerned about other possible unknowns and must expand their monitoring to include radioactivity and corrosive materials. Hazardous material responders use a defined order of detection and atmospheric monitoring to protect responders and the monitoring equipment.

The following monitoring priorities should be considered when responding to an incident with unknown conditions:

- Gamma-radiation hazards need to be ruled out prior to entering the warm zone. Gamma rays, which travel farther than alpha particles or beta particles, are the primary hazard of exposure.
- Corrosive atmospheres and conditions should be ruled out to protect responders and detection equipment. Corrosive atmospheres can adversely affect a monitor's sensors and even destroy them.
- Oxygen levels affect how materials burn and can cause false LEL readings. Low oxygen levels pose dangers to human life and can be lethal.
- Flammable gases and vapors require monitoring to prevent responders from working in flammable environments. A flash fire in a confined space can result in an explosion, and depending on the forces present, it can completely destroy the area.
- Toxic gases and vapors are monitored to prevent poisoning.

Radiation sources are not likely to be present in confined spaces; however, certain industries may call for a raised level of concern—for example, medical and research facilities, mines, and certain manufacturing facilities may have onsite low-dose radioactive sources. Proper preplanning and size-up (both covered in Chapter 2 in more detail) should alert you to the potential dangers.

Incident Access Control

The purpose of any confined-space program is either to make an area safe for people to enter or to keep everyone out—a process known as **access control**. The people who will enter the space to complete an assigned job have roles and responsibilities.

FIGURE 1-13 A scene perimeter at a confined-space rescue scene (a trench).

Courtesy of Ron Zawlocki.

Procedural development must include access control measures. Once a **scene perimeter**—the area surrounding a rescue scene that must be cordoned off to protect onlookers, rescuers, and victims—has been established, access control measures must take place. Only authorized personnel should be present within a scene perimeter. Dealing with small landings, limited access points, and the situation at hand is difficult enough without well-meaning bystanders, friends, family members, or co-workers potentially interfering with the operation (**FIGURE 1-13**).

The Confined-Space Entry Permit

So far, we have discussed confined spaces and permit-required confined spaces, both of which are addressed in OSHA regulations. ANSI standard Z117.1-2003, *Safety Requirements for Confined Spaces* (American Society of Safety Engineers 2003), defines a confined space in terms identical to those used by OSHA, but it better differentiates them as either non-permit-required confined spaces or permit-required confined spaces, as previously discussed. The difference is simple: A non-permit-required confined space has been previously evaluated, and the potential hazards are either not present or have been eliminated through engineering controls.

Knowledge of this differentiation is intended to protect workers who enter a confined space and perform their work—it is not intended for rescuers. Rescuers responding to a confined-space accident must initially treat all confined spaces as permit-required spaces.

SAFETY TIP

To help ensure that the hazards (both atmospheric and physical) are reviewed prior to entry, certain requirements must be met for a basic confined-space entry permit.

Prior to the issuance of a confined-space entry permit, numerous safety concerns must be reviewed. This review includes knowledge of the following:

- The location of the space
- The reason for entry
- Atmospheric monitoring (inside and in proximity to the space)
- PPE requirements
- Lockout/tagout requirements
- Other potential hazards that the space may contain

The Confined-Space Entry Checklist

Some hazards may not exist within or near the space. Some can be eliminated. Still others will be present, and they must be controlled, or the entrant must be protected from them. The only way to effectively review and manage these numerous items is to develop and use a standardized checklist at the confined space, as shown in **Appendix C: Sample Confined-Space Entry Checklist**.

This checklist is the basis of the confined-space entry permit. To help ensure that hazards, both atmospheric and physical, are reviewed prior to entry, certain requirements must be met and recorded on the permit. These requirements can range from simple to technical, but the problems presented by simple ones are no less significant. Correct identification of a confined space may seem simple, but if the identification is incorrect, everything else will be as well. If rescuers take simple items for granted, they may miss the simple problems.

Written documentation in the form of a confined-space entry permit is what you, as a rescuer, should ask for and review as soon as you arrive at a confined-space emergency. If a permit is present, you can begin by verifying the information it contains. A missing permit is a worst-case scenario because it means that no confined-space entry program exists and that you must determine all the information you need to make a rescue on the fly.

Model confined-space entry permits can be found in the 29 C.F.R. 1910.146 OSHA regulations ("Permit-Required Confined Spaces" 1993) and ANSI standard Z117.1-2003, *Safety Requirements for Confined Spaces* (American Society of Safety Engineers 2003). Other examples of confined-space entry permits can be found online, but you must be certain that any examples you use for reference meet the minimum requirements of the OSHA regulations. *NFPA 1006: Standard for Technical Rescue Personnel Professional Qualifications*, 2021 Edition (NFPA 2021), contains sample confined-space entry permits for use by the rescue team.

Determining whether a confined-space entry permit meets the intent of the OSHA regulations requires the following actions:

- Verify by checking that the permit has been filled out properly, that all tests specified by the permit have been conducted, and that all procedures and equipment required by the permit are in place before allowing entry to begin (see **Appendix B: Sample Confined-Space Entry Permit**).
- Terminate the entry and cancel the permit as required.
- Verify that rescue services are available and that there is a means to summon them.
- Remove unauthorized individuals who enter or who attempt to enter the permit space.
- Determine that entry operations remain consistent with the terms of the entry permit and that acceptable entry conditions are maintained at all times.

The Permit-Issuance Period

The entry checklist should note the date and time the permit was issued, as well as the date and time the permit expires. No permit should be issued for an indefinite time period. Conditions change over time, and changes can originate inside or outside the space. An outdated permit invites disaster, giving workers a false sense of security and a terrible result: work in an unsafe environment. Some permits are issued for as long as a year, but a time frame that long must be based on an evaluation of that specific space, its potential hazards, and the establishment that there is a very low potential for hazards to be introduced into the space.

The duration of the permit may not exceed the time required to complete the assigned task or job identified on the permit. The entry supervisor shall terminate entry and cancel the entry permit when the entry operations covered by the entry permit have been completed or when a condition that is not allowed under the entry permit arises in or near the permit space.

Who and What Should Be in the Space

The checklist must accurately identify both the job site or space and the supervisor, as well as the attendant and entrants. Entrants, attendants, and other workers at the job site must know who belongs there and what is expected of them. Employees must know that they are entering the correct space and that they have properly identified the hazards of the space they are entering. The supervisor must also be identified so that workers can get answers to any questions they may have about the space, including potential hazards and the signs and symptoms of exposure.

TIP

The term *workers* is used to describe entrants frequently in this text. Although most confined-space incidents are workplace incidents, that is certainly not always the case. For example, a young child crawling into a storm drain could be a confined-space incident.

The checklist must also identify the equipment to be worked on and the type of work to be performed. This is critical to ensure the space is protected from unexpected and deadly hazards, such as accidental start-ups of equipment, the accidental opening of a valve, the application of a combustible coating inside the space, and so on. By knowing what type of work is to be performed and where it is to be done, as well as what work has not been planned for and what the space was not evaluated for, workers can protect themselves and others from injury or death. A good practice when asking what work was being performed or the purpose of entry is to also ask what the workers carried with them to complete the task because items that workers bring into spaces could be causes of injury or accident. Examples could include corded or battery-operated tools, specific chemical products, or cleaning solvents.

Atmospheric Hazards

Atmospheric checks must be made before and during entry. Atmospheric hazards, the leading cause of death in confined spaces, are among the most critical items evaluated prior to and during a confined-space entry. The entry permit requires that readings be taken for oxygen levels, flammable gases (also known as *explosive gases*), and toxic gases based on the known or suspected hazards in the space.

It is essential to check the operation of monitoring equipment before use and to begin taking readings outside the space. Equally critical is to take readings inside the space either simultaneously or in the following order: oxygen, flammable/explosive limit, and toxicity. The tester must also sign or initial the permit to confirm that all readings have been taken and recorded.

SAFETY TIP

Atmospheric hazards are the most common risk to people in confined spaces.

It may be necessary to provide mechanical ventilation for the space. When the only means of controlling an atmospheric hazard is by using forced ventilation, the space cannot be reclassified, and mechanical ventilation must be in place (**FIGURE 1-14**).

When either isolation or ventilation is deployed, thorough checks must be performed to ensure that the

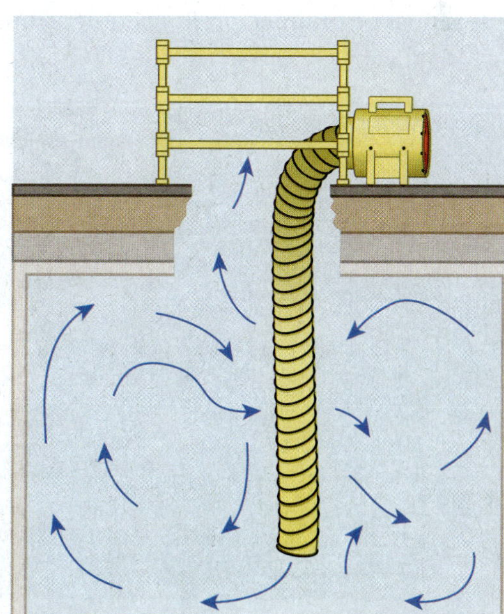

FIGURE 1-14 An illustration of mechanical ventilation in operation: A power ventilator circulating air inside a confined space.

ventilation and isolation measures are effective. After a confined space has been isolated or vented, meter readings for oxygen, combustible gases, and toxic gases must be taken to determine the efficacy of the isolation or ventilation.

Acceptable levels of each atmospheric hazard must be identified on the permit. For oxygen, the level must be between 19.5 percent and 23.5 percent. For combustible gases, levels must be below 10 percent of the LFL for each gas. Toxic gases may not exceed a level greater than the PEL for each gas.

Engulfment

Atmospheric hazards are the leading cause of death in confined spaces, and engulfment is the second-leading cause. Preventing engulfment requires eliminating and securing sources of potential physical or chemical hazards by isolating them from the space and/or controlling flowable materials in the space. This procedure is called *lockout/tagout*, which is covered in detail in Chapter 8. Lockout/tagout procedures include items such as the elimination of the material from the space, the shutdown of pumps, the capping of product lines that enter the confined space (also called **blinding**; see **FIGURE 1-15**), and disconnection or blockage of the lines.

The Importance of Clear Communication

Communication between the attendant and entrant(s) is critical to confined-space safety. Communication procedures must be outlined on the permit so that the attendant can check with the entrants. This becomes especially important if entrants perform work out of sight of the attendant.

The attendant must not enter the confined space for any reason unless another trained person takes over the role of attendant. According to OSHA, an entry is considered to have occurred *as soon as any part of the entrant's body breaks the plane of an opening into the space.* This means, for example, that if an attendant leans into the opening or peers into the opening to maintain contact with an entrant, that attendant has made entry into the space. The attendant is no longer considered an attendant and is instead an entrant. Effective communication procedures protect both the attendant and entrants.

FIGURE 1-15 A blind flange on a confined-space pipe ensures no flow of gases or liquids into the space during maintenance or rescue operations.

Personal Protective Equipment for Confined-Space Rescues

Confined-space rescue teams must be outfitted with appropriate PPE. The PPE and equipment needed for confined-space incidents is seldom necessary for firefighting. Generally, only a small percentage of the equipment carried on a typical fire engine is useful at a confined-space incident.

For the sake of simplicity, confined-space rescue PPE can be classified into two main categories:

- **Personal-issue PPE:** The garments and equipment that each member receives and keeps with them, likely in a personal bag
- **Team-issue PPE:** The PPE that is kept on the response vehicle and is shared (i.e., breathing apparatus)

PPE is not rocket science—it is more like common sense. The rescue personnel on the scene of a confined-space incident need a variety of protective clothing to minimize the effects of the weather and any related trauma caused by working in and around machinery and tools. At a minimum, confined-space rescue team members need an ensemble element consisting of protection for the upper and lower torso, arms, legs, head, hands, feet, and eyes. Other items, depending on the environment expected, could include hearing protection and foul weather gear. *NFPA 1951: Standard on Protective Ensembles for Technical Rescue Incidents* (NFPA 2020), does not address visibility concerns. In areas where visibility is important, team members outside of the actual confined space should be issued reflective vests or belts to be worn. PPE is covered in detail in Chapter 3.

The Importance of Proper Equipment

At every incident, the safety of personnel depends on their level of skill, as developed during both training and real-time incidents. Such training and experience may give personnel an initial advantage in the rescue effort; however, if they do not have the proper PPE or operational equipment and are subsequently injured in the rescue attempt, the operation will not be considered a success overall. In other words, all training and experience can be rendered ineffective if personnel do not have the proper PPE and equipment to give them reasonable protection during the rescue. NFPA 1951 sets the benchmark for technical rescue PPE (NFPA 2020). PPE for confined-space rescue team members should comply with that standard.

By necessity, firefighters carry what they use most, so their standard PPE provides protection for only the most common incidents, such as motor vehicle accidents and fires. This could lead to firefighters performing the initial actions at confined-space rescue incidents without a complete cache of NFPA technical-rescue-compliant PPE. All fire departments should consider the PPE required for confined-space rescue and determine if it is feasible to add to their fire apparatus. Another important element is that all battery-powered or electrical equipment that is used—regardless of whether it is personal equipment or team equipment—should be **intrinsically safe**, which means that the equipment is safe to operate in hazardous areas because it emits levels of electrical or thermal energy that are too low to cause ignition. Just as is true for incidents involving hazardous materials, the less that is known about a situation involving a confined space, the higher the level of PPE that should be used.

Navigating Local Policies for Access to Confined Spaces

All response agencies should have internal working protocols and procedures for emergency response operations, including those with the capacity to respond to a confined-space emergency. The policies, procedures, or protocols of any given agency or response team are usually very similar to those with confined-space response capabilities.

OSHA (2021) established a system known as the **Hierarchy of Controls** to determine which measures are most appropriate for reducing or removing hazards. The purpose of the Hierarchy of Controls is to provide a means of determining ways to implement systems or controls, from most effective to least effective, that protect workers from injuries, illnesses, and fatalities. Further discussion of the Hierarchy of Controls can be found in Chapter 5.

Technical References for Confined-Space Rescue

NFPA Standards

Although NFPA standards do not carry the weight of law, they are used by agencies all over the world as a baseline for achieving industry best practice. NFPA standards are developed through a consensus process that brings together volunteers representing varied viewpoints and interests. This helps to ensure a balanced perspective and the viability of the information represented as truly being best practice.

NFPA standards ensure that confined-space rescue operations are conducted with a focus on safety, compliance, and effectiveness. By following NFPA guidelines, rescue teams can enhance their preparedness, minimize risks, and improve overall rescue outcomes.

The NFPA standards relating to confined spaces include the following:

- ***NFPA 350: Guide for Safe Confined Space Entry and Work* (NFPA 2022a):** This standard offers comprehensive guidance on safe practices for entry, work, and rescue operations in confined spaces. It covers risk assessment, hazard evaluation, and emergency procedures.
- ***NFPA 1006: Standard for Technical Rescue Personnel Professional Qualifications* (NFPA 2021):** This standard outlines the competencies and training requirements for technical rescue personnel, including confined-space rescue technicians.
- ***NFPA 2500: Standard for Operations and Training for Technical Search and Rescue Incidents and Life Safety Rope and Equipment for Emergency Services (2022b) that includes Chapter 4: General Requirements* (NFPA 1670):** Although not specific to confined spaces, this standard provides general guidelines for technical rescue operations, which can be applied to confined-space incidents.

In NFPA terminology, *technical rescue* refers to more than just confined-space rescues. In fact, NFPA 1006 (NFPA 2021) and NFPA 2500 (1670) (NFPA 2022b) identify 20 different, unique, and/or complex search and rescue situations as technical rescue specialty disciplines:

- Tower rescue
- Rope rescue
- Structural-collapse rescue
- Confined-space rescue
- Common-passenger-vehicle rescue
- Heavy-vehicle rescue
- Animal technical rescue
- Wilderness search and rescue
- Trench rescue
- Machinery rescue
- Cave rescue
- Mine-and-tunnel rescue
- Helicopter rescue
- Surface-water rescue
- Swiftwater rescue
- Dive rescue
- Ice rescue
- Surf rescue
- Watercraft rescue
- Floodwater rescue

Confined-space rescue is but one of these, and as a discipline, it rarely exists in isolation. Technical rescue disciplines are not mutually exclusive, and two or more may be intertwined in any given incident. Because of the nature and specific knowledge and skills required during a technical rescue incident, it is not realistic for any one person, or even a single organization, to maintain expertise in all disciplines. Instead, the **authority having jurisdiction (AHJ)**—which is the organization, office, or individual responsible for enforcement of the requirements of a code or standard or for approving equipment, materials, installation, or procedures—is advised to establish written guidelines for each discipline and identify the level of operational response capability desired, based on the probability of incidence, frequency, and risk within their response area.

NFPA 2500 (1670) provides guidelines to assist the AHJ in assessing technical search and rescue hazards within their response area, identifying the level of operational capability for the organization/team, and establishing operational criteria for each discipline (NFPA 2022b). The levels of organizational response capability are defined as follows:

- **Awareness level.** This level represents the minimum capability of organizations that provide response to technical search and rescue incidents. Responders at this level are expected to know just enough about all disciplines to prevent further injury or harm, perform initial size-up, and activate appropriate response resources.
- **Operations level.** At this level, organizations may respond to technical search and rescue incidents. At the incident, they may identify hazards, support technician level rescuers with specific techniques, and assist in search and rescue.
- **Technician level.** This level represents the capability of organizations to respond to technical search and rescue incidents and identify hazards, use equipment, and apply advanced techniques as necessary to coordinate, perform, and supervise technical search and rescue incidents. As it applies to confined-space rescue, responders at the technician level are expected to be capable of everything an operations level responder does, plus performing other duties.

TIP

NFPA 1006: Standard for Technical Rescue Personnel Professional Qualifications (NFPA 2021) outlines the requirements and guidelines for the professional qualifications of personnel involved in technical rescue incidents, including confined-space rescue operations.

NFPA 1006 covers various aspects of technical rescue, including rope rescue, confined-space rescue, vehicle rescue, trench rescue, structural collapse rescue, and water rescue. It specifies the knowledge, skills, and abilities required for personnel involved in these rescue disciplines, as well as the training and certification requirements.

For confined-space rescue specifically, NFPA 1006 provides guidelines on the qualifications, training, equipment, and operational protocols necessary for personnel engaged in such operations. It ensures that rescue teams are adequately prepared to respond to confined-space emergencies safely and effectively, mitigating risks and ensuring the successful extrication of victims.

Overall, NFPA 1006 serves as a crucial benchmark for the professional qualifications and capabilities of technical rescue personnel, including those involved in confined-space rescue, with the ultimate goal of enhancing public safety and reducing the risk of injury or loss of life during rescue operations.

OSHA Regulations

OSHA, a federal agency that sets and enforces workplace safety and health regulations in the United States, provides regulations that have specific requirements for employers and workers to ensure the safety of individuals working in and around confined spaces. These include the following:

- **29 C.F.R. 1910.146, "Permit-Required Confined Spaces" (1993):** This OSHA regulation defines requirements for identifying and classifying permit-required confined spaces, as well as procedures for entry, rescue, and employee training.
- **29 C.F.R. 1926, Subpart AA, "Confined Spaces in Construction" (1979):** OSHA's construction industry regulation for confined spaces outlines requirements specific to construction activities, including rescue procedures and training.

Compliance with OSHA regulations is essential for ensuring the safety and health of workers engaged in confined-space activities. By following OSHA regulations, employers can create a safer work environment and reduce the risk of accidents or injuries during confined-space rescue operations.

Other Technical References for Confined-Space Rescue

Ready access to relevant technical references and resources is crucial for ensuring the safety and effectiveness of emergency response teams. Essential technical references for confined-space rescue operations include the *Emergency Response Guidebook* (ERG) and SDSs. Confined-space rescue operations also require adherence to specific standards and guidelines set forth by regulatory bodies such as the NFPA and OSHA.

The *Emergency Response Guidebook*

The ERG (**FIGURE 1-16**) was developed by the U.S. Department of Transportation (DOT) in collaboration with Transport Canada and the Secretariat of Communications and Transportation of Mexico (U.S. DOT 2024). It provides guidance for first responders and emergency personnel on how to handle hazardous materials incidents, including those that may occur within confined spaces.

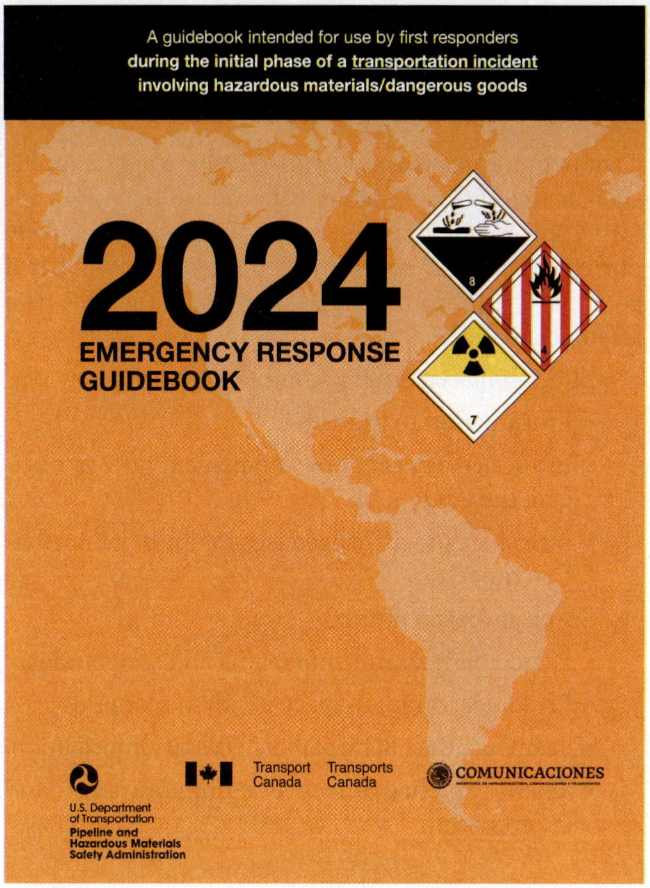

FIGURE 1-16 The ERG can be used for safe and effective confined-space emergency response operations.

Courtesy of U.S. Department of Transportation.

During a confined-space rescue operation involving a chemical spill or exposure, rescue teams can refer to the ERG to quickly assess the hazards and determine the appropriate response actions to ensure the safety of both the victims and responders. The guide helps responders quickly identify hazards associated with specific materials commonly found in confined spaces, such as gases, liquids, and solids. It provides step-by-step instructions on initial actions to be taken during a hazardous materials incident, including evacuation, containment, and protective measures. It also offers safety recommendations for responders, including appropriate PPE and evacuation distances based on the type of material involved and information on placards and labels for hazardous materials that can help responders in their efforts to identify potential risks.

Safety Data Sheets (SDSs)

SDSs (see **Appendix D: Sample Safety Data Sheet [SDS]**) are comprehensive documents prepared by chemical manufacturers, importers, or distributors that provide detailed information about the properties, hazards, handling, storage, and emergency procedures for specific substances. In a confined-space rescue scenario that involves hazardous chemicals, SDS documentation can quickly provide critical information to responders that permits them to assess risks, determine appropriate protective measures, and plan effective rescue operations while minimizing exposure risks.

Using Reference Materials

Conducting routine training using reference materials such as the ERG can help ensure that all necessary information is obtained and communicated effectively during a confined-space incident. The following is a good outline for a training session using the ERG as a reference material:

- ERG familiarization (45 minutes)
  - Introduction to the ERG
  - Introduce key sections and explain how to navigate the guidebook.
  - Introduce practical exercises for finding information quickly.
- Scenario-based training (2 hours)
  - Divide participants into teams and assign roles.
  - Present a confined-space incident scenario.
  - Teams use the ERG to gather necessary information.
  - Teams practice effective communication and coordination.
  - Afterward, teams debrief and discuss the scenario.
- Information gathering and communication workshop (1 hour)
  - Identify critical information needed in confined-space incidents.
  - Practice using the ERG for information retrieval.
  - Develop communication protocols for incident response.
  - Engage in role-playing exercises to practice communication strategies.
- Review and assessment (45 minutes)
  - Review key points from the training.
  - Assess participants' understanding and skills.
  - Provide feedback and engage in a question-and-answer session

In this example, all participants must have access to the ERG and be familiar with its layout before the training begins. The training must use realistic and varied scenarios to cover different types of confined-space incidents, and active participation should be encouraged.

CHAPTER FEATURE: AFTER-ACTION

IN SUMMARY

- For a space to be defined as a confined space, it must have all three of the following characteristics: it must be large enough for a person to enter to perform a specific job, it must have limited means of access and a restricted entrance or exit, and it must not be designed for people to spend significant time within or for continuous occupancy.
- An awareness level responder is responsible for recognition, scene control, and size-up. An operations level responder is responsible for initiating entry into or rescue in areas that are immediately visible. A technician level responder is responsible for entry into or rescue in areas that are not immediately visible.
- Hazards in confined spaces are generally grouped into two categories: atmospheric and physical. Atmospheric hazards may include toxic gases, flammable gases, or oxygen deficiency. Physical hazards include electrical equipment, agitators or paddles, fire-suppression equipment, sharp edges, areas that are difficult to access because of height or other limitations, and the physical state of the material present in the space, as well as other considerations.
- Incident access control involves making the area safe for qualified people to enter or keeping unqualified people out. A confined-space entry checklist provides guidance on controlling access to the scene.
- All response agencies should have internal working protocols and procedures for emergency response operations, including those with the capacity to respond to a confined-space emergency.
- At every incident, the safety of personnel depends on their level of skill and the proper PPE and operational equipment.
- Common technical references include NFPA standards, OSHA regulations, the ERG, and SDSs.

KEY TERMS

access control The processes of making an area safe for qualified people to enter or keeping unqualified people out.

atmospheric hazard A hazard present in the air of a confined space, including toxic gases, flammable gases, or oxygen deficiency.

atmospheric monitoring A method of evaluating the ambient atmosphere of a space, including but not limited to its oxygen content, flammability, and toxicity.

attendant An individual who is stationed outside one or more permit-required confined spaces and monitors the people within and the conditions in the area around the space.

attendant-based rescue A rescue attempt in which the attendant participates as part of the rescue service in a nonentry capacity once entry rescue procedures have been initiated, provided that the attendant performs no duties that might interfere with the attendant's primary duty to effectively monitor and protect all of the authorized entrants for all of the permit spaces they are attending and for whom they are responsible.

authority having jurisdiction (AHJ) An organization, office, or individual responsible for enforcement of the requirements of a code or standard or for approving equipment, materials, installations, or procedures.

authorized entrant A term used to describe a U.S. federally regulated industrial worker designated to enter confined spaces who meets specified training requirements for each specific space they enter.

awareness level responder A first responder who understands the basics of confined-space safety and knows how to safeguard personnel from fire, explosion, unqualified rescue attempts, and other health hazards that are uniquely associated with confined spaces.

blinding The capping of product lines that enter a confined space.

calibrated Carefully assessed, set, or adjusted.

***Code of Federal Regulations* (C.F.R.)** The codification of the general and permanent rules published in the *Federal Register* by the departments and agencies of the U.S. government.

communications plan A plan for communications to flow between rescuers, attendants, and all other parties involved in a confined-space rescue.

confined space A space with all three of the following characteristics: It must be large enough for a person to enter to perform a specific job, have limited means of access and a restricted entrance or exit, and not be designed for people to spend significant time within or for continuous occupancy.

confined-space entry program An employer's overall program for controlling and protecting employees from permit space hazards and for regulating employee entry into a permit space.

confined-space rescue technician A skilled individual trained to operate at both the operations and technician levels within confined spaces.

confined-space supervisor An individual who supervises an operation in a confined space; is responsible for knowing the hazards that may be faced during entry, including the modes, signs (or symptoms), and consequences of the exposure; and is responsible for informing the attendant and entrant(s) of these hazards.

engulfment The surrounding and effective capture of a person by a liquid or finely divided (flowable) solid substance that can be aspirated to cause death by filling or plugging the respiratory system or that can exert enough force on the body to cause death by strangulation, constriction, or crushing.

explosive range The range between the lower exposure limit and the upper exposure limit for a particular gas or vapor.

flammable range The range of concentrations of a mixture of flammable gas or vapor in air over which a flash will occur or a flame will travel upon ignition.

flash point The lowest liquid temperature at which, under certain conditions, a liquid emits vapor that is flammable.

Hierarchy of Controls A system developed by the Occupational Safety and Health Administration (OSHA) that provides a means of determining ways to implement systems or controls, from most effective to least effective, that protect workers from injuries, illnesses, and fatalities.

hot work Work involving electric or gas welding, cutting, brazing, or similar flame- or spark-producing operations.

immediately dangerous to life or health (IDLH) The maximum level to which one could be exposed to a toxic substance or situation and still escape without experiencing any effects that may impair escape or cause irreversible health effects.

KEY TERMS CONTINUED

intrinsically safe A term used to describe battery-powered or electrical equipment that is safe to operate in hazardous areas because it emits levels of electrical or thermal energy that are too low to cause ignition.

lockout/tagout The process of eliminating and securing sources of potential physical or chemical hazards by isolating them from a confined space and/or controlling flowable materials in the space.

lower flammable limit (LFL) or lower explosive limit (LEL) The lower end of the concentration range over which a flammable mixture of gas or vapor in air can be ignited at a given temperature and pressure.

National Institute for Occupational Safety and Health (NIOSH) The agency within the U.S. Department of Health and Human Services that identifies work-related diseases and injuries and the potential hazards of new work-related technologies and practices.

non-permit-required confined space A confined space that does not contain—or in the case of atmospheric hazards, does not have the potential to contain—any hazard capable of causing death or serious physical harm.

Occupational Safety and Health Administration (OSHA) A U.S. government regulatory agency with a mission of ensuring that workers have safe and healthful working conditions free from unlawful retaliation.

operations level responder A first responder who understands rescuer safety, team evaluation, preplanning, atmospheric monitoring, and the proper use of personal protective equipment and is also capable of assembling a portable anchor system for application of a high-point attachment, following lockout/tagout procedures, preparing for entry into a vertically oriented confined space, and entering horizontally and vertically oriented confined spaces. The operations level also covers techniques for patient packaging and recovery.

permissible exposure limit (PEL) A regulatory limit on the amount or concentration of a substance in the air.

permit-required confined space A confined space that is designated as being known to contain—or in the case of atmospheric hazards, has the potential to contain—any hazard capable of causing death or serious physical harm.

personal protective equipment (PPE) The equipment provided to shield or isolate a person from the chemical, physical, or thermal hazards that can be encountered at a specific rescue incident.

physical hazard A hazard present in a confined space, including electrical equipment, agitators or paddles, fire-suppression equipment, sharp edges, areas that are difficult to access because of height or other limitations, and the physical state of the material present in the space, as well as other considerations.

recovery The process of retrieving the body of a deceased person from a confined space.

rescue The process of safely extricating a living person from a confined space.

rescue attendant An individual who is stationed outside a confined space during a rescue attempt who monitors the people within and the conditions in the area around the space.

rescue entrant An individual who is qualified to enter a confined space and assist in a rescue attempt.

scene perimeter The area surrounding a rescue scene that must be cordoned off to protect onlookers, rescuers, and victims.

self-contained breathing apparatus (SCBA) A device worn to provide breathable air in environments with toxic atmospheric hazards.

size-up The ongoing evaluation of problems confronted in a confined-space rescue operation.

standby assistance Ensuring the safety of the people within a confined space and maintaining communication with the workers in the space.

supplied-air breathing apparatus (SABA) A combination of a supplied-air respirator and an escape self-contained breathing apparatus.

supplied-air respirator (SAR) A low-profile respirator worn to provide breathable air in environments with toxic atmospheric hazards.

technical rescue The application of special knowledge, skills, and equipment to resolve unique and/or complex rescue situations.

technician level responder A first responder who is capable of preplanning a confined-space incident; the responder must apply and use a supplied-air respirator or supplied-air breathing apparatus as a rescue entrant and must be well versed in using tripods and other artificial high anchor points, as well as lockout/tagout procedures. These responders are skilled in using rope systems specifically designed for confined-space entry and rescue scenarios and using communications systems.

tripod A portable anchor system device typically meant to support people as they are raised or lowered into and out of a space.

upper flammable limit (UFL) or upper explosive limit (UEL) The upper end of the concentration range over which a flammable mixture of gas or vapor in air can be ignited at a given temperature and pressure.

REFERENCES

American Society of Safety Engineers. 2003. *Safety Requirements for Confined Spaces.* ANSI/ASSE Z117.1-2003. Des Plaines, IL: American Society of Safety Engineers.

Centers for Disease Control and Prevention. 1986. "Preventing Occupational Fatalities in Confined Spaces." https://www.cdc.gov/niosh/docs/86-110/default.html#print.

"Confined and Enclosed Spaces and Other Dangerous Atmospheres in Shipyard Employment." 1994. 29 C.F.R. 1915, Subpart B. https://www.ecfr.gov/current/title-29/subtitle-B/chapter-XVII/part-1915/subpart-B.

"Confined Spaces in Construction." 1979. 29 C.F.R. 1926, Subpart AA. https://www.ecfr.gov/current/title-29/subtitle-B/chapter-XVII/part-1926.

National Fire Protection Association. 2020. *NFPA 1951: Standard on Protective Ensembles for Technical Rescue Incidents, 2020 Edition.* Quincy, MA: National Fire Protection Association.

National Fire Protection Association. 2021. *NFPA 1006: Standard on Technical Rescue Personnel Professional Qualifications, 2021 Edition.* Quincy, MA: National Fire Protection Association.

National Fire Protection Association. 2022a. *NFPA 350: Guide for Safe Confined Space Entry and Work, 2022 Edition.* Quincy, MA: National Fire Protection Association.

National Fire Protection Association. 2022b. *NFPA 2500: Standard for Operations and Training for Technical Search and Rescue Incidents and Life Safety Rope and Equipment for Emergency Services, 2022 Edition* that includes Chapter 4: General Requirements (NFPA 1670). Quincy, MA: National Fire Protection Association.

Occupational Safety and Health Administration. 2021. "Identifying Hazard Control Options: The Hierarchy of Controls." https://www.osha.gov/sites/default/files/Hierarchy_of_Controls_02.01.23_form_508_2.pdf.

"Permit-Required Confined Spaces." 1993. 29 C.F.R. 1910.146. https://www.ecfr.gov/current/title-29/part-1910/section-1910.146.

U.S. Department of Transportation, Pipeline and Hazardous Materials Safety Administration. 2024. *2024 Emergency Response Guidebook: A Guidebook for First Responders During the Initial Phase of a Dangerous Goods/Hazardous Materials Transportation Incident.* Washington, DC: U.S. Department of Transportation, Pipeline and Hazardous Materials Safety Administration.

CHAPTER QUESTIONS

1. Which of the following is a critical element that affects confined-space operations?
   - **a.** atmospheric hazards
   - **b.** fall hazards
   - **c.** entry points
   - **d.** energy sources
2. According to the requirements of a confined-space entry program, one individual must be placed in charge of a program during permit-required entries. OSHA identifies this person as the
   - **a.** attendant.
   - **b.** local fire chief.
   - **c.** entry supervisor.
   - **d.** rescue officer.
3. The process of making an area safe for people to enter or to keep everyone out is called
   - **a.** confined-space control.
   - **b.** rescue scene control.
   - **c.** hazard control.
   - **d.** access control.
4. Which of the following is a characteristic that would define a permit-required confined space?
   - **a.** being poorly ventilated
   - **b.** having a number of openings
   - **c.** having a hazardous atmosphere
   - **d.** being underwater

CHAPTER FEATURE: ON SCENE

TEAMWORK SAVES LIVES AT A CHEMICAL SPILL INCIDENT

Confined-space rescue events are unpredictable and often require cooperation among different agencies and teams. That was certainly the case at a chemical spill event that occurred inside a confined space—in this case, a wood pulp vat—at a paper mill in the spring of 2020.

In this incident, a tank containing bleach ruptured, flooding the confined space with hazardous materials. Workers at the mill were exposed to toxic sulfur dioxide (SO_2) gas, and the danger grew as the hazardous chemicals spread throughout the confined area.

The Emergency Response

Local emergency services, including hazardous materials (HazMat) teams from surrounding communities, responded immediately to the scene. The responders quickly established an exclusion zone around the affected area to prevent further exposure and contamination.

The Rescue Operations

The HazMat teams donned their specialized PPE, including chemical-resistant suits, respirators, and eye protection, before entering the confined space. Decontamination procedures were implemented to ensure the safety of responders and prevent the spread of chemical contaminants. Air-monitoring equipment was used to assess the air quality inside the confined space and determine the presence of toxic fumes.

The Extrication and Treatment

Rescuers carefully accessed the confined space and assisted affected workers as they evacuated the area. Emergency medical personnel provided immediate medical care, including decontamination and treatment for chemical exposure, to affected individuals. Some workers were transported to nearby hospitals for further evaluation and medical treatment.

The Outcome

The coordinated efforts of the HazMat teams and emergency responders resulted in the successful evacuation and treatment of workers affected by the chemical spill. This incident highlights the importance of rapid response, effective communication, and adherence to safety protocols during chemical spill emergencies in confined spaces. Following the incident, an investigation was conducted to determine the cause of the chemical release and identify measures to prevent similar incidents in the future.

This real-life chemical spill incident response underscores the need for HazMat teams and emergency responders to work in sync to mitigate hazards, protect lives, and ensure the safe evacuation and treatment of individuals exposed to toxic substances inside confined spaces.

The positive outcome of this situation shines a spotlight on the importance of continuous training, preparedness, and equipment maintenance for responding to chemical emergencies in industrial settings. Clearly, there is a significant overlap of disciplines in the emergency response world; in this case, what began as a perceived hazardous materials incident quickly morphed into a confined-space rescue requiring skilled technical rescue personnel. The success of this rescue illustrates the value of having a background in hazardous materials when working in the technical rescue discipline.

1. What was the primary hazard workers were exposed to during the chemical spill incident at the paper mill?
   - a. carbon monoxide gas
   - b. sulfur dioxide gas (SO_2)
   - c. methane gas
   - d. nitrogen dioxide gas
2. Which of the following actions was one of the first steps taken by the emergency responders at the scene of the chemical spill?
   - a. initiating confined-space entry
   - b. establishing an exclusion zone
   - c. evacuating the entire mill
   - d. sending workers home for safety

3. What type of specialized PPE did HazMat teams wear during the response to the chemical spill?
   - **a.** fire-resistant suits and helmets
   - **b.** chemical-resistant suits and respirators
   - **c.** water-resistant suits and gloves
   - **d.** heat-resistant suits and goggles

4. What was one key factor that contributed to the successful outcome of the rescue operation at the paper mill?
   - **a.** the use of drones to monitor the scene
   - **b.** the presence of an onsite medical team
   - **c.** coordination between HazMat and rescue teams
   - **d.** immediate evacuation of the mill by workers

CHAPTER

2

Awareness

Sizing Up and Controlling the Incident Site

KNOWLEDGE OBJECTIVES

After studying this chapter, you should be able to:

- Understand the organizational structure and roles and responsibilities of Incident Command Systems (ICSs).
- Understand the concept and value of the size-up.
- Interpret and document information appropriately. (**NFPA 1006: 7.1.3**, pp. 38–42)
- Describe isolation procedures for a confined-space incident. (**NFPA 1006: 7.1.1**, pp. 39–43)
- Explain how to initiate different isolation methods. (**NFPA 1006: 7.1.1**, pp. 39–43)
- List the steps for securing a confined-space incident. (**NFPA 1006: 7.1.6**, pp. 43–48)
- Identify search parameters at a confined-space incident. (**NFPA 1006: 7.1.2**, pp. 48–49; 50–52)
- Describe the process for communicating with confined-space victims. (**NFPA 1006: 7.1.3**, pp. 49; 53)
- Explain the process for establishing a victim profile. (**NFPA 1006: 7.1.2**, pp. 53–55)
- Describe the process for interviewing bystanders. (**NFPA 1006: 7.1.2**, pp. 55–56)

SKILL OBJECTIVES

- Predict the potential hazards associated with a confined-space incident.
- Document all aspects of a confined-space incident, including victim transfer information. (**NFPA 1006: 7.1.3**, pp. 38–42)
- Demonstrate the process of securing and mitigating access to a confined-space incident. (**NFPA 1006: 7.1.1**, pp. 43–48)
- Choose entrance paths.
- Utilize interview techniques. (**NFPA 1006: 7.1.5**, pp. 55–56)

Introduction

Effective **incident command**, or incident management, is crucial for coordinating and managing rescue efforts to ensure the safety and successful outcome of a confined-space rescue operation. This chapter outlines the key responsibilities, strategies, and considerations for incident command personnel in accordance with National Fire Protection Association (NFPA) 1006 standards (NFPA 2021) specific to incident command for confined-space rescue operations (including competency requirements and operational guidelines) and other relevant NFPA requirements for **Incident Command System (ICS)** implementation. The ICS is a system for providing management of personnel, resources, and activities during emergency operations by defining roles and responsibilities and **standard operating procedures (SOPs)**, which are written guidelines for handling specific situations.

Just as important is performing an accurate and in-depth size-up of emergencies—and preplanning for potential emergencies—which will help you to focus on the most critical issues up front. A size-up involves identifying the hazards of a confined space, how you will control those hazards, and the resources you will use during your response.

In this chapter, you will learn the protocols for isolating and securing a confined-space incident site to minimize risks, maintain scene integrity, and ensure the safety of everyone involved. You will also learn about victim profiles, which are detailed summaries of individuals affected by a confined-space incident, including their identity, role or function, health status, and other special considerations about the individuals.

Incident Command for Confined-Space Rescue Operations

All emergency responders have varying levels of training in the art of incident command. Ensuring that all responders within the technical rescue team have updated training levels and updated or expanded knowledge within the subject matter is crucial. The principles of ICS emphasize clarification of a chain of command, a span of control, and the roles and responsibilities of key positions. The National Incident Management System (NIMS) ICS 100, 200, and 700 training courses, which are offered through the Emergency Management Institute of the U.S. **Federal Emergency Management Agency (FEMA)**, are requirements for incident command. FEMA is an agency of the U.S. federal government with a primary purpose of coordinating responses to disasters that overwhelm the resources of local and state authorities.

Preincident Planning and Preparedness

The first step in the ICS process occurs before an incident takes place with the development of **incident action plans (IAPs)**. An IAP is a plan that establishes goals and objectives for the emergency, identifies the resources to be used, and provides the means to accomplish the goals and objectives. IAPs are built in a collaborative manner: Relevant stakeholders work together to develop comprehensive incident action plans tailored to confined-space rescue scenarios. IAPs include such details as objectives, strategies, tactics, and resource allocations. IAPs are covered in detail in Chapter 4.

The next step is to identify and preplan the resources that will be needed for a confined-space rescue operation, including specialized equipment, personnel, and support services. Training is key: All personnel designated for a confined-space rescue operation must be well trained and familiar with relevant Occupational Safety and Health Administration (OSHA) regulations, NFPA standards (e.g., NFPA 350 [NFPA 2022a], NFPA 2500 [NFPA 2022b]), and other local or national standards governing confined-space operations.

In the best case, rescue teams have advance access to all permit spaces from which rescue may be necessary so that they can develop effective rescue plans and practice rescue operations. This allows them to determine whether they have the capability to reach victims within a time frame that is appropriate for the hazards present in the permit space. See **TABLE 2-1** for details and examples of a confined-space preincident plan.

By organizing the confined-space preincident plan in a chart format, it becomes easier to communicate essential information and ensure that all aspects of confined-space work are covered comprehensively. Regular review, updates, and training are essential for maintaining a safe and compliant confined-space program within any organization.

Incident Command Organizational Structure

Effective incident command is essential for orchestrating complex confined-space rescue operations in accordance with NFPA standards and ICS principles. By implementing proactive planning, clear communication, resource management, and safety measures,

TABLE 2-1 The Confined-Space Preincident Plan

Plan Element	Details	Example
Confined-space inventory	Make a list of all confined spaces in the facility/ worksite, including location, type, and associated hazards	▪ Tank #1 (Chemical Storage): Location—West Storage Area, Type—Vertical cylindrical tank, Hazards—Toxic gases, limited entry/exit ▪ Maintenance Pit: Location—Maintenance Workshop, Type—Below-ground pit, Hazards—Mechanical hazards, limited ventilation
Hazard assessment	Identify and assess potential hazards (atmospheric, physical, biological) associated with each confined space	▪ Tank #1: Atmospheric testing required before entry because of potential toxic gas accumulation ▪ Maintenance Pit: Confined-space entry permit required for all personnel because of limited ventilation and mechanical hazards
Entry requirements	Determine when confined-space entry is necessary and establish criteria for authorized personnel	▪ Conduct atmospheric testing using atmospheric monitors before entry ▪ Complete confined-space entry permit with required authorizations and procedures ▪ Wear appropriate personal protective equipment (PPE), including harnesses, helmets, and respiratory protection ▪ Maintain constant communication with attendants/supervisors via two-way radios
Emergency response plan	Define rescue and evacuation procedures specific to confined-space incidents, including coordination with local emergency services	▪ Establish rescue procedures using retrieval systems (tripods, winches) ▪ Coordinate with local emergency services for confined-space rescue training and mutual aid
Equipment and tools	List required equipment for confined-space entry and rescue operations (ventilation, atmospheric monitors, retrieval systems, PPE)	▪ Ventilation fans ▪ Gas-detection monitors ▪ Confined-space tripod with winch ▪ PPE
Training and competency	Provide training for personnel involved in confined-space work (awareness, entry supervision, rescue operations)	▪ Confined-space entry training for authorized personnel ▪ Rescue team training for designated rescue personnel ▪ Yearly entry to maintain proficiency
Entry procedures	Detail step-by-step procedures for safe confined-space entry, including pre-entry checks and communication protocols	
Communication protocols	Establish communication methods and protocols between confined-space entrants, attendants, and supervisors	
Regular review and updates	Schedule periodic reviews of the preincident plan and update procedures, equipment, and training as needed	
Documentation	Maintain detailed records of confined-space inventory, hazard assessments, entry permits, training, and incidents	

incident command personnel can mitigate risks, optimize response efforts, and achieve successful outcomes in confined-space rescue incidents. Regular training, ongoing preparedness, and adherence to established protocols are paramount for maintaining readiness and effectiveness in incident command roles.

Before the Incident

Before an incident occurs, a clear and effective incident command structure must be established. Qualified personnel must be designated into the key positions of an ICS command structure: the **incident commander (IC)**, the operations section chief, the safety officer, the logistics officer, and the public information officer (PIO). The IC is the individual responsible for overseeing the entire incident response, making critical decisions, developing objectives, and ensuring the safety and effectiveness of the operation. Each role must have clear responsibilities assigned to ensure clear lines of communication, coordination, and decision-making.

During the Response

The first step is to conduct a thorough size-up of the confined-space incident, which involves an assessment of all hazards, victim status, the space's structural integrity, and resource requirements. Incident objectives must be defined based on the findings from the size-up. The objectives should prioritize life safety first, incident stabilization second, and property conservation third.

Following the size-up, resources should be allocated efficiently based on incident needs and operational priorities. This process may require coordination with external agencies and mutual aid partners.

Standardized protocols and frequencies are necessary for effective communications management within the incident command structure. Likewise, it is necessary to set up comprehensive safety protocols, including risk assessments, hazard mitigation strategies, and personal protective measures for responders.

As the response evolves, incident command must continuously evaluate and manage risks as they arise, adjusting strategies and tactics as needed to ensure responder safety.

After the Incident

After the incident response concludes, incident command is responsible for accurate documentation of all incident actions, decisions, and outcomes, including incident logs, resource records, and incident reports. This documentation is necessary for postincident debriefings and evaluations, which incident command uses to identify lessons learned, areas for improvement, and recommendations for future incidents.

Establishing an ICS Command Structure

An effective ICS must have a clear command structure for managing emergency incidents efficiently and ensuring coordinated response efforts. The NIMS provides guidelines and principles for establishing command structures that enhance interoperability and collaboration across agencies and jurisdictions. This section outlines the process of establishing a command structure within the ICS according to NIMS principles. See **FIGURE 2-1** for a visual of the hierarchical structure of an ICS command, and see **TABLE 2-2** for a summary of the key roles and responsibilities of each command staff position.

The Incident Commander (IC)

The IC is the highest level in the ICS chain of command. This individual, who should be the most qualified and experienced member of the team, is responsible for the overall management of the incident and has the authority to make decisions and allocate resources effectively. The IC makes all strategic decisions, sets incident objectives, and ensures coordination among all functional areas.

Command Staff

Below the IC in the ICS command structure are the command staff: the PIO, who manages incident information and media relations and ensures accurate and timely public communications; the safety officer, who monitors incident operations for safety hazards and advises the IC on mitigation measures to protect responders; and the liaison officer, who establishes and maintains coordination with external agencies and organizations involved in the incident response.

General Staff

Below the command staff in the ICS hierarchy comes the general staff. Key roles at this level include the operations section chief, who directs and coordinates all incident tactical operations, ensuring resources are utilized effectively to achieve incident objectives; the planning section chief, who develops and maintains the IAP, gathers and analyzes information, and forecasts future incident conditions; the logistics section chief, who provides facilities, services, personnel, and resources to support incident operations, including procurement, communications, and medical support;

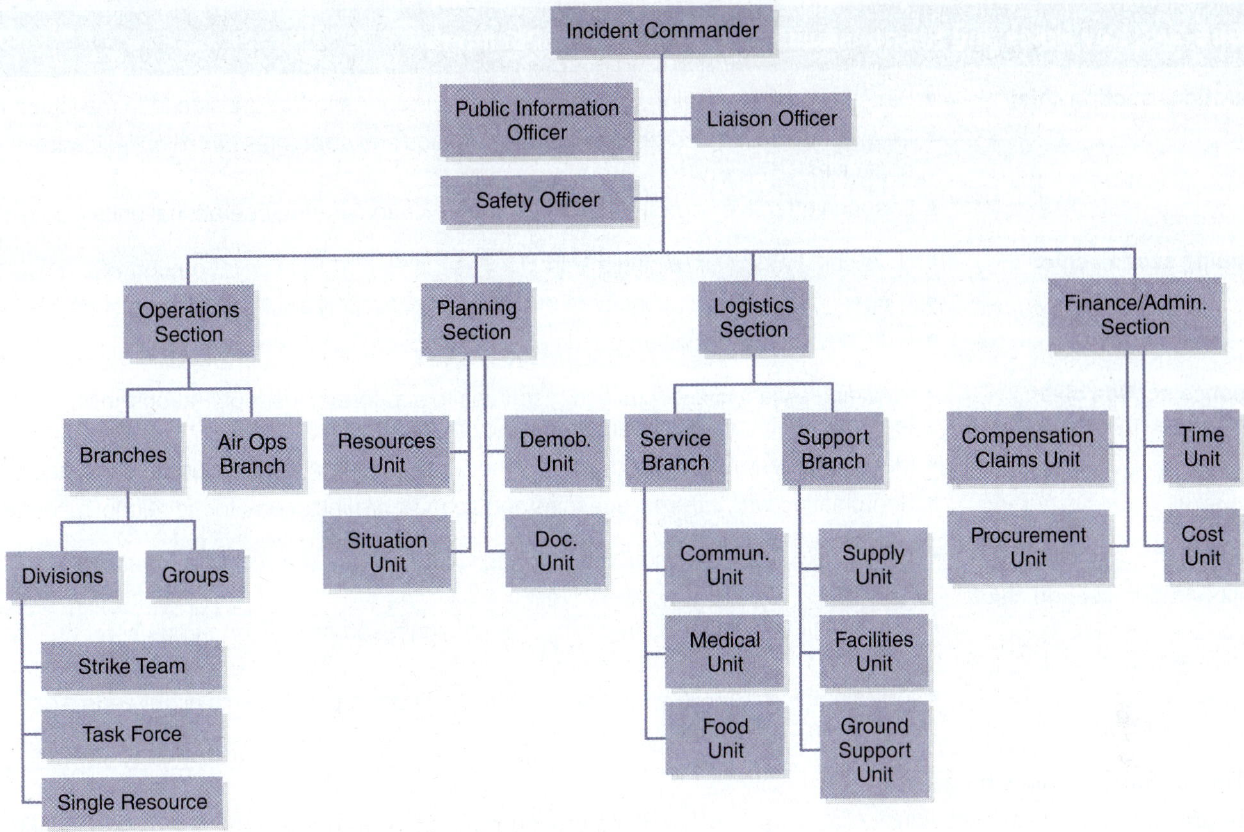

FIGURE 2-1 The hierarchical structure of an ICS command.

Reproduced from Federal Emergency Management Agency (March 2018). ICS Review Document. Accessed at https://training.fema.gov/emiweb/is/icsresource/assets/ics%20review%20document.pdf

TABLE 2-2 Key Roles and Responsibilities of Command Staff Positions

Incident commander (IC)	▪ Overall incident management and decision-making authority ▪ Setting incident objectives and priorities ▪ Directing and coordinating all incident activities
Public information officer (PIO)	▪ Managing public information and media relations ▪ Coordinating with the media and preparing incident updates ▪ Ensuring accurate and timely dissemination of information to the public
Safety officer	▪ Monitoring incident operations for safety hazards and risks ▪ Advising the IC on safety measures and mitigation strategies ▪ Conducting safety briefings and ensuring compliance with safety protocols
Liaison officer	▪ Establishing and maintaining coordination with external agencies and stakeholders ▪ Facilitating information sharing and mutual aid support ▪ Representing the incident management team in interagency meetings

Additional considerations for command staff:

- **Collaboration:** They must work collaboratively to ensure effective incident management and response
- **Communication:** Clear and consistent communication is essential among the command staff members and with their operational units
- **Adaptability:** They must be adaptable and able to adjust strategies based on changing incident conditions and requirements

TABLE 2-3 Key Roles and Responsibilities of General Staff Positions

Position	Responsibilities
Operations section chief	▪ Directing and overseeing all tactical operations and resources assigned to the incident ▪ Developing and implementing operational plans and strategies based on incident objectives ▪ Ensuring effective coordination and communication among operational units
Planning section chief	▪ Developing and maintaining the incident action plan (IAP) in collaboration with the IC ▪ Gathering and analyzing incident information to support planning and decision-making ▪ Forecasting future incident conditions and resource requirements
Logistics section chief	▪ Providing essential services and resources to support incident operations, including facilities, equipment, and personnel support ▪ Managing communications systems, medical services, and transportation logistics ▪ Coordinating with external agencies and support organizations for additional resources
Finance/ administration section chief	▪ Managing financial aspects of the incident, including cost tracking, procurement, and fiscal accountability ▪ Documenting incident-related expenditures and resource use for reimbursement and reporting purposes ▪ Coordinating with finance and administrative support personnel to ensure efficient financial management

Additional considerations for general staff:

- **Integration:** General staff positions collaborate closely with command staff and operational units to achieve incident objectives
- **Resource management:** Each section chief is responsible for resource allocation and coordination within their respective functional area
- **Documentation:** Accurate documentation and record-keeping are essential for tracking incident progress and resource utilization

and the finance/administration section chief, who manages the financial and administrative aspects of the incident, including cost tracking, procurement, and personnel support.

TABLE 2-3 outlines the specific roles and responsibilities of general staff positions within the ICS according to NIMS guidelines. Clear understanding and implementation of these roles contribute to effective incident management, coordination, and response across emergency incidents and jurisdictions.

The IAP

The planning section of an ICS is responsible for developing the IAP based on incident objectives, strategies, and resource needs. The IAP guides operational activities and resource allocation. The IC is responsible for reviewing and approving the IAP, as well as ensuring its alignment with incident objectives and operational priorities.

TIP

In complex incidents involving multiple jurisdictions or agencies, a Unified Command (UC) structure may be established. This allows for coordinated decision-making and resource sharing among participating agencies.

Span of Control

In an optimal ICS command structure, each supervisor or manager should oversee no more than three to seven subordinates (with five being optimal), depending on the supervisor's area of expertise. Limiting the number of subordinates helps ensure effective supervision and communication. To effectively manage the span of control, consider two key factors: the nature and size of the incident.

Transition and Demobilization

As the incident progresses or is resolved, ICS transition planning ensures a smooth handoff of responsibilities and resources. During the demobilization phase of an ICS, resources are released and returned to normal operations once the incident is fully managed and under control.

Preplanning and Size-Up

Before an incident ever takes place, preplanning for potential emergencies helps you identify the basic strategic elements that you will be faced with during an actual emergency, including the hazards of the space, a means of controlling those hazards, and the resources you will need to have available. Preplanning allows you to focus on these key elements without a sense of urgency.

Among those hazards and methods of control should be potential energy sources, other risks and threats, the methods used to control them, and how to confirm that the energy has been isolated. You should be able to identify the action options needed for the anticipated emergencies and craft a basic action plan to initiate your rescue operations.

A word of caution is appropriate here. Even with a comprehensive preplan, you cannot take for granted that conditions in the space have not changed since your initial plan. Even if your last preplan inspection is only hours old, conditions within or near the space may have changed. Your preplan should also be considered as a checklist you can use to compare what you know about the facility with the conditions that exist at the time of the emergency.

Preplanning is clearly a best-case scenario, but what if there is no preplan for a confined-space incident you have been called to? Whether or not preplanning has taken place, you must first conduct a size-up of the incident. The size-up is intended to identify the limiting or strategic factors related to incident conditions. Obviously, a preplan should have already identified those factors, but with or without a preplan, you must be prepared to conduct a **hazard and risk assessment** by determining what could possibly cause harm in the space and the degree of likelihood that harm will be caused. Where do you start? Following is a quick list:

- Develop a method to determine the accuracy of the information you are receiving.
- Identify potential energy sources.
- Determine which energy sources require lockout, as shown in **FIGURE 2-2**.

FIGURE 2-2 Product stored in this grain car is intended to be released through the gates at the bottom (circled). The source of the energy moving this product is gravity, and it must be thought of as stored energy.

- Identify and control sources of residual or stored energy, such as a compressed spring or a charged electrical capacitor.
- Identify the site manager, maintenance supervisor, or most knowledgeable individual at the site for securing all potential items needing lockout/tagout.

Hazard and Risk Assessment

A hazard and risk assessment should start with three simple questions. The first two are: What has happened? What is happening now? The third is based on the answers you gather after asking the first two: What can you predict will happen as this incident plays out?

When you arrive at any emergency, you must determine what has happened. Emergency medical services (EMS) personnel look for the mechanism of injury at an accident scene. Firefighters need to locate the seat of a fire. As a member of a confined-space rescue team, you need to see the whole picture. That includes the victim, the confined space, and the area surrounding it. Seek out the confined-space entry permit, and then find the entry supervisor, the attendant, and any other person who has witnessed or has knowledge of the incident. These are all pieces of the puzzle, and you must figure out how they fit together. To determine what has happened, you should consider the energy sources present in the space, as well as all mechanical aspects of the situation, and answer the following questions:

- Are energy sources or mechanical equipment involved or potentially involved in the incident?

- What effect will the energy source have on your rescue operation?
- Can you control the energy source, or is it adequately controlled already?
- If a victim is trapped in the equipment or energy source, do you have the necessary skills, training, and equipment to rescue them?
- Will you need to call for additional help from a team specifically trained in machinery rescue?

The second question—What is happening now?—is critical to the size-up. If an energy source is involved in the incident, you must determine the following:

- Has the equipment been de-energized? Does it still contain stored energy? Or perhaps no one knows for sure?
- Is this a **beneficial energy source** that must be kept in service to support the rescue? A beneficial energy source is one that does not pose a hazard or for which the hazards are controlled; it can be used for assistance during an emergency (electricity for lights and ventilation equipment, natural gas for generators, etc.).
- Can the energy sources be controlled from outside the space? Or is entry required?
- How is the energy source affecting the victim or the potential for rescue?
- Can you verify that the energy source has been controlled or de-energized and how the control is to be implemented?
- Is this a rescue or a victim recovery?

If it is obvious that the victim is dead from contact with an energy source (e.g., crushing, electrocution, massive thermal burns from steam), your entire operation changes from a rescue to a recovery. Protecting the emergency responders must become paramount, and you may need to take additional steps to protect them. You must compare what is being risked to what you can expect to accomplish. If little can be accomplished (recover), little should be risked. If much is to be accomplished (rescue), how much are you willing to risk?

So what can you predict will happen as this incident plays out? It may seem a little like fortune-telling, but it is not. Your experience and the depth of your knowledge will help you make fairly sound predictions.

If an open electric circuit is present in the confined space, the space is narrow, and your rescuers must enter wearing a self-contained breathing apparatus (SCBA), you can predict, with some certainty, that someone will likely make contact with the circuit and be electrocuted. Similarly, you can expect to find crushing injuries during an assessment of a victim trapped under a heavy object. These are simple predictions based on knowledge and experience, but what happens when you have little or no information? At times, your decisions and course of action may be based on a best guess as to the outcome. You might find that the information you have regarding the emergency leads you to anticipate one outcome, only to have something entirely different transpire.

Interpreting and Documenting Information Appropriately

Interpreting Information Appropriately

In confined-space rescue operations, the ability to interpret information accurately is essential for making informed decisions and ensuring the safety and effectiveness of rescue efforts. This section will discuss the importance of interpreting information appropriately during three key phases of a confined-space incident response: the size-up of the incident, the establishment of a victim profile, and the execution of rescue operations.

Incident Size-Up

The initial size-up of a confined-space incident sets the foundation for the entire rescue operation and begins with the dispatch information (i.e., location, occupancy, etc.). The first question to ask yourself during a size-up is, "What do you see?" During this phase, responders gather critical information to assess the situation and develop a strategic plan. Proper interpretation of available information is crucial for determining the level of risk, identifying hazards, and establishing priorities. The following are the best practices for appropriately interpreting the information gathered during the size-up:

1. **Closely observe the scene.** Responders must carefully observe the confined-space environment and note any visible hazards such as hazardous gases, lack of ventilation, or physical obstructions.
2. **Utilize monitoring equipment.** Deploy air-monitoring equipment to assess atmospheric conditions inside the confined space and interpret the readings the equipment provides to determine the presence of toxic gases, oxygen levels, and potential flammable substances.

3. **Gather an incident history.** Obtain and carefully review information about the nature of the incident, the number of individuals involved, and any previous rescue attempts. This context helps in the comprehension of the urgency and complexity of the situation.
4. **Assess what resources will be needed.** During size-up, responders can determine at least partially what will be needed for the response in terms of human resources and equipment.

The Victim Profile

The creation of a **victim profile**, which is covered in greater depth later in this chapter, involves gathering specific details about the individuals trapped or affected by a confined-space incident. This information assists in tailoring rescue strategies and optimizing resource allocation. The following are the best practices for appropriately interpreting information gathered during the victim profiling process:

1. **Identify victim characteristics.** Determine and assess the number of victims, their locations within the confined space, and their physical condition to help prioritize rescue efforts and allocate resources effectively.
2. **Assess any medical needs.** Gather information on any preexisting medical conditions, injuries, or potential hazards affecting the victims. This influences the choice of rescue techniques and equipment.
3. **Communicate with witnesses.** By interviewing witnesses or colleagues, you can gather and evaluate valuable information about victims' last known whereabouts or activities before the incident.

Rescue Operations

During the execution of rescue operations, accurate interpretation of real-time information is critical for ensuring the safety of both responders and victims. By continuously assessing and adapting the rescue plan based on emergent information, the team is in a better position to navigate evolving challenges. The following are the best practices for interpreting information gathered during the evaluation of rescue operations:

1. **Monitor conditions with vigilance.** By continuously monitoring atmospheric conditions, structural integrity, and any changes in the confined-space environment and interpreting the data yielded, the team can better anticipate potential hazards or obstacles.
2. **Keep communication lines open.** Effective communication channels between incident command and team members will help process incoming information from various sources, coordinate rescue efforts accordingly, and adjust strategies as needed.
3. **Respond to challenges.** Through interpretation of cues from victim responses, careful analysis of feedback from deployed rescue tactics, and adjustment of approaches based on observed outcomes, the team can tailor rescue techniques and provide appropriate support no matter what challenges arise.

Documenting Information Appropriately

Properly securing a confined space starts long before the actual emergency. As previously discussed in this text, having detailed response plans is a crucial element, as is having proper documentation. Part of the documentation includes having a proper confined-space entry plan. This detailed plan and documentation will ensure the safety of individuals entering and working within such spaces.

FIGURE 2-3 shows a template for a confined-space entry plan to secure a confined space. It can be customized to fit the specific needs and requirements of a particular organization and any applicable regulatory standards. All personnel involved in confined-space entry operations must be familiar with this plan and receive appropriate training.

Documenting a Confined-Space Incident

Documenting confined-space incidents can be challenging. **FIGURE 2-4** is a useful guide for documenting all aspects of a confined-space incident.

Methods for Isolating a Confined-Space Incident

Access control—the act of limiting access to a confined-space incident scene—is crucial to ensure the safety of responders, bystanders, and victims. Limiting scene access during a confined-space incident requires a comprehensive approach that combines physical barriers (such as tape or cones) with effective enforcement and coordination with law enforcement agencies. By implementing these access control measures, responders can minimize risks, maintain scene integrity, and ensure the safety of everyone involved in the incident response effort. Implementing effective access control measures involves utilizing lockout/tagout procedures,

Confined-Space Entry Plan

Document Prepared By: [Your Company/Organization]
Date of Preparation: [Date]
Purpose: The purpose of this Confined-Space Entry Plan is to establish guidelines and procedures to ensure the safety of personnel working in or around confined spaces during entry, work, and exit operations.
Scope: This plan applies to all confined spaces within the [Your Company/Organization] premises and outlines the responsibilities, procedures, and precautions necessary for safe entry and work within these spaces.

Key Definitions

Confined Space: A space not designed for continuous occupancy that is large enough and so configured that an employee can bodily enter and perform assigned work, but with limited or restricted means for entry or exit.

- **Open:** There are no obstacles, barriers, or obstructions within the space.
- **Obstructed:** The permit space contains some type of obstruction that a rescuer would need to maneuver around.
- **Elevated:** A permit space where the entrance portal or opening is above grade by 4 feet or more.
- **Nonelevated:** A permit space where the entrance portal or opening is located less than 4 feet above grade.
- **Restricted:** A portal of 24 inches or less in the least dimension.
- **Unrestricted:** A portal size greater than 24 inches in the least dimension.

Authorized Entrant: An employee who is authorized by an employer to enter a confined space

Attendant (or Entry Supervisor): An individual stationed outside a confined space who monitors entrants and ensures their safety

Permit-Required Confined Space: A confined space with one or more of the following characteristics:

- Contains or has the potential to contain a hazardous atmosphere
- Contains a material that has the potential for engulfing an entrant
- Has an internal configuration such that an entrant could be trapped or asphyxiated

Responsibilities of Key Stakeholders

Employer/Management:

- Provide training on confined-space hazards and procedures
- Implement the Confined-Space Entry Plan
- Ensure that appropriate entry permits are issued and followed
- Provide necessary equipment and resources for safe entry

Supervisors:

- Oversee confined-space entry operations
- Ensure that entry procedures are followed
- Conduct pre-entry briefings and post-entry debriefings

Authorized Entrants:

- Follow all entry procedures and safety guidelines
- Use provided PPE and safety devices

Attendants:

- Monitor entrants during entry and work
- Implement emergency procedures if needed
- Maintain communication with entrants and remain at the space (entrance)

Confined-Space Entry Procedures

Pre-Entry Procedures:

- Identify and evaluate confined spaces
- Obtain necessary permits
- Conduct atmospheric testing
- Provide required equipment and PPE
- Lockout/tagout as necessary
- Entry Supervisor to conduct safety check of PPE, communications, and air supply of entrants prior to entry

Entry Operations:

- Perform work as planned
- Maintain constant communication
- Monitor conditions inside the space

Emergency Procedures:

- Establish emergency rescue procedures
- Implement rescue drills and training

Training and Documentation

- Provide confined-space training to all employees involved in entry operations
- Maintain records of training, permits, atmospheric tests, and entry operations

Review and Revision

This Confined-Space Entry Plan will be reviewed annually and updated as necessary to reflect changes in equipment, procedures, regulations, or personnel.

Signatures

I acknowledge that I have read and understood the requirements outlined in this Confined-Space Entry Plan.

Signature: ______________________________

Date: ______________________________

FIGURE 2-3 A sample confined-space entry plan.

Category	Details	Information
Incident Information	Incident Date/Time	[Exact date and time the incident occurred]
	Location	[Address, facility name, etc.]
	Space Description	[Description of the confined space (type, size, entry points, etc.)]
	Incident Description	[Detailed account of the incident (what happened, how it happened, etc.)]
		[Identification of hazards involved (chemical, physical, atmospheric, etc.)]
		[Initial observations upon arrival]
Personnel Information	Incident Commander	[Name and contact information]
		[Role and responsibilities]
	Rescue Team Members	[Names and contact information of all team members]
		[Specific roles and responsibilities during the incident]
	Safety Officer	[Name and contact information]
		[Role and responsibilities]
Pre-Incident Information	Confined-Space Permit	[Copy of the entry permit, if applicable]
		[Details of pre-entry procedures and safety checks]
	Risk Assessment	[Summary of risk assessment conducted prior to entry]
		[Identified hazards and mitigation measures]
Incident Response Actions	Initial Response	[Time of notification and arrival on scene]
		[Initial actions taken (e.g., hazard assessment, securing the area, etc.)]
	Rescue Operations	[Description of rescue procedures used]
		[Timeline of rescue actions (start and end times)]
		[Equipment used during the rescue]
	Atmospheric Monitoring	[Results of atmospheric testing (e.g., oxygen levels, toxic gases, etc.)]
		[Times and locations of readings]
Victim Information	Identification	[Names and identification details of victims]
	Condition of Victims	[Initial condition upon discovery]
		[Description of injuries or symptoms]
	Medical Treatment	[First aid or medical treatment provided on-site]
		[Time and details of treatment]
Victim Transfer Information	Transfer Details	[Time of transfer from confined space to medical facility]
		[Method of transfer (e.g., ambulance, airlift, etc.)]
		[Name and contact information of medical personnel involved]
	Receiving Facility	[Name and location of the medical facility where victims were taken]
		[Contact information of the facility]
		[Admission details and initial assessment by medical personnel]
Postincident Actions	Debriefing	[Summary of debriefing session with rescue team]
		[Key points discussed and lessons learned]

FIGURE 2-4 A confined-space incident documentation guide.

Postincident Actions	**Incident Investigation**	*[Steps taken for incident investigation]*
		[Findings and conclusions]
		[Recommendations for preventing future incidents]
	Reporting	*[Completion and submission of required incident reports]*
		[Notifications made to relevant authorities (e.g., OSHA, local regulatory bodies)]
Additional Documentation	**Photographs and Videos**	*[Visual documentation of the incident scene]*
		[Photos of confined-space entry points, hazards, and rescue operations]
	Witness Statements	*[Statements from witnesses and involved personnel]*
	Communication Logs	*[Records of all communications during the incident (radio, phone, etc.)]*
	Equipment Logs	*[Documentation of equipment used, including maintenance and calibration records]*
Conclusion	**Summary of Incident**	*[Overall summary of the incident and response actions]*
	Evaluation	*[Evaluation of the response's effectiveness]*
		[Recommendations for improvements and training needs]

FIGURE 2-4 *(Continued)*

TABLE 2-4 Best Practices for Access Control

Communication channels	Establish clear communication channels with all involved parties, including emergency responders, incident commanders (ICs), and law enforcement personnel
Continuous monitoring	Regularly monitor the scene perimeter to ensure that barriers remain intact and unauthorized access is prevented
Flexibility first	Adapt access control measures based on changing conditions or evolving incident dynamics
Clear signage	Supplement physical barriers with visible signage that communicates the reason for restricted access and the potential dangers associated with the confined-space incident

TIP

Identify all potential energy sources, determine whether they are hazardous energy sources or beneficial energy sources, and take control of them. Beneficial energy sources, such as this electrical box, should not be allowed to pose a hazard during the incident. It may be necessary to take steps to control the hazards so that the energy can be safely used.

© BanksPhotos/iStock/Getty Images Plus/Getty Images

barrier tape, and cones and potentially involving law enforcement personnel as well. See **TABLE 2-4** for a summary of best practices for access control.

Lockout/Tagout

Lockout/tagout is the process of eliminating and securing sources of potential physical or chemical hazards by isolating them from a confined space and/or controlling flowable materials in the space (**FIGURE 2-5**), as noted in Chapter 1. You will learn more about lockout/tagout in Chapter 8.

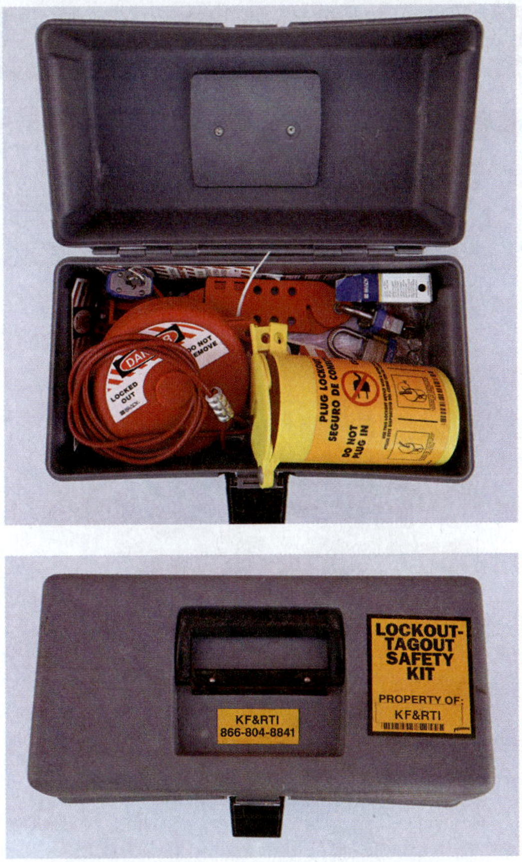

FIGURE 2-5 A lockout/tagout kit is utilized to ensure safety during confined-space operations by preventing accidental equipment activation.

> **SAFETY TIP**
>
> As part of your confined-space response equipment, you should carry basic lockout/tagout equipment.

Barrier Tape

Barrier tape is a highly visible and effective tool for cordoning off restricted areas (**FIGURE 2-6**). It serves as a physical and visual barrier, warning individuals to stay away from an incident scene. Barrier tape can do the following:

- **Define the perimeter.** If the affected area is surrounded with barrier tape, the boundaries of the restricted zone will be clearly marked.
- **Communicate danger.** Tape can be customized with appropriate warning messages (e.g., "Danger—Confined-Space Incident") to convey the seriousness of the situation.
- **Prevent unauthorized entry.** Barrier tape securely attached to stable objects (e.g., cones, fences, or equipment) is an effective method of deterring unauthorized access.

FIGURE 2-6 Barrier tape can be utilized for marking off hazardous areas during confined-space operations to ensure safety and restricted access.

Cones

Conventional traffic cones are another effective means of controlling access and directing traffic away from hazardous areas while guiding its flow. Cones should be positioned to redirect pedestrian or vehicular traffic away from the scene and at a safe distance.

Cones can create a physical barrier when they are placed strategically along a scene perimeter, coupled with barrier tape, to physically block access routes. Cones should be brightly colored, providing a visual cue to attract attention and signal that an area is off-limits.

Law Enforcement

In situations where additional enforcement of access control is necessary, law enforcement personnel may be needed to assist with crowd management and enforcement of restricted access. Collaboration with law enforcement agencies can provide much-needed support in coordinating access control efforts and ensuring compliance with safety directives.

Officers can be deployed to monitor the scene perimeter and prevent unauthorized individuals from entering a restricted area. They can also direct vehicular and pedestrian traffic and maintain order around an incident scene—support that is particularly important in cases where the scene is located in a busy or congested area.

Steps for Securing a Confined-Space Incident

Securing a confined space and managing an incident within it requires a well-defined set of steps to ensure the safety of the individuals involved. To better understand the process of securing and mitigating access to a confined-space incident, follow the steps in **SKILL DRILL 2-1**.

SKILL DRILL 2-1

Demonstrate Securing and Mitigating Access to a Confined-Space Incident

Step 1: Evaluate the Situation

Perform an initial assessment of the scene, including examining the area for hazards.

Step 2: Establish Communication

Use a radio to communicate with control or other on-site teams.

Step 3: Implement Rescue Procedures

Start the rescue process by gathering the necessary equipment.

Step 4: Ensure Safety Precautions

Carefully check and use appropriate PPE for each team member.

SKILL DRILL 2-1 CONTINUED

Demonstrate Securing and Mitigating Access to a Confined-Space Incident

Step 5: Coordinate Evacuation and Assistance

Have a medical bag and team member(s) on standby near the evacuation area.

Step 6: Conduct a Postincident Assessment

Evaluate the scene postincident for any overlooked hazards or equipment damage.

Document observations and complete all reports.

Continues.

SKILL DRILL 2-1 CONTINUED

Demonstrate Securing and Mitigating Access to a Confined-Space Incident

Step 7: Review and Update Procedures

Meet with the team to discuss and review incident procedures and identify areas for improvement.

During the documentation phase, make updates to standard procedures based on lessons learned from the incident.

Step 8: Communicate Findings

Hold a debrief session with involved personnel, gathering feedback and discussing improvements.

SKILL DRILL 2-1 CONTINUED

Demonstrate Securing and Mitigating Access to a Confined-Space Incident

Step 9: Ensure Compliance with Regulations

The safety officer should verify that all actions taken during the incident comply with safety regulations.

Step 1: Evaluate the Situation

- **Assess the nature of the incident.** Determine the type of emergency or incident within the confined space (e.g., hazardous atmosphere, trapped personnel, or equipment failure).
- **Consider the risks and hazards involved in the confined space.** Identify potential dangers such as toxic gases, lack of oxygen, or physical obstructions.
- Consider the risk-versus-benefit and recovery-versus-rescue equations.

Step 2: Establish Communication

- **Ensure clear communication with personnel involved in the incident.** Use established communication protocols such as radios, signaling systems, or visual cues.
- **Establish contact with emergency services if needed.** Provide them with specific details about the confined space and the location and nature of the incident.
- **Use alternate forms of signaling if needed.** In many circumstances, verbal communication is not possible or is severely hampered by noise or other factors. One form of alternate communication is made via tugging on a rope; this method is commonly referred to as **OATH** (O for okay = one tug; A for advance = two tugs; T for take up slack = three tugs; H for help = four tugs). Another form, using air horns as signals, is commonly referred to by the acronym **SUDRH** (S for stop = one short blast; U for up = two short blasts; D for down = three short blasts; R for rope free = four short blasts; and H for help = one long blast).

Step 3: Implement Rescue Procedures

- **Activate confined-space rescue procedures.** Follow established protocols for confined-space rescue, including deploying trained rescue personnel.
- **Use appropriate equipment for rescue operations.** Ensure rescue teams have access to all necessary tools, such as harnesses, ropes, and breathing apparatus.

Step 4: Ensure Safety Precautions

- **Closely monitor atmospheric conditions.** Continuously monitor air quality within the confined space to ensure that it remains safe for entry and rescue operations.
- **Use personal protective equipment (PPE) diligently.** Ensure all personnel involved have appropriate PPE for the specific hazards present.

Step 5: Coordinate Evacuation and Assistance

- **Evacuate all nonessential personnel from the area.** Clear the surrounding area of unnecessary personnel to facilitate rescue operations.
- **Coordinate with medical responders if injuries are involved.** Provide medical information and prepare for immediate medical assistance upon completion of the rescue.

Step 6: Conduct a Postincident Assessment

- **Evaluate the effectiveness of the response.** Review the incident to identify any areas for improvement in procedures, training, or equipment.
- **Document the incident.** Maintain detailed records of the incident, including actions taken, personnel involved, and outcomes. Indicate the response mode (rescue or recovery) and the strategy (entry or nonentry) involved.

Step 7: Review and Update Procedures

- **Review the confined-space incident response plan.** Identify lessons learned and update procedures accordingly to enhance future response capabilities.
- **Provide additional training and resources.** Address any gaps in training or equipment needs identified during the incident.

Step 8: Communicate Findings

- **Share findings and recommendations with relevant stakeholders.** Inform management, safety teams, and personnel about the incident and the resulting improvements made.
- **Conduct debrief sessions.** Gather feedback from involved personnel to enhance response procedures and ensure continuous improvement.

Step 9: Ensure Compliance with Regulations

- **Verify compliance with relevant safety regulations and standards.** Ensure that all actions taken during the incident are aligned with legal requirements and industry best practices.

Search Parameters for a Confined-Space Incident

Search parameters for an incident should be defined before a confined-space technical rescue team enters the confined space. When establishing the search parameters, a logical sequence of steps should be followed. Each step will be covered a bit more in-depth in the following discussion, but the most critical steps are the safety briefing, the assessment of entry and access points, and the performance of a survivability profile.

Step 1: Conduct a Safety Briefing

During a **safety briefing**, you will go over the safety procedures for entry into the confined space. You will ensure that all involved understand the standard safety protocols and procedures required before entering.

To complete the safety briefing, you will go over the confined space's entry checklist (see Chapter 1) and review training resources and materials related to confined-space safety, including hazard identification and emergency response.

Step 2: Assessment of Entry and Access Points

Careful assessment of the entry and access points for a confined space is critical to managing safety during an incident response. For this process, you will investigate optimal designs for entry points to ensure easy access to and emergency egress from the confined space; discuss and decide upon which equipment will be necessary for safe entry and egress, such as harnesses (**FIGURE 2-7A**), atmospheric monitors (**FIGURE 2-7B**), and communication devices; and review all relevant permits used to control entry, including authorization processes and permit requirements (29 C.F.R. 1910.146 1993). To better understand the process of choosing entrance paths, follow the steps in **SKILL DRILL 2-2**.

Step 3: Perform a Survivability Profile

To perform a survivability profile of the confined space, you will need to assess the methods and devices you will use for monitoring air quality, including detection of toxic gases and oxygen levels' flammability; determine how temperature and humidity levels will affect survivability within the space; decide upon the tools and techniques you will use to identify and assess potential hazards; determine which strategies and protocols for

A. B.

FIGURE 2-7 Equipment that might be necessary for safe entry into and egress from a confined space can include **(A)** a Class 3 rescue harness or **(B)** a four-gas atmospheric monitor.

A: Courtesy of CMC Rescue, Inc; B: © Jones & Bartlett Learning. Courtesy of MIEMSS.

emergency response and rescue operations you will deploy; identify which key factors will contribute to survivability or pose risks to individuals; and go over all necessary industry standards and guidelines.

TIP

A **survivability profile** is an attempt to gain a clear understanding of the threats to health and safety within a confined space, followed by an analysis of what procedures and actions can be made to mitigate the threats and, in effect, make the confined-space entry survivable.

Factors affecting survivability in a confined space include environmental and physical hazards, air quality, availability of resources, temperature and humidity levels, and vulnerabilities in the space's structure and its entry and exit points.

Step 4: Consider Case Studies and Best Practices

If time allows, it is best to review real-life incidents involving confined spaces, including lessons learned and best practices, and pause to consider the recommended practices and strategies for ensuring safety during confined-space entry and operations.

Communicating with Confined-Space Incident Victims

In any confined-space incident, effective communication is critical for ensuring the safety and well-being of both conscious and unconscious victims. When dealing with victims who are conscious, the rescuer's approach must be empathetic, clear, and focused on gaining essential information—all while providing reassurance.

Communicating with other rescue or job-site personnel when victims are unconscious requires coordination, clarity, and a structured approach to ensure a coordinated rescue effort. In the industrial setting, communication by voice or radio could be hampered by the ambient noise of the space. Alternative communication procedures, such as hand signals or hard-wired communication systems, may be necessary. Read on to discover some best practices for communication in both situations.

Communication with Conscious Victims

First, you must assess the situation. Before initiating communication with conscious victims in the space, it is critical to ensure your own safety and that of the other rescuers and assess the immediate surroundings. Identify any hazards that could endanger you or the victim.

SKILL DRILL 2-2

Choose Entrance Paths

Step 1: Conduct an Initial Assessment

Assess the entry area and surrounding environment for overall risks and determine the site conditions.

Step 2: Identify Hazards

Identify any specific hazards at the scene, such as unstable structures, the presence of toxic gases, or sharp objects.

Identify and mark any hazards (such as loose debris or particularly confined areas of the space) for awareness.

SKILL DRILL 2-2 CONTINUED

Choose Entrance Paths

Step 3: Evaluate Entry Points

Evaluate potential entry points for their structural stability, accessibility, and any obstacles.

Step 4: Select Primary and Secondary Paths

Map out primary and secondary paths for entry and evacuation.

Discuss secondary (alternative) entry paths and mark them for easy identification.

Continues.

SKILL DRILL 2-2 CONTINUED

Choose Entrance Paths

Step 5: Prepare for Entry

Prepare entry equipment, checking ropes, lighting, and PPE.

Don safety gear and check to make sure fellow team members' equipment is properly set up.

Step 6: Monitor and Adjust Entry Paths

Monitor the entry as the team begins the rescue to ensure all paths are clear and safe.

Stay in continuous communication, making adjustments to entry paths as necessary based on real-time assessments.

Next, you must verbally approach conscious victims with caution. Introduce yourself calmly and clearly. Speak slowly and avoid using technical jargon. Maintaining a respectful and nonthreatening demeanor is essential to build trust.

As you speak with victims, be sure to provide constant and consistent reassurance. Victims in confined spaces may be frightened or disoriented. Reassuring them that help is on the way and that you are there to assist will help encourage them to stay calm and conserve their energy.

While speaking with victims, carefully gather essential information by asking them simple, direct questions. Examples include the following:

- "Can you tell me your name?"
- "Do you know what happened?"
- "Are you injured?"
- "Do you feel any pain or discomfort?"

After you have gathered the most essential information, carefully explain the rescue process and your rescue plan. Describe the steps being taken to rescue them in detail. Let them know what they can expect next and approximate time frames for each element of the rescue. Reiterate constantly that their safety is your top priority.

As the rescue progresses, take care to maintain open communication. Keep victims informed of any developments or changes in the rescue plan. Listen actively to their concerns and address them promptly.

Last, always remember to perform reassessments of the situation regularly and consistently. This includes close monitoring of the victim's condition and may involve adjusting your approach accordingly. Providing conscious victims with regular updates on your progress will help keep them calm and will work to maintain both their continued cooperation and a sense of hope.

Communication with Other Personnel on Behalf of Unconscious Victims

A first step when organizing your communication strategy with other onsite personnel when an incident's victims are unconscious is to establish clear protocols. Ensure that a predefined communication plan is in place for dealing with unconscious victims, and designate a team leader who will be responsible for coordinating communications.

A big part of effective communication with other onsite personnel is the efficient coordination of resources. If all available rescue resources—including medical teams, technical rescue teams, and incident command—are coordinated effectively through continuous communication, the incident response will go much more smoothly.

FIGURE 2-8 A visual representation of the communication chain, illustrating the flow of information and coordination between team members during operations. If one link in the chain is broken, then a communication breakdown can occur.

When communicating with other personnel, never assume that they will be well versed in jargon or other language associated with rescue. Use standardized, clear terminology when conveying critical information to others. Keep your messages concise and focused on essential details. Following are a few examples of clear messaging:

- "Victim located. Unconscious."
- "Standby for medical assessment information."
- "Patient is unconscious but breathing on their own."
- "No further injuries noted."
- "Patient is packaged and ready for retrieval."

As you communicate with others involved in the rescue operation, take time to confirm that they understand you clearly, and always remember: Communication is a two-way street (**FIGURE 2-8**). For example, when you are ready to start entry into the space, you could say, "Entry team, confirm readiness." A simple message like "Medical team, do you copy?" when communicating with medical support provides assurance that your messages are heard and understood.

As the rescue operation progresses, provide frequent updates on victim status, rescue progress, and any changes in conditions within the confined space to the rest of the team. Use established visual or audible signals that you know will be understood to convey urgent messages or any changes to the rescue plan.

Be sure to maintain detailed records of all communications related to an incident because this documentation will be crucial for postincident analysis and improvement of rescue protocols.

Establishing a Victim Profile

Creating victim profiles is an essential way to gain a full understanding of the individuals involved, their roles, and their potential needs. This section explores the concept of victim profiles and describes when and how to develop them, as well as how the information they contain can aid in the rescue process.

Understanding Victim Profiles

In the context of a confined-space incident, a *victim profile* refers to a detailed summary of individuals affected by the incident. It includes pertinent information such as their identity, role or function within the confined space, health status, and any special considerations or vulnerabilities.

The victim's identity information can include their name, age, contact information, and identifying characteristics. Role information may include the victim's job title and responsibilities, as well as the tasks being performed at the time of the incident. Regarding health status, preexisting medical conditions, allergies, or special health considerations are all important factors. Special considerations or vulnerabilities could include lists of potential language barriers, mobility issues, or other factors that may affect rescue efforts.

When to Develop Victim Profiles and Put Them to Use

Victim profiles should be initiated as soon as information about the incident becomes available. Although they are particularly useful during the initial phases of rescue planning, they are also vital throughout a rescue operation. Early in the incident response, profiles provide critical information that can speed up or otherwise aid the process. As the situation evolves, profiles are continually updated and used to help with decision-making, resource allocation, and communication with rescue teams.

Potential Information Sources for Victim Profiles

Information about victims can be obtained from various sources, including the following:

- **Emergency contacts.** Family members, employers, or colleagues are all excellent sources for gathering background information on victims.
- **Work records.** Employment records and personnel files may contain details about job roles and qualifications.
- **Medical records.** Health records, including preexisting conditions, allergies, or medical considerations, may be accessible by close contacts of victims.
- **Colleague interviews.** Interviews with co-workers or individuals familiar with the victim's work and responsibilities can provide information that will be critical to a successful rescue.

Utilizing Victim Profiles in Rescue Operations

Victim profiles serve as valuable tools for informing rescue strategies and operations. Profiles help prioritize resources and personnel based on the specific needs of individuals involved, and they inform communication strategies, ensuring that relevant information is shared with appropriate stakeholders. Health-related information in victim profiles guides medical interventions and treatment plans. Last, profiles highlight any special considerations or hazards associated with individual victims, enhancing safety protocols during rescue operations.

How HIPAA Laws Affect the Medical Information of Victims

In the context of confined-space incidents and emergency response, the **Health Insurance Portability and Accountability Act of 1996 (HIPAA)** plays a crucial role in protecting the privacy and confidentiality of individuals' medical information, including victims involved in such incidents. HIPAA is a federal law that sets standards for the protection of individuals' health information, known as protected health information (PHI). The law applies to covered entities, such as healthcare providers, health plans, and healthcare clearinghouses, as well as their business associates who have access to PHI. This section provides important information about HIPAA laws and how they apply to the handling of medical information related to victims of confined-space incidents.

In the context of a confined-space incident, the following HIPAA rules are particularly relevant:

- **Privacy rule.** This rule establishes national standards to protect individuals' medical records and other PHI. It regulates the use and disclosure of PHI by covered entities.
- **Security rule.** This rule sets standards for the security of electronic protected health information (ePHI) and requires covered entities to implement safeguards to protect the confidentiality, integrity, and availability of PHI.

When dealing with medical information of victims in confined-space incidents, it is important to adhere to HIPAA regulations. Generally, covered entities (individuals and organizations that are subject to following HIPAA rules) must obtain written authorization from the individual or their authorized representative before disclosing PHI. HIPAA permits covered entities to disclose PHI to first responders, including emergency medical personnel and law enforcement, without authorization if necessary to provide treatment or protect public health and safety. Covered entities should only disclose the minimum amount of PHI necessary for the intended purpose.

Compliance with HIPAA in Confined-Space Rescues

Emergency responders may need access to limited medical information (e.g., known allergies, medical conditions) to provide appropriate treatment during rescue operations. This information can often be found on file at the place of employment. Rescuers must maintain confidentiality and security when handling medical information, ensuring that PHI is not improperly accessed, used, or disclosed.

Failure to comply with HIPAA regulations can result in significant penalties, including fines and legal consequences. It is essential for organizations and individuals involved in confined-space rescues to understand and adhere to HIPAA requirements when handling the medical information of victims. The full rescue team must be respectful of victims' rights and maintain privacy and confidentiality when collecting and utilizing information for victim profiles. This includes obtaining prior consent and ensuring that sensitive information is handled with care and discretion.

Interviewing Bystanders

Conducting interviews with bystanders who have witnessed or were present during a confined-space incident is a critical component of the information-gathering process and will help you gain insight into the incident. Bystander interviews can provide firsthand accounts, observations, and details that will directly contribute to your understanding of the sequence of events, identification of potential hazards, and improvement of your rescue procedures.

Bystander interviews play a vital role in incident investigation and analysis, too, because the perspectives they offer will complement your formal reports and documentation. Key reasons to conduct bystander interviews include the following:

- **Obtaining immediate accounts.** Bystanders may provide real-time details and observations that are fresh and accurate.
- **Identifying critical information.** Bystanders may have witnessed specific actions, conditions, or events leading up to the incident that will be valuable for understanding the incident and planning the rescue.
- **Identifying potential hazards.** Bystander accounts can help identify underlying hazards or risks associated with the confined-space and rescue operations.
- **Improving rescue procedures.** Insights bystanders can provide may inform improvements in your rescue protocols, training, and preparedness.
- **Capture contact information.** Be sure to get bystanders' contact information so that you can reach out for follow-up questions or ask if they recall additional details.

The following are key considerations and guidelines for conducting effective interviews with bystanders during confined-space incidents.

Guidelines for Conducting Interviews

When conducting interviews with bystanders of a confined-space rescue event, the following guidelines will help you ensure both effectiveness in your information gathering and sensitivity to the subjects you are interviewing:

- **Establish a rapport.** Approach the bystanders respectfully and work to establish a rapport with them. The goal should be to create a comfortable environment for the mutual sharing of information.
- **Ask open-ended questions.** Open-ended questions generally provide much more detailed responses. Prompt witnesses to describe what they saw, heard, and experienced, and remind them to be as detailed as possible.
- **Be an active listener.** Pay close attention to bystanders' accounts without interrupting. By allowing them to talk freely and without interruption, you will benefit from more cogent and understandable narratives.
- **Carefully clarify any details.** Seek clarification on specific details that will be of significant benefit to the rescue operation, such as the timing of the incident, a record of all actions taken to that point, and any observations made during the incident.
- **Meticulously document the interview.** Take thorough notes to accurately capture details and statements. Recording interviews is always a good idea,

but you must be sure to request and receive consent before recording as dictated by your state or the authority having jurisdiction (AHJ).

- **Stay respectful.** Remember that bystanders' responses to your questions are likely to be emotional. Express concern about witnesses' privacy and describe your sensitivity to it when discussing the incident with them. Avoid any unnecessary probing or intrusive questions.

Key Interview Questions

Your interview questions should be tailored to gather only relevant information from bystanders and to deter information overload—bystanders are likely to be excited and upset and may have a tendency to provide more information than is necessary. As a rescue team, you can develop a set of interview questions to have ready when an incident occurs. Sample questions may include the following:

- Can you describe what you observed leading up to the rescue operation?
- What actions did you witness during the incident?
- Did you notice any hazards or safety concerns in or around the confined space?
- Is there any additional information or detail you believe is important for us to know about the incident or the space?

Analyzing and Using Bystander Information

After conducting interviews, it is important to take time to analyze the information you have gathered to extract key insights and information that might be critical to the rescue operation. After the incident and rescue have concluded, it is important to rapidly integrate information from bystander accounts into incident reports, debriefings, and postincident reviews. This input will help inform improvements in rescue procedures and training.

Always prioritize ethical considerations when interviewing bystanders. It must be clear throughout the interview process that you are careful to respect their privacy, emotional well-being, and rights. In addition to obtaining consent for recording interviews, be sure to reassure your interview subjects that you will handle sensitive information they provide with confidentiality and discretion—and follow through on those reassurances.

CHAPTER FEATURE: AFTER-ACTION

IN SUMMARY

- It is essential to understand the key responsibilities, strategies, and considerations for incident command personnel in accordance with NFPA 1006 standards specific to incident command for confined-space rescue operations (including competency requirements and operational guidelines) and other relevant NFPA requirements for ICS implementation.
- An accurate and in-depth size-up will help you to focus on the most critical issues upfront. A size-up involves identifying the hazards of a confined space, how you will control those hazards, and the resources you will use during your response.
- Interpreting and documenting information appropriately is essential for making informed decisions and ensuring the safety and effectiveness of rescue efforts.
- Methods for isolating a confined-space incident include lockout/tagout procedures, barrier tape, cones, and engagement of law enforcement.
- The nine steps for securing a confined-space incident are as follows:
  - Evaluate the situation.
  - Establish communication.
  - Implement rescue procedures.
  - Ensure safety precautions.
  - Coordinate evacuation and assistance.
  - Conduct a postincident assessment.
  - Review and update procedures.
  - Communicate findings.
  - Ensure compliance with regulations.
- The four steps to establish search parameters for a confined-space incident are as follows:
  - Conduct a safety briefing.
  - Assess entry and access points.
  - Perform a survivability profile.
  - Consider case studies and best practices.

- Isolating and securing a confined-space incident site is necessary to minimize risks, maintain scene integrity, and ensure the safety of everyone involved.
- Victim profiles are detailed summaries of individuals affected by a confined-space incident, including their identity, role or function, health status, and other special considerations about the individuals.
- Bystander interviews can provide firsthand accounts, observations, and details that will directly contribute to your understanding of the sequence of events, identification of potential hazards, and improvement of your rescue procedures.

KEY TERMS

beneficial energy source An energy source, the presence of which does not pose a hazard or the hazards are controlled, that is of assistance during an emergency (electricity for lights and ventilation equipment, natural gas for generators, etc.).

Federal Emergency Management Agency (FEMA) An agency of the U.S. federal government with a primary purpose of coordinating responses to disasters that overwhelm the resources of local and state authorities.

hazard and risk assessment An assessment that determines what could possibly cause harm and the degree of likelihood that harm will be caused.

Health Insurance Portability and Accountability Act of 1996 (HIPAA) A federal law that sets standards for the protection of individuals' health information, known as *protected health information* (PHI).

incident action plan (IAP) A plan developed by the incident commander (IC) that establishes goals and objectives for the emergency, identifies the resources to be used, and provides the means to accomplish the goals and objectives.

incident command The overall system or framework used to manage emergency response operations, ensuring effective coordination, communication, and resource management.

Incident Command System (ICS) A recognized system for providing management of personnel, resources, and activities during emergency operations by defining roles and responsibilities and standard operating procedures.

incident commander (IC) The individual responsible for overseeing the entire incident response, making critical decisions, developing objectives, and ensuring the safety and effectiveness of the operation.

OATH A method of silent communication involving tugs on a rope (O for okay = one tug; A for advance = two tugs; T for take up slack = three tugs; H for help = four tugs).

safety briefing A briefing in which the safety procedures for entry into a confined space are discussed.

standard operating procedures (SOPs) Written guidelines for handling a specific situation.

SUDRH A method of communication involving signals using air horns (S for stop = one short blast; U for up = two short blasts; D for down = three short blasts; R for rope free = four short blasts; and H for help = one long blast).

survivability profile An attempt to gain a clear understanding of the threats to health and safety within a confined space, followed by an analysis of what procedures and actions can be made to mitigate the threats and, in effect, make the confined-space entry survivable.

victim profile Detailed summaries of individuals affected by a confined-space incident, including their identity, role or function, health status, and other special considerations.

REFERENCES

Health Insurance Portability and Accountability Act. Pub. L. No. 104-191, § 264, 110 Stat. 1936.

National Fire Protection Association. 2021. *NFPA 1006: Standard on Technical Rescue Personnel Professional Qualifications, 2021 Edition.* Quincy, MA: National Fire Protection Association.

National Fire Protection Association. 2022a. *NFPA 350: Guide for Safe Confined Space Entry and Work, 2022 Edition.* Quincy, MA: National Fire Protection Association.

National Fire Protection Association. 2022b. *NFPA 2500: Standard for Operations and Training for Technical Search and Rescue Incidents and Life Safety Rope and Equipment for Emergency Services, 2022 Edition.* Quincy, MA: National Fire Protection Association.

"Permit-Required Confined Spaces." 1993. 29 C.F.R. 1910.146. https://www.ecfr.gov/current/title-29/part-1910/section-1910.146.

CHAPTER QUESTIONS

1. What does an incident size-up begin with?
   a. the type of confined space
   b. the number of resources required
   c. the dispatch information
   d. the rescue or recovery decision
2. Which of the following is a method for isolating a confined-space incident?
   a. using barrier tape and cones
   b. shutting down access roads
   c. restricting local law enforcement access
   d. sending out a public service announcement
3. Which one of the following is a potential information source for a victim profile?
   a. talking to emergency contacts
   b. doing an internet search
   c. checking the victim's phone
   d. talking to command staff
4. It is important to integrate information from bystander accounts into
   a. incident reports, IAPs, and postincident reviews.
   b. incident reports, debriefings, and postincident logs.
   c. incident reports, debriefings, and personal notes.
   d. incident reports, IAPs, and personal notes.

CHAPTER FEATURE: ON SCENE

A TANK RESCUE

At a confined-space rescue event, you have to be prepared for anything. Responders to a water tank rescue that occurred in the summer of 2016 were ready for the challenge.

In this incident, a worker employed by a contractor was performing maintenance inside a large storage tank at a water pollution control plant. The worker entered the tank to conduct routine maintenance activities but soon collapsed and lost consciousness as a result of oxygen deficiency.

The Emergency Response

Fellow workers immediately noticed that the victim was in distress and called for help. The fire department responded to the emergency with specialized confined-space rescue teams. Upon arrival, the rescue teams assessed the situation and determined that the victim was unconscious inside the oxygen-deficient tank.

The Rescue Operations

To safely extract the worker, the fire department utilized specialized retrieval systems, including harnesses and rope systems, to lower rescue personnel into the tank. Ventilation equipment was deployed to improve the air quality inside the confined space and restore oxygen levels.

The rescuers carefully secured and stabilized the unconscious worker using safety harnesses and retrieval lines. Additional equipment, such as atmospheric monitoring devices, was used to continuously assess the air quality inside the tank during the rescue operation.

The Extrication and Treatment

Once the worker was successfully extracted from the tank, emergency medical personnel provided immediate medical attention onsite. The worker was transported to a nearby hospital for further evaluation and treatment.

The Outcome

Despite challenging conditions, the coordinated efforts of the fire department's confined-space rescue teams resulted in the successful rescue of the worker. This incident highlights the importance of proper training, equipment, and adherence to safety protocols in confined-space rescue operations.

This real-life tank entry rescue also underscores the inherent risks associated with confined-space work and emphasizes the critical role of emergency responders in ensuring the safety and well-being of workers in such environments. The incident also served as a learning opportunity for improving safety practices and emergency response procedures in similar scenarios.

1. What was the primary cause of the worker's collapse inside the storage tank?
   - **a.** toxic fumes
   - **b.** oxygen deficiency
   - **c.** overheating
   - **d.** dehydration
2. What was the first action taken by the worker's colleagues after noticing the victim was in distress?
   - **a.** attempting a rescue with proper equipment
   - **b.** administering first aid
   - **c.** calling for help from the fire department
   - **d.** ventilating the tank immediately
3. What specialized equipment did the fire department use to lower rescue personnel into the tank?
   - **a.** cranes and pulleys
   - **b.** harnesses and rope systems
   - **c.** ladders and safety nets
   - **d.** scaffolding and hoists
4. What was the primary purpose of deploying ventilation equipment during the rescue operation?
   - **a.** to cool the interior of the tank
   - **b.** to remove toxic gases from the environment
   - **c.** to improve air quality by restoring oxygen levels
   - **d.** to dry out the tank's interior
5. Which type of equipment was used during the rescue operation to continuously monitor the air quality inside the tank?
   - **a.** a carbon monoxide detector
   - **b.** an atmospheric monitoring device
   - **c.** a retrieval line
   - **d.** an oxygen deficiency monitor

CHAPTER

3

Awareness

Safety Measures and Victim Retrieval

KNOWLEDGE OBJECTIVES

After studying this chapter, you should be able to:

- Indicate common safety measures used in confined-space incidents. (**NFPA 1006: 7.1.6**, pp. 62–82)
- Identify confined-space protocols related to safety, communication, and medical procedures. (**NFPA 1006: 7.1.4**, pp. 62–82)
- Discuss the need for rehabilitation during confined-space incidents. (**NFPA 1006: 7.1.7**, pp. 82–83)
- Describe fall prevention techniques. (**NFPA 1006: 7.1.4**, pp. 84–87)
- Explain the process used for retrieving a victim. (**NFPA 1006: 7.1.4**, pp. 87–89)

SKILL OBJECTIVES

- Establish safety measures. (**NFPA 1006: 7.1.6**, pp. 62–82)
- Implement all needed safety and communication protocols. (**NFPA 1006: 7.1.4**, pp. 62–82)
- Don applicable personal protective equipment (PPE). (**NFPA 1006: 7.1.4**, pp. 66–69)
- Construct an escape plan from the area of a confined-space incident. (**NFPA 1006: 7.1.2**, pp. 63–65)
- Choose communication devices that can be utilized anywhere in a confined-space incident. (**NFPA 1006: 7.1.3**, pp. 69–71)
- Demonstrate the ability to operate in and exit adjacent areas to a confined space. (**NFPA 1006: 7.1.2**, pp. 80–82)
- Demonstrate fall prevention methods. (**NFPA 1006: 7.1.4**, pp. 84–87)
- Utilize nonentry retrieval systems. (**NFPA 1006: 7.1.4**, pp. 87–89)
- Demonstrate methods for securing a portal for victim passage. (**NFPA 1006: 7.1.4**, pp. 87–89)

Introduction

As emergency responders, we understand that there is a certain degree of risk associated with fire, police, hazardous materials (HazMat), and emergency medical services (EMS) work. All too often, we hear of an emergency worker being injured or, worse yet, killed in the line of duty. The National Fallen Firefighters Foundation has started a safety initiative called Everyone Goes Home®. This program is based on the 16 Firefighter Life Safety Initiatives, which are designed to bring safety to the forefront and make it an integral part of the culture of the fire service.

This organizational safety culture should not be restricted to firefighters. It should be part of every emergency organization. We do what we do not because of the thrill or the adrenaline rush but because we seek to render assistance to others and make a difference in their lives. We should not be blind risk-takers but professionals who know how to manage risks. If we cannot protect ourselves, how can we protect others?

Safety, Communication, and Medical Protocols

Confined-space rescue is not so much about the confined space as it is about the hazards unique to the space. In previous chapters, we have covered atmospheric and physical hazards, Occupational Safety and Health Administration (OSHA) requirements, roles of personnel, air monitoring, the incident command system, and basic rescue size-up, among other topics. In this chapter, we cover common safety and communication measures and protocols, such as hazard avoidance, personal protective equipment (PPE), communication methods, and temperature stresses.

Keeping the Focus on Safety

The first priority of incident response is life safety—that of the rescuers and victims. If you do not manage the risks to the rescuers, how can you help the victims? Managing the risks means that you must weigh the risks against the benefits to be gained: If little is to be gained, little should be risked.

Emergency operations are dynamic. At times, they seem to change minute by minute, and the information can be conflicting or may even contradict earlier information you received. As the incident commander (IC), it is your job to sort through the information and scrutinize it to get to the factual basis.

But to do that, you must keep a balanced approach to the risks. You cannot get tunnel vision or work on preconceived ideas of what the problem is, and you must observe everything and sort through it rapidly. Focusing only on the hazards associated with confined spaces can let something as simple as rescuers' heat injuries go unanticipated until you are suddenly confronted with them, and they are now the emergency. Dramatic changes at an emergency, such as a wall collapse or injuries to rescue personnel, can cause your entire operation to fall apart immediately. You must then work to regain control of an emergency within an emergency. It is better to prevent the second emergency from occurring in the first place.

Hazard Avoidance

All confined spaces must meet the three requirements to be classified as such:

- Be large enough to enter and work in
- Have a limited means of entry and egress
- Not be designed for continuous employee occupancy

Permit-required confined spaces must have only one of the following hazardous conditions to be classified as such:

- The potential to contain a hazardous atmosphere
- Contain a material with an engulfment potential
- Have an internal configuration that might cause entrapment or asphyxiation
- Contain any other recognized safety or health hazard

The difference between a confined space and a permit-required confined space is that when you find a hazard or hazardous condition, you must identify it, analyze it to see how it affects your operations, and control or eliminate it. That is a big challenge, and it may not be possible to identify and analyze all the hazards present at the emergency.

That doesn't mean you shouldn't look for hazards. It means you will have to sort them out and assign some of them precedence over others. Some of the hazards can be quickly resolved, whereas others will take so much time and effort to eliminate that you will have no chance of rescuing the victim.

The use of respiratory protection during a confined-space rescue in an oxygen-deficient atmosphere is one example of this. Let's say you make the decision to get to the victim quickly, but you also need to protect the rescuers; you cannot provide enough ventilation to eliminate the atmospheric hazard, so you provide the rescuers with protection. The situation is what it is, and you cannot effectively change it. You have no choice but to work within the specific limits set by the low-oxygen hazard, and you can take only a few extra steps to eliminate the obstacle to the rescue.

At other times, the problem can be simply fixed and the hazard eliminated, but first, you must recognize it as a risk. Loose tools or equipment lying around the vertical opening to a confined space present risk. These items can be kicked into the opening and strike people who are in the space. Eliminate or control the objects near the opening, and you will reduce the chances of them falling into the space.

These are two black-and-white examples, but unfortunately, controlling risk isn't that cut-and-dried. Just because you have controlled some objects that have the potential to fall into the space or provided respiratory protection doesn't mean you have eliminated the need for head protection or continuous monitoring.

If you had a confined-space incident during which you were able to control a hazardous atmosphere or your monitoring showed that no atmospheric hazard existed, how comfortable would you feel, as the IC, about having rescuers enter without respiratory protection? There is a difference between controlled hazards and no identified hazards. If you had made the decision to ventilate a space to control an identified atmospheric hazard, but you know that without the ventilation, the hazardous atmosphere would reoccur, would you require respiratory protective equipment?

What do you do in a situation where no respiratory hazard is present? For example, if the oxygen content is between 19.5 percent and 23.5 percent, no toxic or flammable vapors or gases are present to worry about (because you performed your size-up well and continuously monitored the space), and no dust that can become airborne is in the space. Based on your size-up, you might feel comfortable that a problem is unlikely to occur within the space.

These are ideal conditions, but emergencies rarely produce ideal conditions. You are there because something went wrong, and the conditions are not ideal—for the victim or for the responders. Either way, you control the conditions, or the conditions control you. Your available PPE is either adequate for the conditions or not. If the PPE is adequate, one problem is solved. If it is not adequate, you must ask why.

On a hot summer day, wearing a self-contained breathing apparatus (SCBA) and protective clothing intended to prevent exposure to chemicals is adequate for respiratory and chemical exposures, but what effect will it have on the emergency services personnel wearing it? Safety is more than the obvious hazards associated with the confined space. Wearing such gear on a hot day during a rescue will quickly drain rescuers' energy, and it may even endanger their lives—if attention is not paid to the operational limitations it creates. The SCBA and protective clothing may be adequate for the hazard, but how long can people work effectively in this gear? How long can they wear it while standing around waiting for orders or as part of a backup team? Something as simple as high ambient temperatures can double resource needs, and the resource you will need most will likely be more people.

In some cases, hazards become insurmountable, and escape is necessary. To better understand the process of constructing an escape plan, follow the steps in **SKILL DRILL 3-1**.

SKILL DRILL 3-1

Constructing an Escape Plan

Step 1: Locate Exit Points

Clearly mark exit points with signage or flagging for easy recognition.

Step 2: Assess Risks

Assess any risks around the exit points, such as potential obstructions, narrow passageways, or nearby hazardous materials.

Continues.

SKILL DRILL 3-1 CONTINUED

Constructing an Escape Plan

Step 3: Ensure Clear Communication

Conduct a communication check to ensure radios and other devices are working. The Incident Commander should brief the team on the escape plan and communication protocols.

Step 4: Post Rescuer at Exit Route

Post a rescuer or rescuers on standby at the exit routes.

Step 5: Prepare for Medical Emergencies

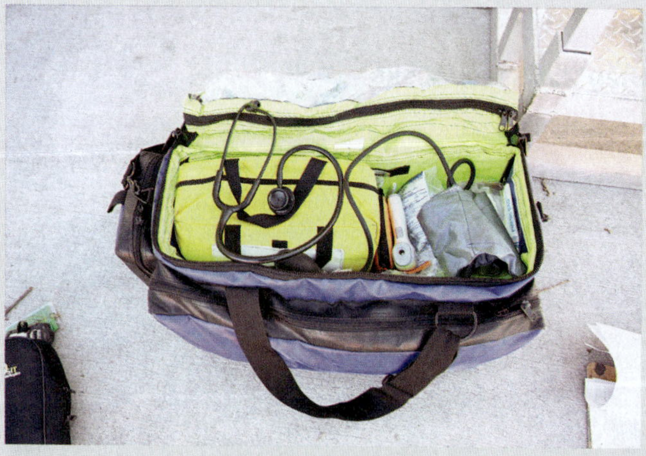

Prepare medical kits and emergency supplies and keep them accessible near entry and exit points. Medical personnel should be briefed on the escape plan and their roles.

SKILL DRILL 3-1 CONTINUED

Constructing an Escape Plan

Step 6: Practice the Escape Plan

Practice the escape route by moving through designated exit paths in a controlled drill.

Step 7: Execute the Escape Plan

Simulate an escape while moving swiftly and efficiently, exiting through marked points.

Step 8: Review and Document

Review the escape plan, document feedback, and discuss improvements for future drills.

Step 7: Courtesy of Illinois Fire Service Institute.

PPE's Role in the Development of a Safety Culture

Selecting and using the proper PPE is just one component in providing a safe way for rescue personnel to operate on the scene of a confined-space incident. Instilling a safety mindset into the culture of your team is also critical and does not happen overnight. Rather, the development of a safety culture is the result of many hours of training and the discipline that comes from everyone on the team being accountable. Such a culture is established because individual team members demand it from each other.

When an injury brings down a team member, everyone suffers the loss. There is nothing wrong with being safe or demanding that the team maintain a positive attitude about maintaining safety. Ultimately, safety culture is something to be proud of, and it should be emulated by other teams that deliver specialized rescue services.

Personal-Issue PPE

Personal-issue PPE is equipment that is given to every member of a confined-space rescue team. This equipment is typically stored in backpacks. Personal PPE should include coats, trousers or coveralls, helmets, gloves, boots, and safety glasses. Although this equipment is not shared, it should be properly cleaned after each use. Proper cleaning removes contaminants, protects the equipment, ensures that it is ready for the next response and that the equipment does not fail prematurely, and provides an opportunity for additional detailed inspection.

Torso, Arm, and Leg Protection (Coat and Trousers or Coverall). Garments provide protection for the upper and lower torso, arms, and legs. Garments include coats, trousers, and coveralls. Shorts or short-sleeved shirts should not be considered. If it is hot, provide for frequent rehabilitation (rehab) breaks, and set up misting and other ventilation fans to cool the area. If the weather is a major consideration, set up a makeshift tent (tarp or pop-up tent) over the rescue area to shield it from the sun, rain, or snow. Make sure that you wear the proper protection for the environment in which you are working.

The National Fire Protection Association (NFPA) code 1951 (NFPA 2020) garment (coat and trousers or coverall) is a best practice for confined-space rescue operations (**FIGURE 3-1**). One disadvantage of a coverall is that it retains heat because of its one-piece design. Also, a suit that is too large or too small or has the wrong proportions can cause discomfort and limit the ease of movement. Firefighting gear is very bulky and not recommended as standard protection for confined-space rescue. Ideally, rescuers should be as comfortable as possible so that they can concentrate on the efforts at hand and not their clothing. This consideration alone makes a case for providing each member of the rescue team with his or her own PPE ensemble.

Head Protection (Helmet). The most critical piece of protective equipment in a confined-space rescue is the helmet. Head trauma is an injury from which a victim may not easily rebound and, in fact, could prove to be a career- or life-ending incident. Appropriate helmets must be included in special confined-space rescue gear.

Considering that the most likely sources of head injuries on a confined-space rescue scene are from materials falling in the confined space, a good choice for a confined-space emergency is an NFPA 1951–compliant technical rescue helmet. It is lightweight and balanced, and it provides impact protection (**FIGURE 3-2**). Many varieties of this type of helmet are available commercially from rescue gear manufacturers.

FIGURE 3-1 An NFPA 1951 garment.

> **TIP**
>
> The large profile of a fire helmet—particularly the brim—makes it poorly suited for use in the tight confines of a confined space.

Another option is the **American National Standards Institute (ANSI)** Type 2, Class G helmet, which is used by many technical rescue teams. (ANSI is a nonprofit organization that administers and coordinates the U.S. voluntary standards and conformity assessment system.) Never use helmets that are not rated by the NFPA or ANSI. Helmets must be issued or approved by the authority having jurisdiction (AHJ).

Hand Protection (Gloves). The glove is often the undoing of the rescuer. We know that gloves must be worn, but performing any rescue-related function is difficult when wearing them. When rescue personnel are standing around talking, they may have their gloves on—but all too often, the first time they are required

FIGURE 3-2 An NFPA code 1951–compliant technical rescue helmet.

to do something, *off* come the gloves. Gloves provide zero protection to your hands when stored in your back pockets.

NFPA 1951 requires design and performance compliance for cut resistance, puncture resistance, heat/flame resistance, abrasion resistance, hand function, grip, and ease of donning (NFPA 2020). Thankfully, the days of using leather work gloves and Nomex flight gloves to provide a safe and efficient glove for technical rescue operations are gone. NFPA-compliant technical rescue and extrication gloves are now available through a number of quality manufacturers.

Firefighting gloves are designed for fighting fires. They are bulky, hot, and stiff, and they do not provide the dexterity needed for rescue operations. Conversely, if your sole task is to move equipment and lumber, the firefighting glove will protect your hands from slivers and abrasions.

A better choice in a confined-space rescue incident for support personnel who are performing tasks like moving equipment and lumber is the standard leather glove. These gloves are sturdy enough to keep abrasions and splinters to a minimum and still comfortable enough that rescuers will generally keep them on while working (**FIGURE 3-3**).

Foot Protection (Boots). All confined-space rescue team personnel should wear boots that protect their feet from the hazards that are commonly found in a confined-space area (**FIGURE 3-4**). Not only are there many opportunities to drop something on your foot, but nails and other sharp debris are also typically found all around the site. In addition, it is a good idea to have a high-top boot. This will provide a measure of support in the event you step on a piece of lumber or other equipment and turn your ankle. NFPA 1951–compliant boots require protection from the following: impact, cuts, abrasion, heat, punctures, bending, electrical conduction, and slipping (NFPA 2020).

FIGURE 3-3 A rescuer wearing standard leather gloves.

FIGURE 3-4 NFPA 1951–compliant boots for technical rescue operations are designed to provide protection, durability, and comfort while meeting the specific safety requirements set by the NFPA.

TIP

The NFPA sets the minimum requirements for technical rescue ensemble elements. As defined by the standard, ensemble elements provide protection to the upper and lower torso, arms, legs, head, hands, and feet. The elements that are tested and certified will have a label stating that they are certified. In addition to providing the body protection described in the standard, garments that also provide protection from bloodborne pathogens will have a label stating that they protect the wearer from bloodborne pathogens.

Eye Protection (Safety Glasses). A standard pair of safety glasses approved by ANSI Z87.1 (International Safety Equipment Association [ISEA] 2020) is more than satisfactory for confined-space rescue (**FIGURE 3-5**). Eye injuries are most likely to occur during a confined-space rescue when rescuers are cutting and nailing or when soil is blowing or falling.

ANSI Z87.1 (ISEA 2021), referenced by OSHA and in NFPA 1500 (NFPA 2021a), is the standard that covers protection for faces and eyes. Measurable characteristics of lens thickness, impact resistance, penetration by projectiles, and optical quality are detailed in Sections 9 and 10 of the standard. Eyewear used by technical rescue members should be marked with *Z87*, indicating that the product meets the ANSI standards. Most safety glasses can be purchased with or without sun or glare protection, and they will stay in position when the rescuer's head moves. In addition, they will rarely fog because the outside of the lenses is ventilated. Make sure that any safety glasses used for confined-space rescue are approved by ANSI for that purpose.

Full-face goggles, although they provide the most protection, will usually fog when the rescuer gets hot. When that happens, what comes next? The goggles end up hanging around the neck of the rescuer and not covering the eyes.

One last point about eyewear: Rescue personnel should never wear reflective sunglasses because rescue personnel frequently need to confirm movement by eye contact.

FIGURE 3-5 A pair of safety glasses approved by ANSI.

TIP

Although they may not exactly qualify as pieces of equipment, sunblock and insect repellent with DEET (chemical name N, N-diethyl-meta-toluamide) should be part of every confined-space rescuer's PPE pack. Technical rescue incidents are often of prolonged duration by nature, and a rescuer should be prepared for possible challenges, including those related to weather, clothing, and hydration.

Firefighting Turnout Gear. For most confined-space rescue incidents, the use of turnout gear is not a best practice. It is bulky, and the insulation can overheat a rescuer who is performing rescue functions. However, firefighters often will arrive on the scene and begin work in turnout gear. In addition, when hazards such as a broken natural gas line are present, full firefighting PPE is essential for all fire and rescue personnel in the hot zone until the hazard is controlled and removed.

Team-Issue PPE

Team-issue PPE is equipment that is stored in the confined-space rescue team's response unit and available for use by any member who needs to use it. This equipment is either shared or disposable. Shared PPE must be properly sanitized after each use. This is important because shared equipment that has been stowed without sanitation may contaminate the next user. Additionally, dirty PPE will likely not be used by the next member who needs the protection.

Respiratory Protection. Respiratory protection should always be readily available at the scene of a confined-space rescue incident. Whenever there is any indication that an immediately dangerous to life and health (IDLH) atmospheric problem might exist, an atmospheric monitor should be utilized, in addition to SCBA or supplied-air respirator (SAR), before entry into the environment (see Figure 2-7b in Chapter 2). In dry and windy conditions with flying dust and dirt, an air-purifying respirator (APR) should be used. Whereas particulate respirators can be disposable, SCBA and SAR respirators must be sanitized after each use in accordance with manufacturers' recommended procedures.

Because of the respiratory hazards posed by confined spaces, all members of the confined-space rescue team must be compliant with *Code of Federal Regulations* (C.F.R.) training and annual fit-testing requirements as outlined in Title 29 of the C.F.R., which covers labor regulations ("Permit-Required Confined Spaces" 1993; "Respiratory Protection" 1998). An adequate supply of respirator face pieces in various sizes must be available. Remember that if rescuers are using respiratory protection, all live victims will need the same level of protection. Rescuers must evaluate the head size and shape of the victim and make their best guess on the size of the mask needed.

Hearing Protection. Hearing protection is a good idea for anyone who is working around compressors or saws. Choose a level of protection that will shield the rescuer's ears from high-noise frequencies but will not block out all communication. It is usually more dangerous for the rescuer to be unable to hear anything than it is for them to be exposed to loud noises.

Communication Protocols

Protocols must be determined and implemented to guarantee clear communication with all personnel involved in the incident, as covered in Chapter 2 ("Steps for Securing a Confined-Space Incident"). Communication protocols, such as radios, signaling systems, or visual cues, must be planned and executed throughout a response. To better understand the process of selecting communication devices, follow the steps in **SKILL DRILL 3-2**.

Mental Health Protocols

A comprehensive mental health program helps responders develop resilience, enabling them to cope better with the high-stress situations inherent in confined-space incidents. Regular mental health support and stress management techniques reduce the risk of burnout, ensuring that responders remain effective and capable of performing their duties. Responders who are mentally healthy are more focused, make better decisions, and work more cohesively as a team, leading to more successful rescue operations.

A formal mental health program normalizes the conversation around mental health, encouraging

SKILL DRILL 3-2

Selecting Communication Devices

Step 1: Assess the Environment

Assess the surrounding environment for hazards.

Step 2: Select Devices

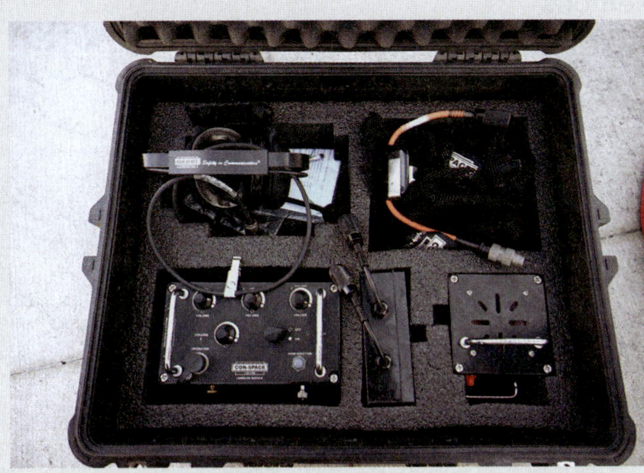

Select appropriate communication devices based on environmental conditions, distance, and terrain considerations.

Continues.

SKILL DRILL 3-2 CONTINUED

Selecting Communication Devices

Step 3: Test Devices

Check the functionality of devices, including their volume, range, and clarity in different environmental areas.

Step 4: Implement Devices

Deploy communication devices at designated checkpoints within the operational area.

Step 5: Prepare Alternate Communication Methods

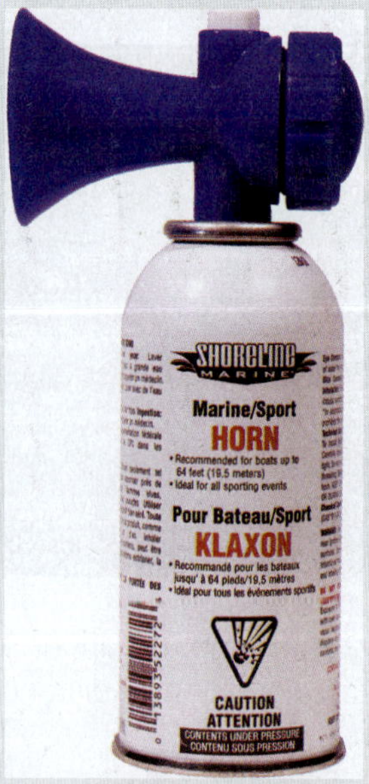

Prepare backup communication methods, such as air horns, hand signals, or physical markers, for use if devices fail or become unreliable.

Step 6: Monitor Communication

Monitor device channels and confirm that signals are being received without interference.

SKILL DRILL 3-2 CONTINUED

Selecting Communication Devices

Step 7: Check Equipment and Document Performance

Complete a final check on all communication equipment and document operational performance for future reference and improvement.

Step 1: Courtesy of Illinois Fire Service Institute.

responders to seek help without fear of stigma or judgment. Providing continuous mental health support ensures that responders can recover from traumatic experiences and maintain their mental well-being throughout their careers. It contributes to higher job satisfaction and morale, aiding in the retention of experienced and skilled responders.

The integration of mental health protocols and programs for responders to confined-space incidents is critical. These protocols not only support the immediate well-being of responders during and after incidents but also contribute to their long-term mental health and job satisfaction. By prioritizing mental health, emergency services can maintain a resilient, effective, and dedicated workforce capable of handling the demanding nature of confined-space rescues.

Medical Monitoring Protocols

Medical monitoring is the process of performing a medical evaluation of emergency response personnel by determining and recording basic vital signs such as blood pressure, respirations per minute, and pulse. Medical monitoring of rescue personnel is not just an intervention; it is *prevention*. It may provide the first indicator of a developing medical emergency, or it may identify an emergency worker who should not be assigned certain tasks.

Medical monitoring does not replace managing the risks at an incident; it only helps you do that. In addition to having EMS on the scene for the victims, you should have EMS personnel on the scene who are assigned to stand by to treat rescue personnel. These EMS resources are dedicated to the rescue workers.

Hydration

In confined-space rescue operations, maintaining proper hydration is essential for the health, safety, and performance of rescue teams. Following is information on hydration, including the signs of dehydration, effective hydration strategies, and practical tips to ensure confined-space responders stay properly hydrated during incidents. **TABLE 3-1** shows signs and symptoms of hydration deficiencies and strategies for their management.

Proper hydration is crucial for confined-space rescue teams because of the physical exertion, elevated temperatures, and potential exposure to heat stress during operations. Adequate hydration does the following:

- Supports overall physical and cognitive performance
- Helps regulate body temperature and prevent heat-related illnesses
- Enhances endurance and reduces the risk of fatigue and muscle cramps

TABLE 3-1 Hydration Deficiencies

Signs and Symptoms	Strategies for Hydration Management
▪ Thirst ▪ Dry or sticky mouth ▪ Dark urine or infrequent urination ▪ Fatigue or dizziness ▪ Headache ▪ Rapid heartbeat ▪ Dry skin ▪ Irritability or confusion	▪ **Pre-hydration:** Encourage team members to start hydrating well before deployment ▪ **Water accessibility:** Provide ample access to clean, cool drinking water throughout the incident ▪ **Hydration schedule:** Establish regular hydration breaks during operations, especially in hot or strenuous conditions ▪ **Electrolyte replacement:** Offer electrolyte drinks or sports drinks to replenish lost minerals and maintain electrolyte balance ▪ **Individual hydration plans:** Tailor hydration plans based on individual needs and fluid preferences

TIP

Implementing effective hydration strategies can help ensure that confined-space responders maintain optimal hydration levels during incidents. These strategies include the following:

- **Drink before thirst.** Encourage responders to drink water before feeling thirsty to prevent dehydration.
- **Use hydration packs.** Equip team members with hydration packs or water bottles for easy access during operations.
- **Monitor urine color.** Encourage checking urine color as a simple indicator of hydration status (pale yellow to clear indicates adequate hydration).
- **Avoid caffeine and alcohol.** Discourage consumption of caffeinated or alcoholic beverages because they can contribute to dehydration.
- **Eat hydrating foods.** Include hydrating foods, such as fruits and vegetables with high water content, in your meals or snacks.

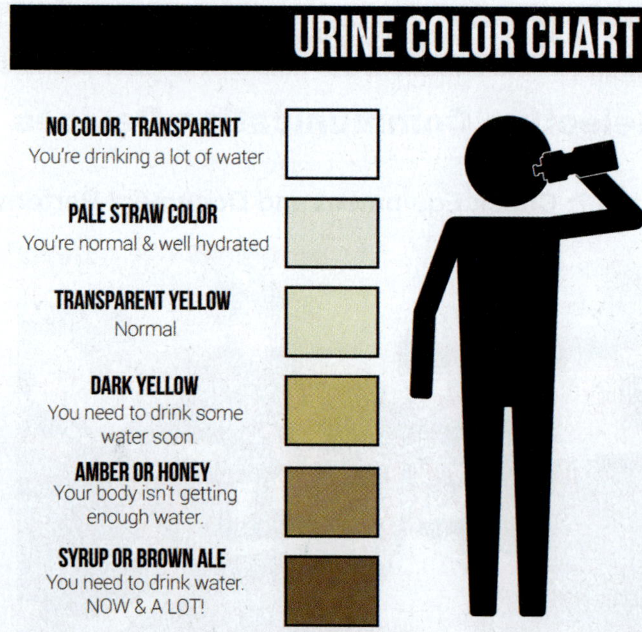

FIGURE 3-6 This hydration chart shows how a person's urine color corresponds to their hydration level.

Hydration deficiencies can be detected by monitoring the color of an individual's urine (**FIGURE 3-6**). The lighter the color of the urine, the more hydrated an individual is.

Temperature Stress

The ambient temperature conditions during rescue operations can cause injuries. If you are working outside, you must consider the air temperature as well as conditions that can affect the air temperature. Inside the confined space, you must also be concerned with conditions that can affect the temperature within the space, as shown in **FIGURE 3-7**. Cooling or heating equipment can create high- or low-temperature conditions unique to the space. If you have ever gone into a walk-in freezer on a hot summer day, it doesn't seem too bad at first, but it quickly becomes too cold to stay in for very long. The temperature can be more than 100°F outside the freezer, but once you are in the freezer, it may as well be winter.

No matter what causes the air temperature to go up or down, you must think about what it will do to the rescue personnel and how it will affect your rescue operation. Exposure to high or low temperatures can affect people's ability to aid in the rescue. Rescuers can also become victims in need of medical treatment or rescue. Whether people are inside or outside of the confined space, they can be affected by temperature. It is a matter of how long they are exposed, the severity of the temperature conditions, and how prepared they are to deal with the conditions.

Air temperatures as low as 80°F combined with high humidity can make the air temperature feel the equivalent of 90°F and above. This is known as the *apparent temperature*, and the body reacts as though the temperature really was 90°F or above. (Another term that may be more familiar is *heat index*; see **FIGURE 3-8**.) Include this with the other factors listed earlier (age, physical condition, impervious clothing, etc.), and you'll see that heat stress injuries can occur before the potential hazard is recognized. Elevated temperatures expose rescue personnel to the risk of **heat cramps**, a condition in which painful muscle spasms occur during activity; **heat exhaustion**, a condition in which the body responds to excessive losses of water and salt, usually through excessive sweating; or **heat stroke**, a condition in which the body's temperature rises rapidly, the sweating mechanism fails, and the body is unable to cool down. Preventing heat-related injuries is the best method of dealing with heat stress.

FIGURE 3-7 Confined spaces can be above or below the ambient temperature. Shown here is a steam coil for heating the product normally stored in this tank.

© pidjoe/iStock/Getty Images Plus/Getty Images

To avoid heat stress, you must take steps to reduce the impact that heat has on the human body. These preventive steps include the following:

- Changing the ambient conditions within the confined space by heating or cooling the atmosphere within the space

Temperature (°F)

Relative Humidity (%)	80	81	84	86	88	90	92	94	96	98	100	102	104	106	108	110
40	80	81	83	85	88	91	94	97	101	105	109	114	119	124	130	136
45	80	82	84	87	89	93	96	100	104	109	114	119	124	130	137	
50	81	83	85	88	91	95	99	103	108	113	118	124	131	137		
55	81	84	86	89	93	97	101	106	112	117	124	130	137			
60	82	84	88	91	95	100	105	110	116	123	129	137				
65	82	85	89	93	98	103	108	114	121	126	130					
70	83	86	90	95	100	105	112	119	126	134						
75	84	88	92	97	103	109	116	124	132							
80	84	89	94	100	106	113	121	129								
85	85	90	96	102	110	117	126	135								
90	86	91	98	105	113	122	131									
95	86	93	100	108	117	127										
100	87	95	103	112	121	132										

Likelihood of Heat Disorders with Prolonged Exposure or Strenuous Activity

Caution | Extreme Caution | Danger | Extreme Danger

FIGURE 3-8 This heat index chart illustrates the perceived temperature by combining air temperature and humidity levels, helping to assess the risk of heat-related illnesses during outdoor activities.

Data from National Weather Service/National Oceanic and Atmospheric Administration.

- Limiting the number of people exposed to the heat (including direct sunlight) by keeping people out of the heat
- Using fans to blow air across people to cool them
- Having entry team members drink liquids both before and after entry into the confined space to replace lost fluids
- Managing the amount of time that personnel must wear impervious protective clothing, like the HazMat suit shown in **FIGURE 3-9**. The rescue team may have to be prepared to have multiple backups or second-entry teams on hand based on the environmental factors at work in the space.
- Implementing an action plan that has people pace their work to match conditions
- Medical monitoring of rescue personnel as conditions dictate

Mechanical ventilation can provide cooling. Blowing fresh, clean, cooler air into the confined space assists the victim and the rescuers, and it can have an ancillary benefit because it also cools people within the space. Just remember that if you are ventilating to cool the space, you still must take the precautions needed for any ventilation of a confined space. Know the source of your supply air, know where the air will exhaust, and evaluate the effects within the space. Determine whether you can—or want to—exchange the air within the space.

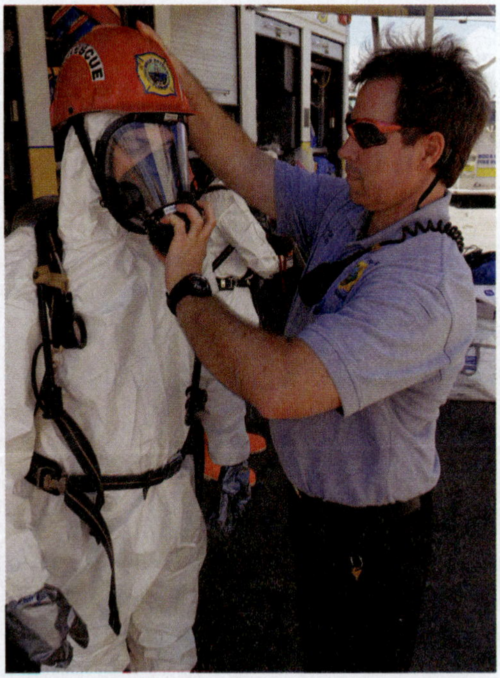

FIGURE 3-9 Protective clothing can increase the potential for heat stress to rescuers. These Level B suits can trap moisture from sweat and slow down the evaporative cooling effect.

TIP

Changing the ambient conditions within the confined space to improve rescuer safety also increases the chance of survival for victims.

In summer months, both the inside and outside temperatures of a confined space may be high. In that case, all you might be doing is moving hot air. If you have a tank with limited entry and ventilation openings, effective air movement for cooling may be difficult. Similarly, it may be difficult to cool a large tank. In such cases, you will need to be resourceful.

Directing the cooling air so that it is blowing across the rescuers within the space may be the most efficient way to provide cooling. Keep in mind the safety of the victim and the condition of the atmosphere within the space. If you are blowing so much air into the space that it creates airborne aerosols and dust, you may create more problems than you solve. An even simpler limitation may be that you can use only one fan, and its use is limited to providing ventilation and fresh air to the victim.

Keeping people out of the heat sounds so simple that it almost doesn't need to be mentioned. Despite that simplicity, support personnel (anyone not directly involved in the rescue) can become victims of heat stress. Manage where you set up the command post or where you stage people.

Staying out of the heat and out of direct sunlight should be a concern in hot weather. If time and equipment permit, set up some type of shade covering. If it is difficult to provide shade or the temperatures are very high, use a fan to provide air movement. We sweat to cool our bodies; sweat provides evaporative cooling. Air movement increases evaporation, which is why you feel so much cooler in a breeze when you are sweating.

Remember that backup personnel waiting in the sun for their chance to work are being exposed to the heat. This preheating reduces the length of time they can function effectively when they are called to work. A balanced approach to managing the risk of heat injuries includes staying out of the heat to maximize your available work time.

Stay Hydrated

Hydrate, hydrate, hydrate. Sweating reduces the volume of liquid within the body. You must do something to replace those liquids, as shown in **FIGURE 3-10**. Reduce the volume of liquids in the body too much, and the body will take steps to protect itself. Heat cramps, heat exhaustion, and heat stroke can follow.

To prevent heat injuries, begin with the basics. Have entry team members drink water or a sports drink to replace electrolytes, both before and after entry. Base the amount of liquid on the conditions you are faced with. You must consider not only the actual temperature but also the apparent temperature to avoid heat injuries. When hydrating, be aware that it can take 20 minutes or more for the liquid to be available for the body to use. Drinking liquids before entry prepares you so that you will have the liquids available while you are working. Drinking liquids after leaving the confined space replaces lost liquids. Have the entry team (and anyone else working under similar conditions) hydrate even if they state that they are not thirsty. Thirst may not be a reliable indicator of dehydration, and if you wait until you are thirsty, you may be well on your way to heat stress.

Dress Appropriately

Most emergency workers wear some type of protective clothing. Body armor, turnout gear, and other types of PPE are designed for a specific hazard or hazards. While fighting a fire, firefighters need to protect their skin from heat and steam. Unfortunately, as good as that equipment is at protecting you from external heat and moisture, it is equally as good at holding in body heat and preventing sweat from evaporating. Other emergency workers may be wearing clothing that is designed to keep out the rain or cold, and it can be impervious, just like firefighters' protective clothing.

FIGURE 3-10 As you sweat, you lose fluids. You must replace those fluids to help prevent heat-related injuries.

Although this equipment may be good at protecting you from hazards that are typically encountered in fire emergencies, what problems could it create in a confined-space rescue? Yes, you will have to wear some type of protective equipment, but is it the right equipment for the hazards you will be facing? Dress appropriately for the incident conditions. Wear what protective clothing is needed, but make sure you consider the effect that the clothing will have on personnel.

High temperatures and impervious clothing create difficult tactical conditions for emergency workers. Their efficiency will decrease, it will take longer to accomplish the assigned tasks, and it will most likely take more people to accomplish those tasks. Consider if it is appropriate to have rescue personnel who are staged outside of the hazard area open up or remove the protective clothing while they are awaiting assignments.

People assigned as the backup rescue team to the entry team will have to continue to wear their protective clothing so that they can be quickly deployed. Keep them out of the heat as much as possible, and dress them only to the point at which they can quickly finish donning the protective equipment and immediately go to work.

Pace Yourself

Pacing the work of the rescue team and having realistic expectations as to what they can accomplish is integral to the safety of the rescue operation. Matching the work to the conditions is always a good idea, regardless of the temperature. The IC should assign work to people only if they can accomplish it safely based on the current working conditions. Those decisions, once made, must be enforced: Never let people overextend themselves.

This does not mean discouraging efficiency; instead, it means that an IC must monitor people to keep them from taking shortcuts or rushing so that they do not injure themselves or others. Sometimes, the span of control may have to be reduced for the sake of safety. The goal of the action plan is to accomplish the needed objectives, prevent injuries, and provide for a flow to the emergency operation so that you maintain the required level of control and thus succeed.

Heat-Related Injuries

Will you have incident-specific hazards that might be best managed through medical monitoring? Yes, you will, and heat stress is one of those hazards that requires medical monitoring both before entry into the confined space and after exiting the space. Comparing pre-entry and post-exit vital signs such as pulse, blood pressure, respiration, and body temperature can indicate the onset of heat injuries. Conducting medical monitoring reduces the risk of heat injuries by giving real-time information as to what is going on within the body. An increase in body temperature between the pre-entry and the post-exit monitoring may indicate the need for follow-up monitoring or immediate medical attention to prevent further heat injuries. Prevention of heat stress injuries by hydration and monitoring is the key factor. See **TABLE 3-2** for a summary of signs, symptoms, and corrective measures for heat injuries.

Left untreated, heat stroke can cause death. Although prevention of heat stress injuries is important, you must also be able to recognize the signs and symptoms of heat stress. Knowing the signs and symptoms lets you evaluate your own physical condition as well as the condition of others around you. The symptoms of heat cramps include severe muscle cramps in the abdomen and legs; they are believed to be caused by loss of fluids from sweating heavily. Heat cramps may be the first sign of heat stress, but depending on how severe the heat injury is, they may not occur.

If your heat injury is severe enough, you may go right past heat cramps and exhibit symptoms of heat exhaustion or heat stroke. Victims of heat exhaustion may feel weak, nauseous, dizzy, and faint and have pale and clammy skin as the blood flows to the skin and away from vital organs in an attempt to cool the body. Heavy perspiration and normal or below-normal body temperature also occur. Heat exhaustion victims rarely lose consciousness, but they may. If not treated, heat exhaustion can lead to heat stroke.

Heat stroke is a true medical emergency and is the most serious heat injury. Without immediate medical attention, people can suffer serious injury to vital organs, and up to 20 percent of heat stroke victims die from it. Heat stroke signs and symptoms include lack of

TABLE 3-2 Signs, Symptoms, and Corrective Measures for Heat Injuries

Condition	Signs and Symptoms	Corrective Measures
Heat cramps	▪ Painful muscle cramps or spasms, typically in the abdomen, arms, or legs ▪ Sweating ▪ Fatigue or weakness	▪ Move to a cool place and rest ▪ Drink water or electrolyte drinks ▪ Gently stretch and massage the affected muscles ▪ Avoid strenuous activity and seek medical attention if cramps persist
Heat exhaustion	▪ Profuse sweating ▪ Fatigue and weakness ▪ Nausea or vomiting ▪ Dizziness or lightheadedness ▪ Headache ▪ Cool, moist skin ▪ Elevated body temperature (but less than 104°F)	▪ Move the affected individual to a cool, shaded area ▪ Loosen clothing and apply cool, wet cloths to the skin ▪ Provide sips of water or electrolyte drinks ▪ Monitor vital signs and seek medical attention if symptoms worsen
Heat stroke	▪ High body temperature (above 104°F) ▪ Lack of sweating (dry, hot skin) ▪ Rapid pulse ▪ Confusion or disorientation ▪ Seizures ▪ Loss of consciousness	▪ Call for immediate medical assistance (heat stroke is a medical emergency) ▪ Move the individual to a cool area and remove excess clothing ▪ Cool the body rapidly with cold-water immersion or ice packs ▪ Monitor vital signs and provide supportive care until help arrives

sweating (even though the skin may be wet from earlier perspiration); dry, hot, and red skin; elevated body temperature; and possible convulsions or unconsciousness. The body can no longer regulate temperature, and the body temperature may rise to 104°F or more.

Cold-Related Injuries

You must also be concerned about injuries from cold temperatures. Acute cold injuries may be less serious than heat injuries, but just as for heat injuries, you must consider the hazards presented by cold temperatures and work to prevent injury. Wind plays a significant role in how the body senses and reacts to cold temperatures. Most of us have heard the term *windchill*, and if you live in a colder climate, windchill information is generally included in winter weather reports (**FIGURE 3-11**). See **TABLE 3-3** for a summary of signs, symptoms, and corrective measures for cold injuries.

Cold conditions may be present because of the weather or from intentional or accidental conditions in the confined space, such as a refrigerated hold on a ship. As the ambient temperature drops, the body attempts to produce heat to keep up with the loss of heat to the cold. When the body cannot produce enough heat to stay warm, the blood supply retreats to the head and torso in an attempt to protect the brain and vital organs. This, in turn, leads to shivering and a loss of muscle coordination.

Medical monitoring during cold-temperature conditions, just as with hot conditions, is intended to recognize and prevent injury. In the case of cold temperatures, injuries include **hypothermia**, or a lowering of the body's core temperature, and **frostbite**, a medical condition in which body tissues—typically the nose, fingers, or toes—are injured as a result of exposure to extreme cold. See **TABLE 3-4** for a summary of monitoring and response protocols for heat- and cold-related injuries.

When the condition lasts long enough that the body's core temperature begins to drop, it leads to hypothermia. If the hypothermia is severe enough, unconsciousness and death can follow. It does not have to be extremely cold for hypothermia to occur. A person wearing wet clothing in 50°F air can suffer hypothermia. Victims of hypothermia may not recognize that they have fallen victim. This failure to recognize the onset of hypothermia may cause them to deny that they are suffering from it. Just a 1-degree drop in core body temperature is enough to initiate hypothermia. Reducing your core temperature from 98.6°F to 96°F may be all it takes.

Temperature (°F)

Wind (mph) / Calm	40	35	30	25	20	15	10	5	0	−5	−10	−15	−20	−25	−30	−35	−40	−45
5	36	31	25	19	13	7	1	−5	−11	−16	−22	−28	−34	−40	−46	−52	−57	−63
10	34	27	21	15	9	3	−4	−10	−16	−22	−28	−35	−41	−47	−53	−59	−66	−72
15	32	25	19	13	6	0	−7	−13	−19	−26	−32	−39	−45	−51	−58	−64	−71	−77
20	30	24	17	11	4	−2	−9	−15	−22	−29	−35	−42	−48	−55	−61	−68	−74	−81
25	29	23	16	9	3	−4	−11	−17	−24	−31	−37	−44	−51	−58	−64	−71	−78	−84
30	28	22	15	8	1	−5	−12	−19	−26	−33	−39	−46	−53	−60	−67	−73	−80	−87
35	28	21	14	7	0	−7	−14	−21	−27	−34	−41	−48	−55	−62	−69	−76	−82	−89
40	27	20	13	6	−1	−8	−15	−22	−29	−36	−43	−50	−57	−64	−71	−78	−84	−91
45	26	19	12	5	−2	−9	−16	−23	−30	−37	−44	−51	−58	−65	−72	−79	−86	−93
50	26	19	12	4	−3	−10	−17	−24	−31	−38	−45	−52	−60	−67	−74	−81	−88	−95
55	25	18	11	4	−3	−11	−18	−25	−32	−39	−46	−54	−61	−68	−75	−82	−89	−97
60	25	17	10	3	−4	−11	−19	−26	−33	−40	−48	−55	−62	−69	−76	−84	−91	−98

Frostbite Times: 30 minutes; 10 minutes; 5 minutes

$$\text{Wind Chill (°F)} = 35.74 + 0.6215T - 35.75 (V^{0.16}) + 0.4275T (V^{0.16})$$

Where, T = Air Temperature (°F) V = Wind Speed (mph)

FIGURE 3-11 This windchill chart shows how cold it feels by factoring in both air temperature and wind speed, providing a guide to the increased risk of frostbite and hypothermia in cold, windy conditions.

Data from National Weather Service/National Oceanic and Atmospheric Administration.

TABLE 3-3 Signs, Symptoms, and Corrective Measures for Cold Injuries

Condition	Signs and Symptoms	Corrective Measures
Hypothermia	▪ Shivering ▪ Confusion or disorientation ▪ Slurred speech ▪ Weak pulse ▪ Slow, shallow breathing ▪ Loss of coordination ▪ Cold, pale skin ▪ Unconsciousness (in severe cases)	▪ Move the affected individual to a warm, dry area ▪ Remove wet clothing and replace with dry, warm layers ▪ Cover the person with blankets or insulated materials ▪ Provide warm, nonalcoholic beverages and monitor vital signs ▪ Seek immediate medical attention for severe cases ▪ Place heat packs in groin area and armpits to assist in warming
Frostbite	▪ Numbness or tingling in **extremities** (arms, hands, legs, and feet) and the ears and nose ▪ Pale, waxy skin ▪ Hard, cold skin ▪ Joint or muscle stiffness ▪ Blistering (in severe cases)	▪ Protect the affected area from further exposure to cold ▪ Gently rewarm frostbitten areas by immersing in warm (not hot) water or using body heat ▪ Avoid rubbing or massaging frostbitten skin ▪ Dress the area with sterile, dry bandages and seek medical attention promptly

TABLE 3-4 Monitoring and Response Protocols for Heat- and Cold-Related Injuries

Pre-deployment assessment	Screen responders for underlying health conditions or medications that may increase susceptibility to heat- or cold-related injuries
Continuous monitoring	Use heat stress monitoring devices to track environmental conditions and responders' physiological responses; in cold environments, use temperature monitoring devices to assess the same
Buddy system	Implement a buddy system to monitor each other for signs of heat- or cold-related distress and provide assistance if needed
Incident command	Ensure that incident commanders are aware of heat- and cold-related risks and are in a position to initiate appropriate interventions

Beyond hypothermia is frostbite, an injury caused by the freezing of the skin and underlying tissues. It typically occurs in extremely cold temperatures and affects areas such as the fingers, toes, nose, ears, cheeks, and chin. Frostbite is a particular risk when rescuers are in direct contact with ice, freezing metals, or very cold liquids. Frostbite injuries require immediate medical attention.

Protection against cold temperatures and cold-induced injuries includes dressing for the temperature to prevent loss of body heat, not allowing your clothing to get wet from your own sweat or the liquids in the confined space, and pacing your work. Recognizing a cold-temperature hazard may be difficult during warmer weather. As pointed out earlier, under the right conditions, 50°F is low enough to cause hypothermia, and on a warm winter day, 50°F feels pretty warm. If you have preplanned the confined space and know a cold-temperature hazard is likely to be present, you can take steps in advance to address the problem. If you have not had the chance to preplan, you may only discover the cold hazard during your size-up.

As with heat, cold affects tactical operations. Extra layers of clothing and gloves limit mobility and dexterity. You must evaluate the cold to determine how much of a strategic factor it produces.

Just as with high temperatures, prevention of the injury should be your goal. As discussed earlier, to prevent heat injuries, fans or blowers can be used to cool people. Now we have the reverse problem: We do not want to cool people; we want to keep them from getting colder and possibly even warm them. This being said, air movement, especially the wind, now becomes your enemy. As air circulates around the body, it draws off heat.

Typically, when we think of wind, we think of the naturally occurring wind in the environment. To recognize windchill and how it can add to the potential for cold injury, you must think of any source of air movement. Moving air from blowers used for mechanical ventilation can produce a windchill effect as it blows across you. If you are from a warmer climate and not accustomed to considering windchill, you may not immediately recognize ventilation equipment as a source that can cause windchill. However, think of it this way: A fan blowing air at a speed of 15 miles per hour (24 km/h) is the same as a 15-mile-per-hour (24-km/h) wind. If that artificial wind is blowing in an area where the temperature is 40°F, it produces a windchill equal to 22°F.

You must manage exposure to cold temperatures in ways somewhat similar to exposure to high temperatures. Although the practices parallel those for preventing heat injuries—except that you are trying to warm the area instead of cool it—you should also follow these procedures:

- Change the temperature by heating the atmosphere within the confined space.
- Minimize the number of people exposed to the cold.
- Minimize the amount of time that personnel must wear impervious protective clothing—in this case, to prevent sweating.
- Have rescue personnel pace their work to match conditions.
- Provide medical monitoring as conditions dictate.

If you attempt to heat the atmosphere within a confined space, be certain it is safe to do so. You do not want the heater to become an ignition source. You also must ensure that the heat introduced into the space does not cause greater vaporization of any materials contained in the space. The vapors generated by the heat may be toxic, flammable, or both, and they may displace the oxygen within the space. If you know these conditions will not occur, you may want to consider if it is in the victim's best interest to warm the confined space or direct warmed air to the victim to protect them from hypothermia. Whatever you do, be realistic about what you can accomplish.

Just as with exposure to high temperatures, you want to minimize the number of people exposed to the cold. Get people out of the wind, frequently relieve people who must remain in the cold to perform their assigned work, and rehabilitate them in heated areas whenever possible. Try to keep both rescuers and victims dry and out of any precipitation that is occurring. When you are wet, your heat loss is increased tenfold. Cold temperatures and hot temperatures will require additional personnel at the emergency scene. Protecting people against cold injuries benefits everyone at the incident.

If the work rescuers are performing causes them to sweat and they cannot slow down the pace to prevent sweating, you must consider how to either rotate the workers to get them out of the cold or how to get dry clothing for them.

As previously mentioned, pacing the work is always a good idea. It keeps people working within reasonable physical limits and can lead to greater efficiency. Pacing your work in cold weather is advantageous because of the layers of clothing you are wearing. Not only do you not produce as much body heat and not sweat as much, but layers of clothing affect movement and dexterity.

There is another advantage to maintaining a modest work pace during cold weather: pacing your rate of breathing as well. Each time you inhale in cold weather, you bring cold air into your lungs. At a normal pace, your body can warm the air and limit the cooling of the blood. When you begin breathing faster during cold weather, your body cannot warm the air in your lungs, which can lead to cooling of the blood and an increased loss of body heat. Match the rate at which you are working with the conditions you face. Accept that you are doing the best that you can under conditions that you may not be able to change.

Training and Awareness for Heat- and Cold-Related Injuries

Heat-related injuries pose significant risks to confined-space responders and can affect the success of rescue operations. Cold-related injuries can pose significant risks to confined-space responders, particularly in environments with low temperatures or exposure to cold surfaces. By understanding the types of heat- and cold-related injuries, recognizing signs and symptoms, and implementing preventive measures and response protocols, fire departments and rescue agencies can protect the health and safety of their teams during confined-space incidents. Ongoing training, monitoring, and adaptation of heat and cold safety practices are essential for maintaining resilience and readiness in challenging environments.

See **TABLE 3-5** for a summary of preventive measures for heat- and cold-related injuries. The following are also important to consider:

- **Safety training for hot and cold weather.** Provide comprehensive training on recognizing, preventing, and responding to heat- and cold-related injuries.
- **Symptom recognition.** Educate responders on the signs and symptoms of heat- and cold-related illnesses and the importance of early intervention.
- **Cultural awareness.** Consider cultural factors that may influence perceptions of heat or stress and adaptation strategies.

Most of all, remember that constant monitoring of vital signs, especially body temperature, will help ensure rapid recognition of heat- or cold-related illnesses and will put your team in a position to provide proper treatment and prevent more serious injury.

As this section on safety, communication, and medical protocols has shown, operating in and around confined spaces poses many risks to responders. To better understand how to safely operate in and exit areas adjacent to a confined space, follow the steps in **SKILL DRILL 3-3**.

TABLE 3-5 Preventive Measures for Heat- and Cold-Related Injuries

Heat-Related Injuries	Cold-Related Injuries
▪ **Hydration:** Encourage regular hydration with water or electrolyte drinks before, during, and after confined-space operations ▪ **Appropriate clothing:** Wear lightweight, breathable clothing and use cooling vests or gear if available ▪ **Limit exposure:** Schedule work in confined spaces during cooler times of the day, if possible ▪ **Rest breaks:** Implement regular rest breaks in shaded or cool areas to prevent overheating	▪ **Proper clothing:** Dress in layers with insulated, moisture-wicking materials ▪ **Head and face protection:** Wear hats, face masks, and goggles to protect exposed skin ▪ **Stay dry:** Keep clothing and footwear dry to minimize heat loss ▪ **Limit exposure:** Schedule work in confined spaces during warmer times of the day, if possible

SKILL DRILL 3-3

Operating In and Exiting Areas Adjacent to a Confined Space

Step 1: Identify Hazards and Assess Risks

Conduct a thorough assessment of the area, identifying hazards and evaluating risk levels.

Identify hazards, document findings, and mark high-risk areas.

SKILL DRILL 3-3 CONTINUED

Operating In and Exiting Areas Adjacent to a Confined Space

Step 2: Define Safe Zones

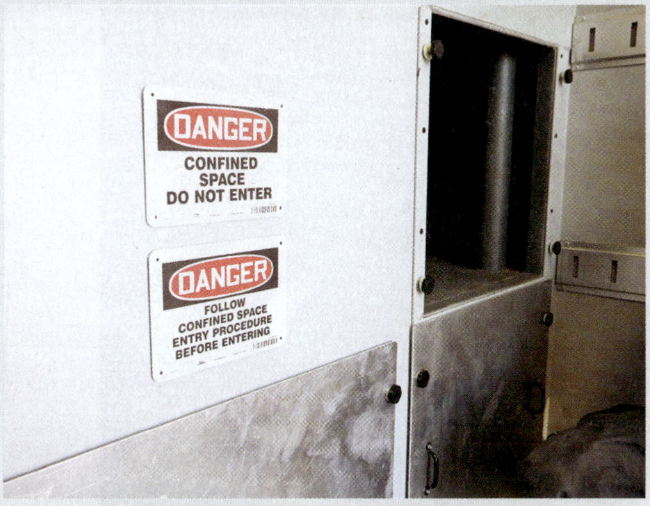

Mark and cordon off designated safe zones.

Step 3: Ensure Clear Communication

The Incident Commander should brief the team on communication protocols and signal use for clear and concise information sharing.

Step 4: Conduct Continuous Monitoring

Actively monitor the environment and communication channels for any changes in conditions or hazards.

Step 5: Ensure Proper PPE Use

Don appropriate PPE, ensuring correct fit and coverage.

Continues.

SKILL DRILL 3-3 CONTINUED

Operating In and Exiting Areas Adjacent to a Confined Space

Step 6: Develop and Practice Exit Strategies

Discuss and practice emergency exit procedures with all team members.

Step 7: Carry Out Operations

Be sure to follow all safety procedures while performing tasks, reassessing conditions as necessary.

Step 8: Demonstrate Proper Exits

Follow all established exit protocols to leave the operational area safely.

Step 8: Courtesy of Illinois Fire Service Institute.

Rehabilitation Needs for Responders

Confined-space rescue operations can be physically demanding and pose significant risks to responders' health and safety. Implementing effective rehab measures is essential to ensure the well-being and readiness of confined-space technician responders.

As you have learned, confined-space technical rescue teams face various physical and physiological challenges during rescue operations, including heat stress, dehydration, exhaustion, extreme cold, or exposure to hazardous substances. Proper rehab is crucial for the following:

- Preventing and addressing fatigue and heat-related illnesses
- Monitoring vital signs and hydration levels
- Restoring physical and mental readiness for continued operations

Designated Rehabilitation Areas

Establishing dedicated areas for responder rehab near the incident site is essential for efficient rehabilitation. **Rehabilitation areas** (or rehab areas) are places where responders can rest, rejuvenate, and receive emotional support during a rescue (**FIGURE 3-12**). A rehab area must also provide some degree of privacy because confidential discussions often take place within them.

At a minimum, rehab areas should provide shelter from environmental elements such as sun, rain, and cold; rest areas that provide seating and shade for hot-weather operations or warm clothing, heated shelter, and hot beverages for cold-weather operations; hydration stations with access to water, nutritious snacks, and drinks containing electrolytes; medical supplies and equipment for monitoring vital signs; and medical evaluation and monitoring spaces. In wet conditions, rehab areas should have nonslip surfaces and access to dry clothing and clean, waterproof gear.

FIGURE 3-12 Team members at a rehab area.

Rehabilitation Equipment

Essential equipment for rehab during confined-space incidents should be placed and ready in rehab areas. Items to have on hand include the following:

- **Portable tents or canopies** to provide shelter and shade in outdoor rehab areas
- **Cooling devices**, such as portable fans, misting systems, or cooling vests, to combat heat stress, or **warming stations** with warm and dry clothing, hot beverages, and portable heaters
- **Hydration stations** featuring water coolers, electrolyte drinks, and hydration packs to maintain responders' fluid balance
- **Medical monitoring equipment**, such as pulse oximeters, blood pressure cuffs, and thermometers, for assessing vital signs
- **Comfort items**, such as folding chairs, blankets, and comfort kits (e.g., snacks, personal care items)
- **Communication tools**, such as radios or mobile phones, which are essential for maintaining contact with incident command and backup teams

Rehabilitation Protocols

Standardized rehab protocols should be in place that outline the proper criteria for assessing the need for rehab, **work/rest cycles** (also known as *rotation schedules*) for confined-space technical rescue teams and their backups (more on that in the next section), procedures for monitoring vital signs and hydration levels, and guidelines for return-to-duty assessments and medical evaluations.

Backup Teams and Rotation

Prolonged confined-space rescue operations are unsustainable without backup personnel. As mentioned earlier, a rotation schedule for confined-space technical rescue teams that allows for seamless deployment of backup personnel will ensure continuous operations without overburdening individual responders.

Backup teams help maintain high levels of situational awareness and team cohesion. If backups are on hand, incident command can promptly address unexpected emergencies or equipment failures. Long-duration events must also have backups for command-level personnel; in these instances, protocols must be set up for proper handoff of control of the operation.

Lifting with Fall Prevention

Many ways exist to lift victims or rescuers out of a confined space. Your equipment should be efficient to use, provide **mechanical advantage** (a force created through mechanical means, including but not limited to a system of levers, gearing, or ropes and pulleys, usually creating an output force greater than the input force) and safety, be simple to operate, prevent falls, and allow you to maintain proficiency with the equipment. Some of the simplest equipment may be mechanical winches designed for confined-space rescue, but the simplest may not always match the circumstances.

An OSHA rule known as the "Six-Foot Rule" states that working surfaces greater than six feet (1.8 m) from the lower level or a work location within six feet (1.8 m) of an unprotected edge must have some sort of fall protection device, such as a guardrail system, a personal fall arrest system, or a travel restraint system. See **TABLE 3-6** for a summary of fall prevention techniques for awareness level responders.

Equipment

Depending on your knowledge and proficiency, rope, pulleys, carabiners, and other types of rope equipment may be best to use. This text is not about using rope and other rope rescue equipment for confined-space rescue—that information is beyond its scope. An additional course of study on rope rescue will be necessary

TABLE 3-6 Fall Prevention Techniques for Awareness Level Responders

Category	Techniques
Hazard recognition	▪ Identify potential fall hazards in locations such as rooftops, scaffolding, ladders, and elevated platforms ▪ Assess worksite for uneven surfaces, slippery conditions, unprotected edges, and inadequate guardrails
Engineering controls	▪ Ensure installation of guardrails, safety nets, or barriers around elevated work areas to prevent falls ▪ Implement measures to secure tools and equipment to prevent objects from falling and causing injuries to workers
Administrative controls	▪ Establish comprehensive fall protection plans with procedures for working at heights and using personal protective equipment (PPE) ▪ Provide thorough training on fall hazards and prevention techniques ▪ Supervise work activities for compliance
PPE	▪ Ensure proper use of harnesses, lanyards, and fall arrest systems when working at heights ▪ Regularly inspect and maintain PPE according to the manufacturer's guidelines
Safe work practices	▪ Maintain good housekeeping to minimize trip hazards and prevent slips and falls ▪ Instruct workers to avoid overreaching and use appropriate equipment (e.g., ladders, scaffolds) for elevated work
Emergency preparedness	▪ Establish rescue procedures and ensure responders are trained and equipped for safe rescue operations
Continuous monitoring and communication	▪ Maintain regular communication with workers and monitor activities for safety compliance ▪ Encourage reporting of unsafe conditions or hazards for prompt corrective action

before attempting a confined space rescue such as a **high-angle rescue** (a form of rescue that occurs in situations where the slope angle exceeds 50 degrees). Rope rescue demands a high level of initial training, and you must be able to maintain proficiency through continuous training.

Retrieval winches are designed for hoisting people and may have stainless steel or galvanized steel cables attached to allow them to be used as lifting devices. The winch provides a mechanical advantage for raising and lowering people. (Please note that not all retrieval winches can be used for controlled lowering—some can be used only for raising.) In addition to the mechanical advantage, other built-in features include fall protection, handle brakes to prevent movement of the load when you release the handle, the ability to adjust the mechanical advantage provided, and a clutch mechanism (shown in **FIGURE 3-13**) to prevent applying the force of the mechanical advantage on an entangled person.

Limitations to the use of retrieval winches include the length of the cables, periodic inspection requirements that mandate that some units be returned to the manufacturer for routine inspection and maintenance, the cost per unit, the need to buy a separate unit for training use, and the manufacturer's requirement that the unit be sent back to the manufacturer for inspection after each rescue use. As with any other equipment you obtain, you must look at the advantages and disadvantages of the equipment and match it to your own needs.

FIGURE 3-13 This retrieval device has a variety of features that make it valuable for confined-space rescue. In addition to being useful in several different positions, it can be used as an untended safety line because it provides fall protection.

A winch with a built-in self-retracting safety line can be a great advantage during confined-space rescue. If you were faced with a limited number of team members and needed to maintain both a safety line and a retrieval line, it could be difficult to maintain the proper amount of slack on the safety line while raising or lowering the person with the retrieval line. The use of a self-retracting safety line allows the line from the winch to freely extend or retract as fast as the wearer moves. If the wearer moves too quickly (such as accelerating during a fall), the line locks and stops the fall. Self-retracting winches limit free fall to 18 inches (45.7 cm) or less. A self-retracting line used during raising or lowering operations is one less line that needs to be tended by a person.

Snap hooks, shown in **FIGURE 3-14**, allow you to connect D-rings and O-rings to the life line. Snap hooks are available with either spring-loaded or double-locking closures, as shown in **FIGURE 3-15**. You should

FIGURE 3-14 A snap hook for connecting O-rings, D-rings, and other equipment to retrieval equipment.

FIGURE 3-15 Match the snap hook to the size of the device to which it is to be connected. The snap hook on the left is a spring-loaded snap hook, and the one on the right is a double-locking snap hook. Double-locking snap hooks are the only type of snap hook that should be used to support human life.

only use double-locking snap hooks for rescue operations, and you must ensure you match the size of the snap hook to the size of the attachment to which it will be connected.

Using a snap hook with a throat opening that is too large can allow an O-ring or D-ring to either side load the gate or slip under the keeper of the gate and slide out of the snap hook. As with any other load, the load that is applied to the snap hook must be along the major axis of the snap hook. Loading the keeper is not recommended because the keeper is tested to withstand only a 350-pound (158.7-kg) side load. Applying a load greater than 350 pounds (158.7 kg) sideways to the keeper of the snap hook can deform the keeper and leave an opening where the keeper was. If the throat opening of a snap hook is greater than 1 inch, a locking feature should be included in the design.

TIPS

- Be vigilant in identifying workers engaging in unsafe practices related to working at heights.
- Encourage a strong safety culture where workers prioritize fall prevention and actively participate in safety initiatives.
- Refer complex fall prevention issues or incidents to qualified safety professionals for further evaluation and intervention.
- Manage slack in retrieval or lowering lines to minimize fall distances.

By adhering to these fall prevention techniques and key points, awareness level responders can effectively contribute to a safer work environment and assist in mitigating fall-related risks and incidents. Awareness of these measures is critical for recognizing hazards, promoting safety practices, and supporting overall safety initiatives at the worksite.

Like all the rest of your equipment, snap hooks must also be inspected and maintained. Follow the manufacturer's instructions. Look for corrosion of the metal; watch for distortion of the snap hook and keeper; and monitor the functionality of the swivels, springs, and other movable parts of the snap hook.

Protect the integrity of your rescue equipment. Using rescue equipment to raise or lower equipment, tools, or materials into or out of a confined space can damage the equipment and should not be allowed. Rescue equipment should be reserved for use only to support human life. Yes, you can use the rescue equipment to move people into or out of a confined space during routine work practices, but raising or lowering equipment or tools can overload the tripod, winch, or other apparatus associated with the rescue equipment. The damage caused by the overload may not be visible or obvious, and the equipment can fail without warning.

Equipment used for rescue purposes must be capable of performing to its design limits during an emergency. The demands placed on the equipment and the need for total reliability mandate the many standards for rescue equipment. There is a certain expectation that equipment used during rescue will be subject to extraordinary loads so that it will not be easily overstressed and fail.

Using hoisting equipment that has been used to lift tools and equipment and may have been overstressed can cause the failure of the equipment and injure or kill both the victims and rescuers. You may see claims by some manufacturers that their tripods and associated equipment can be used to hoist people and equipment. How do you know whether the equipment had too large a load on it? Equipment not dedicated solely to supporting human life can be used for anything. Did someone borrow a tripod or hoist and use it to lift an engine out of a car? After all, it is an equipment lift, too. Not everyone appreciates the importance of this equipment and knows how to use it properly.

Although this problem is difficult to manage, consider the consequences of not managing it. If you plan on managing the risks associated with the rescue problem, you must start before the emergency and make sure that your equipment can be used with confidence.

Victim Retrieval

To meet NFPA 2500 (NFPA 2022), each member of the rescue team must be certified as an emergency medical responder (EMR), have cardiopulmonary resuscitation (CPR) training, and perform a primary survey for any victims in the space (airway, breathing, and circulation). A victim with a compromised airway cannot get air. If a victim is lying face down in a puddle of water, the airway is not maintained. Given enough time, no airway means death. When a victim is breathing, or in cases where you can reestablish the airway to the point where the victim begins breathing on their own, the problem is solved. But what about a victim who is not breathing? How can you provide rescue breathing while wearing respiratory protection in a contaminated atmosphere?

As this text is being written, CPR methods are changing, and rescue breathing may become a thing of the past for first responders. However CPR may change, work to get a rescuer into the confined space as soon as possible. Once the rescuer is in the space, they can provide a self-contained source of oxygen for the victim. You should consider whether it is practical to change the atmosphere in the space as quickly as possible. The ultimate goal is to remove the victim from the contaminated atmosphere as quickly as possible.

Performing compressions during CPR in a confined space may be difficult, if not impossible. Again, getting the victim out of the space as soon as possible resolves that problem, but it may present the issue of how to handle a spinal injury. If your victim dies from the atmosphere in the space, the other injuries will not matter.

When your victim is viable (airway, breathing, and circulation appear good), look for injuries and prepare your patient for transport. Above all, follow local protocols for treating victims. This text is not intended to provide comprehensive information on the treatment of victims.

Victim Stabilization

How was your victim injured? Did they fall? Was there a preexisting medical condition? Or were they exposed to a hazard within the space that caused them to fall or pass out? Evaluate the space and work from the facts that you know or develop. Yes, a fall can produce trauma injuries, and you should treat victims as trauma patients. But what caused the fall? Was it some hazard associated with the confined space, or did the victim simply fall as a result of something unrelated to the confined space?

Despite the precautions rescuers must take to protect themselves, you must consider whether the victim needs to be immobilized. If so, your rescue stretcher must be able to accommodate a backboard, and you must know how to use the equipment.

Victim Removal

Remove the victim in the manner that is best for the victim. Do not let the removal process become more hazardous than letting the victim remain in the space as you work to create a safer way out. Removing a victim horizontally, as shown in **FIGURE 3-16**, is often better than lifting vertically. If your only available removal method is vertical, ensure your equipment is designed to do the job. Rescue stretchers in the vertical position may make your victim more apprehensive and may present problems with the use of a backboard, and you face the risk of the victim falling out of the stretcher (with or without equipment failure). See **TABLE 3-7** for a summary of victim retrieval actions for awareness level responders.

If you must place the stretcher vertically, as shown in **FIGURE 3-17** and **FIGURE 3-18**, make sure that the victim is properly positioned and secured to the stretcher. Double-check, prior to the lift, that you have fastened the victim to the stretcher as needed for a vertical lift. Do not attempt to use a stretcher that is not designed to be used in the vertical position.

Once again, training is the key to your success. The degree of competency needed to package and lift a patient must be 100 percent. Load test the stretcher and equipment before you begin to raise the stretcher out of the space. Bringing the stretcher just off the ground and allowing the victim's full weight and the stretcher to load the equipment is one way to do that. Don't forget to check fasteners, knots, straps, and the like to be sure they are in the proper position and are holding the victim in place. An improperly positioned strap or knot may be uncomfortable for the victim, but it may also cause additional injury. Only after you are satisfied that everything checks out okay should you begin the lift. If conditions are not satisfactory, take the load off the stretcher and fix the problem. As stated earlier, your level of training and competency are invaluable. You will not have a procedural checklist in the space with you, so you must rely solely on your training.

Another reason for packaging a victim, aside from injury, would be to maintain an airway. If a victim must be lifted vertically, hooked to the dorsal attachment of a harness, the victim's airway would be closed for the duration of the lift.

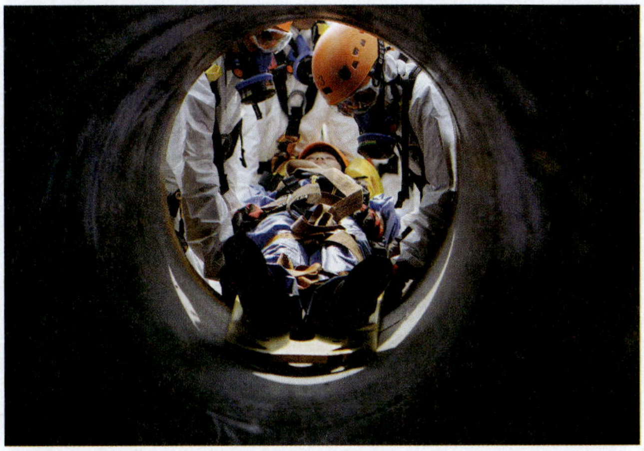

FIGURE 3-16 The location of your victims and your access to them will affect how you can remove them.

TABLE 3-7 Victim Retrieval Actions for Awareness Level Responders

Step	Actions
Recognize hazards	▪ Approach the confined space cautiously from a safe location outside ▪ Assess potential hazards, such as atmospheric (e.g., low oxygen, toxic gases), physical, and biological hazards ▪ Identify common confined-space hazards based on National Fire Protection Association (NFPA) 1006 guidelines (NFPA 2021b)
Initiate emergency response	▪ Immediately notify emergency services and qualified rescue personnel ▪ Provide accurate details about the confined-space incident and the victim's condition ▪ Activate emergency procedures according to established protocols
Establish communication	▪ Attempt to establish communication with the victim using safe methods (e.g., verbal communication, signals) ▪ Maintain continuous communication with the victim and relay information to incoming rescue teams using two-way radios or communication devices
Ensure scene safety	▪ Set up barriers or warning signs to control access to the confined space ▪ Continuously monitor the atmosphere for changes and potential hazards using gas detection equipment if available
Assist incoming rescue teams	▪ Brief incoming rescue team on confined-space hazards and victim's condition (if known) ▪ Provide support to the rescue team with logistics, equipment setup, and coordination for safe and effective rescue operations
Document and report	▪ Record detailed observations, actions taken, and communications during the incident ▪ Prepare a comprehensive incident report documenting circumstances, actions, and outcomes for review and analysis
Safety considerations	▪ Wear appropriate personal protective equipment (PPE), such as helmets, safety vests, and eye protection ▪ Adhere to established procedures and protocols for confined-space emergencies ▪ Continuously monitor the situation and communicate effectively with all parties involved to ensure a coordinated and safe response

FIGURE 3-17 The narrow width of these stairs might make it impossible to carry a victim to the ground, requiring vertical removal.

FIGURE 3-18 The location of this manhole, on a steep hillside, combined with the narrow opening, will require you to remove the victim in a vertical position.

CHAPTER FEATURE: AFTER-ACTION

IN SUMMARY

- Instilling a safety mindset into the culture of your team does not happen overnight but rather is the result of many hours of training and the discipline that comes from everyone on the team being accountable.
- Confined-space rescue PPE can be classified into two main categories:
  - Personal-issue PPE: The garments and equipment that each member receives and keeps with them, likely in a personal bag.
  - Team-issue PPE: Equipment that is kept on the response vehicle and shared (i.e., breathing apparatus).
- At a minimum, confined-space rescue team members need an ensemble element consisting of protection for the upper and lower torso, arms, legs, head, hands, feet, and eyes, as well as respiratory protection. Other items, depending on the environment expected, could include hearing protection and foul-weather gear.
- Communication protocols, such as radios, signaling systems, or visual cues, must be planned and executed throughout a response.
- Medical monitoring is necessary to ensure the safety of rescuers. This includes monitoring rescuers' hydration levels and their exposure to hot or cold environments. Conducting a rescue in extreme temperatures, whether they are hot or cold, can lead to injury. Heat cramps, heat exhaustion, and heat stroke are risks in hot environments, and hypothermia and frostbite are risks in cold environments.
- A comprehensive mental health program helps responders develop resilience, enabling them to cope better with the high-stress situations inherent in confined-space incidents.
- Effective rehab measures are essential to ensure the well-being and readiness of responders.
- Fall prevention equipment must be capable of lifting victims or rescuers out of a confined space, be efficient to use, provide mechanical advantage and safety, be simple to operate, prevent falls, and allow you to maintain proficiency with the equipment.
- Effective retrieval of a victim generally requires the victim to be immobilized and packaged for removal.

KEY TERMS

American National Standards Institute (ANSI) A nonprofit organization that administers and coordinates the U.S. voluntary standards and conformity assessment system.

extremities The arms, hands, legs, and feet.

frostbite A medical condition in which body tissues are injured as a result of exposure to extreme cold, typically affecting the nose, fingers, or toes.

heat cramps A medical condition in which painful muscle spasms occur during activity in hot environments.

heat exhaustion A medical condition in which the body responds to excessive losses of water and salt, usually through excessive sweating.

heat stroke A medical condition in which the body's temperature rises rapidly, the sweating mechanism fails, and the body is unable to cool down.

high-angle rescue A rescue that occurs in situations where the slope angle exceeds 50 degrees.

hypothermia Lowering of the body's core temperature.

mechanical advantage A force created through mechanical means, including but not limited to a system of levers, gearing, or ropes and pulleys, usually creating an output force greater than the input force.

medical monitoring Basic medical evaluation of emergency response personnel by determining and recording basic vital signs such as blood pressure, respirations per minute, and pulse.

rehabilitation area A place where responders can rest, rejuvenate, and receive emotional support during a rescue.

retrieval winch A mechanical device designed to allow the operator to wind up or wind out a rope or cable

KEY TERMS CONTINUED

to support a person being raised or lowered into or out of a confined space.

snap hook A hook with a spring that allows the entrance but prevents the escape of something.

work/rest cycle A rotation schedule for confined-space technical rescue teams and their backups.

REFERENCES

International Safety Equipment Association. 2020. *American National Standard Occupational and Educational Personal Eye and Face Protection Devices.* ANSI/ISEA Z87.1-2020. Arlington, VA: International Safety Equipment Association.

National Fire Protection Association. 2020. *NFPA 1951: Standard on Protective Ensembles for Technical Rescue Incidents, 2020 Edition.* Quincy, MA: National Fire Protection Association.

National Fire Protection Association. 2021a. *NFPA 1500: Standard on Fire Department Occupational Safety, Health, and Wellness Program, 2021 Edition.* Quincy, MA: National Fire Protection Association.

National Fire Protection Association. 2021b. *NFPA 1006: Standard on Technical Rescue Personnel Professional Qualifications, 2021 Edition.* Quincy, MA: National Fire Protection Association.

National Fire Protection Association. 2022. *NFPA 2500: Standard for Operations and Training for Technical Search and Rescue Incidents and Life Safety Rope and Equipment for Emergency Services, 2022 Edition.* Quincy, MA: National Fire Protection Association.

"Permit-Required Confined Spaces." 1993. 29 C.F.R 1910.146. https://www.ecfr.gov/current/title-29/part-1910/section-1910.146.

"Respiratory Protection." 1998. 29 C.F.R. 1910.134. https://www.ecfr.gov/current/title-29/subtitle-B/chapter-XVII/part-1910/subpart-I/section-1910.134.

CHAPTER QUESTIONS

1. What is the first priority of incident response?
   a. dispatch information
   b. size-up
   c. additional resources
   d. life safety
2. A backup rescue team member's equipment should be
   a. located in the staging area.
   b. worn by the member so that they are ready to go.
   c. near the entrance for quick use in an emergency.
   d. stored out of the way until it is needed.
3. The "Six-Foot Rule" states that
   a. working surfaces in a confined space greater than six feet (1.8 m) from the lower level must have fall protection.
   b. working surfaces in a confined space less than six feet (1.8 m) from the lower level must have fall protection.
   c. working surfaces in a confined space must be located six feet (1.8 m) apart.
   d. working surfaces in a confined space require a six-foot (1.8-m) clearance.
4. Rehabilitation protocols include a rotation schedule known as a
   a. work/return cycle.
   b. work/rest cycle.
   c. work/release cycle.
   d. work/reassignment cycle.

CHAPTER FEATURE: ON SCENE

A TRENCH RESCUE

Many may not think of a trench rescue as also being a confined-space rescue event (in fact, OSHA does not classify a trench as a confined space), but the methods and procedures for trench and confined-space rescues go hand in hand, and if a trench meets the three conditions of a confined space, it should be treated as such. Although OSHA addresses trenches and excavations in a separate standard, many of the same hazards identified exist in trench rescue. An event involving a collapsed trench occurred in the summer of 2019.

In this incident, a construction worker became trapped in a collapsed trench at a worksite. The trench collapse occurred while the worker was performing excavation work, resulting in the worker becoming partially buried under soil and debris.

The Emergency Response

Following the collapse, emergency services, including the fire department and an urban search and rescue team, rushed to the scene. Upon arrival, the rescuers found the worker trapped in a trench with limited access for extraction.

The Rescue Operations

The fire department and urban search and rescue teams used shoring and trench rescue techniques to stabilize the collapsed trench and prevent further collapse. Shoring equipment, including pneumatic shores and trench boxes, was strategically deployed to support the trench walls and create a safe workspace for rescuers.

The rescuers worked meticulously to carefully remove soil and debris from around the trapped worker while ensuring their safety and preventing further collapse. Throughout the rescue operation, atmospheric monitoring equipment was used to monitor the air quality inside the trench and ensure the safety of both the rescuers and the trapped worker.

The Extrication and Treatment

After several hours of meticulous work, rescuers successfully extricated the trapped construction worker from the collapsed trench. Emergency medical personnel provided immediate medical attention on-site before transporting the worker to a nearby hospital for further evaluation and treatment.

The Outcome

This successful rescue operation highlights the importance of rapid emergency response, effective coordination among emergency services, and specialized training in trench rescue techniques, along with knowledge of confined-space rescue techniques.

This incident underscores the inherent risks associated with excavation and trenching work and reinforces the critical need for proper safety protocols and equipment to prevent accidents and respond effectively to emergencies in confined spaces.

This real-life collapsed trench rescue event exemplifies the dedication and expertise of emergency responders in safeguarding lives and ensuring the safety of workers in hazardous work environments. It also serves as a valuable case study for improving safety practices and emergency response procedures in construction and excavation operations.

Another notable and important highlight is that technical rescue is often an all-inclusive activity. Having expertise within a specific area is always beneficial in any circumstance, of course, but having a varied knowledge of the entirety of the rescue discipline is essential for emergency response technical rescue teams.

1. What was the primary cause of the emergency in this trench rescue scenario?
   - **a.** worker falling into the trench
   - **b.** equipment malfunction during excavation
   - **c.** trench collapse during excavation
   - **d.** chemical leak inside the trench
2. What was the first action taken by emergency services upon arriving at the scene?
   - **a.** immediate extraction of the trapped worker
   - **b.** stabilizing the trench with shoring equipment
   - **c.** administering first aid to the worker
   - **d.** evacuating all other workers from the site

A TRENCH RESCUE CONTINUED

3. What equipment was used to support the trench walls during the rescue operation?
   - a. cranes and scaffolding
   - b. heavy-duty ropes and harnesses
   - c. hydraulic lifts and platforms
   - d. pneumatic shores and trench boxes

4. Why was atmospheric monitoring equipment used during this trench rescue operation?
   - a. to check for flammable gases
   - b. to ensure air quality and the safety of both rescuers and the trapped worker
   - c. to detect underground water leaks
   - d. to ensure the trapped worker's safety and prevent further collapse

5. What additional insight does this incident provide regarding technical rescue?
   - a. Technical rescue teams benefit from having knowledge across different rescue disciplines.
   - b. Technical rescue teams should have expertise in a single area.
   - c. Trench rescue operations should be handled separately from confined-space rescues.
   - d. Trench rescue specialists should only be involved in technical rescue.

CHAPTER

4

Awareness

Criteria for a Basic Search and the Incident Action Plan

KNOWLEDGE OBJECTIVES

After studying this chapter, you should be able to:

- Describe the criteria for a basic search. (**NFPA 1006: 7.1.5**, pp. 94–95)
- Explain how to determine the need for additional resources. (**NFPA 1006: 7.1.6**, pp. 95–96)
- List the elements of an incident action plan (IAP). (**NFPA 1006: 7.1.5**, pp. 96–103)
- Describe how a size-up is incorporated into an IAP. (**NFPA 1006: 7.1.5,** p. 103)
- Explain how to gain information from witnesses during the incident size-up. (**NFPA 1006: 7.1.5**, pp. 103–104)
- Discuss how operational protocols are determined by the authority having jurisdiction (AHJ). (**NFPA 1006: 7.1.6**, pp. 104–107)

SKILL OBJECTIVES

- Demonstrate the methods used to identify the required resources for a confined-space incident. (**NFPA 1006: 7.1.1**, pp. 95–96)
- Ascertain the needed resources for a confined-space incident. (**NFPA 1006: 7.1.6**, pp. 95–96)
- Communicate needed information while following an IAP while functioning in an Incident Command System (ICS). (**NFPA 1006: 7.1.7**, pp. 98–100)
- Select appropriate planning forms and document hazards within the AHJ. (**NFPA 1006: 7.1.6**, pp. 105–107)

Introduction

This chapter delves into the fundamental aspects of conducting a basic search, determining the necessity for additional resources, and formulating an incident action plan (IAP). In this chapter, you will explore how the initial and continuous size-up assessments may be integrated into an IAP and the establishment of operational protocols by the authority having jurisdiction (AHJ).

When an incident occurs, the primary objective is to locate and assist affected individuals swiftly and safely. A basic search is the foundational process in this endeavor. It involves a set of criteria designed to maximize efficiency while ensuring the safety of search personnel.

Incidents can vary significantly in scope and complexity, often necessitating additional resources beyond what is immediately available. Assessing the situation, evaluating current resources, and identifying gaps that require external support provide a framework for decision-making that ensures adequate resources are deployed to manage the incident effectively.

As you learned in Chapter 2, an IAP is a blueprint for managing an incident that outlines strategies, objectives, and resource allocations, and a size-up is a continuous process of evaluating an incident to gather critical information. An IAP includes incident objectives, tactical plans, safety measures, and communication protocols, and the size-up is a critical element of its creation.

At an incident scene, the AHJ plays a critical role in defining the operational protocols that guide response efforts. The AHJ establishes these protocols based on its legal authority, standard operating procedures, best practices, and available resources. If the AHJ's directives and the incident response activities are not in alignment, the response effort will be compromised.

The Awareness Level Responder's Nonentry Roles

As detailed in National Fire Protection Association (NFPA)1006 Job Performance Requirement (JPR) 7.1.4, awareness level responders are expected to have the training to recognize and respond to emergency situations without engaging in direct rescue operations that require entry into a hazardous environment (NFPA 2021). Vital support roles that awareness level responders can play in performing rescues without entering a hazardous area include operating equipment and systems that can facilitate the rescue of individuals from a safe distance, placing ladders to provide an escape route or assist in the rescue of others, and using retrieval systems designed to extract others. These can include mechanical winches, rope systems, or other retrieval devices.

Awareness level responders must be able to identify potential hazards associated with the rescue scenario and take appropriate precautions to mitigate the risks they present. They must also have proficiency in the setup and operation of retrieval systems, ensuring that they are used effectively to extract individuals without requiring entry into the hazardous environment. Effective communication skills are necessary to coordinate with other responders and provide clear instructions during the rescue operation.

Other areas of support that awareness level responders can provide include the following:

- **Starting ventilation:** The responder should be able to initiate and use ventilation equipment to improve air quality and remove hazardous atmospheres from the incident area.
- **Limiting access:** The responder should implement measures to control and restrict access to the incident site, ensuring only authorized personnel are present.
- **Marking work zones:** The responder should be able to clearly mark and define work zones using appropriate signage, barriers, and markers to ensure a safe and organized working environment.
- **Implementing safety measures and fall prevention strategies:** The responder should be proficient in safety procedures and setting up fall prevention measures such as barriers, warning tape, and other safety devices to protect personnel from fall hazards.

Criteria for a Basic Search

Performing search operations within a confined space requires careful planning, adherence to safety protocols, and specialized training for a rescue team. The criteria for conducting basic search operations in such environments involve several key considerations, including air-monitoring requirements, specialized equipment needs, and training in the use of ropes for rescue purposes.

Before a search can begin, a thorough risk assessment must take place. The risk assessment will identify potential hazards and will help you develop control measures to mitigate risks. Next, you must establish clear emergency procedures and protocols for responding to unforeseen circumstances or incidents during the search operation. Effective teamwork, communication, and leadership are necessary throughout the planning and risk assessment phases, and regular debriefings and feedback sessions will help the team continuously improve its performance.

Air-Monitoring Requirements

Before entering a confined space, you must conduct an initial assessment of its atmospheric conditions. A portable atmospheric monitor will allow you to check oxygen levels and scan the space for the presence of flammable gases and toxic substances.

When monitoring for entries involving a descent into atmospheres that may be stratified, the atmospheric envelope should be tested at a distance of approximately four feet (1.22 m) in the direction of travel and to each side. If a sampling probe is used, the entrant's rate of progress should be slowed to accommodate the sampling speed and detector response.

As the search operation goes on, you must continuously monitor the space's air quality. The atmospheric monitor you use must be capable of providing real-time readings and alarms if hazardous conditions are detected. Rescuers cannot enter a confined space if the air quality is outside safe limits. The team must refer to established occupational safety standards and regulations for acceptable air quality levels.

During a search, the team must always take into consideration the effects of air movement on gas concentrations within a confined space. Continuous monitoring of different levels within the space will help ensure that uniform air quality is maintained, no matter which areas of the space rescuers must enter. A space may be small enough that a few rescuers introduced into the area could lead oxygen levels to drop. If an atmospheric monitor alarms at any time, rescuers must immediately evacuate the space.

Specialized Equipment Needs

A prepared rescue team arrives with the equipment they will need to perform the basic search. This includes personal protective equipment (PPE) and equipment designated for communication, rescue, proper lighting, ventilation, and lockout/tagout.

Appropriate PPE for a basic search includes respiratory protection for immediately dangerous to life or health (IDLH) environments, helmets, gloves, and coveralls. All equipment must be properly fitted and well maintained. Reliable communication devices include two-way radios or other types of intrinsically safe communication systems for lower explosive limit (LEL)/upper explosive limit (UEL) atmospheres that will enable rescuers to maintain contact among team members, both inside and outside the confined space.

Specialized rescue equipment, including harnesses, tripods, hoists, and winches, is required for safe entry and exit of a confined space and the retrieval and evacuation of victims. Adequate lighting sources are required to illuminate a confined space during the search operation, especially in areas with low visibility. Ventilation equipment must be fully set up and in place prior to any entry into a confined space.

Specialized Training in Rope Rescue

Rescue team members must undergo specialized training in technical rope rescue techniques. This training should cover rope systems, knots, anchor points, mechanical advantage systems, and patient packaging, and it must be specific to rope operations within confined spaces. This training must place emphasis on the unique challenges and hazards associated with working at heights or in tight spaces.

Regular practical exercises and simulations are necessary to reinforce rope rescue skills and enhance team coordination. Team members should be encouraged to obtain relevant certifications in rope rescue, such as those offered by recognized organizations or training institutions.

Many confined spaces in industry have top or side entry points that are elevated and accessed by ladder onto a platform or catwalk. An injured or unconscious victim in a space like this must be lowered to ground level using rope systems, so teams must be proficient at using rope-lowering systems.

Determining the Need for Additional Resources

Determining the need for additional resources is a dynamic process that requires careful consideration of various factors, thorough assessment methodologies, and informed decision-making. By prioritizing safety, optimizing efficiency, and maintaining flexibility, responders can effectively address the complexities and challenges inherent in confined-space incidents, ensuring a swift and successful outcome.

The Initial Resource Assessment

Upon initial response to a confined-space incident, the incident commander (IC) must swiftly assess the situation and deploy available resources accordingly. This assessment includes evaluating the nature and severity of the incident, the size and complexity of the confined space, the number and condition of personnel involved, and the existing equipment and expertise onsite.

Factors Influencing Resource Needs

The requirements for additional resources in confined-space incident response are influenced by the complexity of both the incident and the rescue operations, the confined space's size and configuration, the duration of the response, the space's environmental conditions (as well as the environmental conditions outside the space), and the safety of the responders.

Incidents involving multiple hazards, entrapment, or hazardous materials may necessitate specialized equipment and expertise that is not readily available on the scene. Larger or more intricate confined spaces may require additional personnel for effective monitoring, communication, and rescue operations. Prolonged incidents may deplete existing resources, necessitating reinforcements to sustain operations over an extended period.

Harsh environmental factors such as extreme temperatures, toxic atmospheres, or limited access may mandate specialized equipment and personnel trained to operate in such conditions. The need for technical rescue capabilities, such as rope rescue, confined-space entry, or hazardous materials response, may arise depending on the intricacy of the rescue scenario. If existing resources are insufficient to ensure the safety of responders, additional support may be required to mitigate risks effectively.

Specialized teams, equipment, and resources often require significant time to ready and deploy. Plan for early notifications to ensure these resources are available.

Assessment Methodologies

ICs have several assessment methodologies at their disposal to determine the need for additional resources, including risk assessments, resource inventories, consultations with subject-matter experts, and incident progression monitoring.

A comprehensive risk assessment permits the IC to identify potential hazards, assess their severity, and evaluate the adequacy of existing resources to mitigate these risks. A resource inventory allows the IC to take stock of the available personnel (as well as their expertise) and equipment and compare them against the requirements of the incident to identify gaps and deficiencies. Solicitation of input from specialized personnel, such as technical rescue teams, hazardous materials technicians, or medical professionals, will help the IC evaluate whether they will be needed based on the specific circumstances of the incident. Incident progression monitoring allows the IC to reassess resource needs as the situation evolves and adapt to changing conditions and emerging challenges.

The Decision-Making Process

Once ICs assess the factors influencing the resource needs of the response and determine which methodologies will be employed, they must make informed decisions regarding the allocation of additional resources. This decision-making process should be guided by twin overarching goals: prioritizing responder safety and maximizing the effectiveness of the response effort.

The Incident Action Plan

The development of the IAP is the point at which the transition from preparation to action takes place. No one person or piece of equipment can accomplish a confined-space rescue. It takes a coordinated plan to complete the rescue because there is interdependency between the different parts of your plan.

TIP

If using generators to power equipment (e.g., ventilation fans), remember that the generator will produce carbon monoxide that must not blow back into the space. Generators must be placed far enough away from the space to eliminate this risk. Better yet, use battery-powered equipment so that generators are not required in or near the space. Batteries can be replaced and recharged in a staging area.

Before a space can be entered, it may require ventilation, but before ventilation can occur, someone must bring the fan where it is needed. Simply shouting, "Get a fan!" is far from productive. Someone must be tasked with getting the fan. Someone else must bring an extension cord for the fan. Yet another person must find an electrical power source for the extension cord.

ICs develop action plans, and they put them into action by directing others—no matter how complex the emergency may be. Sometimes, putting a plan into action is simple, but other times, a plan is complex and requires the execution of tasks simultaneously to achieve its objectives, as shown in **FIGURE 4-1**.

If the IC doesn't create an IAP, others may be forced to create their own competing plans. It's like the old saying: When you fail to plan, you plan to fail.

The IAP Framework

An IAP outlines the objectives, strategies, and tasks required to mitigate the incident and safely rescue any individuals in distress. It has six areas of focus: the incident overview, the incident objectives, the organizational structure and responsibilities, the operational strategies to be used, the incident tactics, continuity and follow-up, and incident review and debrief. See **FIGURE 4-2** for a sample outline of an IAP.

In the incident overview section, the IAP will feature a full description of the confined-space incident. This includes a brief summary of the incident, its location, the type of confined space the incident occurred within, the number of individuals involved, and any known hazards or risks. It also includes the current status of the incident, including the environmental conditions present, the potential for exposures (e.g., the

FIGURE 4-1 In developing an action plan, identify the sequence of actions needed to complete your operation. Equipment (such as retrieval equipment) may need to be in place prior to entry.

Sample IAP Outline

1. Incident Overview
- Location: *[Specify location]*
- Type of Incident: Confined-space rescue
- Number of Victims: *[Specify number]*
- Hazards/Risks: *[Identify key hazards]*
- Rescue vs. Recovery: *[Identify objective]*
- Entry Rescue vs. Nonentry Rescue: *[Identify rescue type]*

2. Incident Objectives
- Primary Objective: Rescue trapped individuals safely and efficiently.
- Secondary Objectives: Ensure responder safety, coordinate resources, and maintain effective communication.

3. Organization and Responsibilities
- Incident Commander: *[Name]*
- Safety Officer: *[Name]*
- Operations Personnel: *[Assign roles and responsibilities]*

4. Operational Strategies
- Safety Procedures: Use appropriate PPE, conduct air monitoring, and follow confined-space entry protocols.
- Communication Plan: Establish radio channels, designate communication points, and implement accountability systems.
- Rescue Operations: Deploy specialized teams for confined-space entry and victim retrieval.

5. Incident Tactics
- Size-Up and Assessment: Identify hazards, assess risks, and establish control measures.
- Control Measures: Implement lockout/tagout procedures, ventilate the space, and establish access control.
- Rescue Tactics: Use rope rescue techniques, coordinate victim extraction, and provide medical support.

6. Continuity and Follow-Up
- Shift Changes: Plan for operational continuity and rest periods for responders.
- Documentation: Maintain accurate records of incident actions, decisions, and outcomes.

7. Incident Review and Debriefing
- Postincident Evaluation: Conduct a debriefing session to review incident response, identify lessons learned, and document recommendations for improvement.

FIGURE 4-2 A sample IAP outline.

presence of hazardous gases or lack of ventilation), and if known, the condition of any victims in the space.

The incident objectives section clearly states the primary goal of the incident response. This could include rescuing trapped individuals, securing the scene, or mitigating any hazards that are present. It also lists any secondary objectives of the response, such as ensuring responder safety, maintaining clear communication, and effective coordination of resources. To better understand how to communicate information gathered during a size-up in an IAP, follow the steps in **SKILL DRILL 4-1**.

SKILL DRILL 4-1

Communicating Information While Following an IAP

Step 1: Review the IAP

Review the IAP with the team, highlighting key objectives, tasks, and communication strategies.

Step 2: Set Up Communication Protocols

Test all communication equipment. The Incident Commander should brief team members on communication protocols.

Step 3: Use Proper Radio Etiquette

Practice clear, standardized radio etiquette to facilitate effective communication.

SKILL DRILL 4-1 CONTINUED

Communicating Information While Following an IAP

Step 4: Relay Updates

Relay all real-time updates to the command center. As updates stream in, the command center should document them.

Step 5: Communicate with Different Teams

Maintain constant communication between command and field teams, ensuring coordination across different operational areas.

Step 6: Keep Detailed Records

Document communications in a logbook or electronic record for later review and accountability.

Step 7: Review Communication Effectiveness

Conduct a debrief to evaluate the effectiveness of the communication protocols used, lessons learned, and improvements for future incidents.

Steps 4 and 6: Courtesy of Illinois Fire Service Institute.

The organizational structure and responsibilities section contains critical information about the incident command system (ICS) that has been devised for the response. The ICS description lists the designated roles and responsibilities for ICs, safety officers, operations personnel, and all other key positions. This section also details the resources that are available for the response, including personnel, equipment, and all specialized teams, as well as the allocation of those resources based on the needs of the operation.

The operational strategies section details the safety procedures and protocols that will be followed during the response. This includes the use of PPE and procedures for entry and air monitoring of the space. Also within the operational strategies is the full communication plan (inclusive of information flow among responders, ICS, and any involved external agencies) and the overall strategies for the rescue operation, including entry methods, victim retrieval techniques, and medical considerations (specialized rescue strategies, such as rope rescue techniques and victim stabilization methods, are included in the incident tactics section that immediately follows).

In the incident tactics section, procedures are defined for conducting the size-up assessment, hazard identification, risk evaluation, and scene evaluation. This section also contains the control measures planned for the rescue operation, including the isolation of energy sources, ventilation methods, and the establishment of exclusion zones.

The IAP's continuity and follow-up section lists the plan for ongoing operations, such as shift changes, rest periods, and procedures for continuous monitoring of the scene. The follow-up portion contains documentation of all actions taken, emergent developments, and lessons learned.

The last section of the IAP, the incident review and debrief, is a postincident evaluation, so it is completed after the termination of the response. This section requires a debriefing session at which the team will review the actions taken during the incident response and identify strengths and weaknesses the response uncovered. At the session, the team will document key takeaways for future incidents. See **TABLE 4-1** for a summary of the process of building an IAP.

Implement the Action Plan

By this point in the emergency, the IC must know what needs to be done and what is needed for success. Now that the plan is set, others must execute it, as shown in **FIGURE 4-3**. Everyone should know their roles, as well as how and when they are expected to perform them.

Evaluate the Effectiveness of the Action Plan

Review what is happening at the scene. Is it what you expected to happen? Or is it different? You must assess your progress because action plans rarely run exactly

TABLE 4-1 Building an Incident Action Plan (IAP) for a Confined-Space Rescue

Understand incident objectives	Clearly define the primary objectives of the rescue operation, such as victim retrieval, hazard mitigation, and responder safety
Assess available resources	Evaluate the resources needed for the operation, including personnel, equipment, and specialized tools required for confined-space rescue
Develop operational strategies	Outline specific strategies and tactics to achieve the identified objectives, considering the unique challenges of confined-space environments
Establish safety protocols	Prioritize safety measures for responders and victims, including atmospheric monitoring, personal protective equipment, and communication protocols
Assign responsibilities and tasks	Delegate roles and responsibilities to team members based on their expertise and capabilities, ensuring clarity and accountability during the rescue operation
Communicate and coordinate	Establish clear lines of communication and coordination among rescue team members, the incident commander, and support personnel to facilitate effective teamwork
Review and update continuously	Regularly review and update the IAP based on evolving incident conditions, feedback from responders, and changes in operational requirements

as you intended. At times, the plan will be effective, and you will not need to make many adjustments. At other times, nothing will seem to work; you will make constant adjustments, but you will struggle to complete the objectives you created. Evaluating the plan and its progress must be an ongoing process, as shown in **FIGURE 4-4**. Continue your evaluation of the effectiveness of your action plan until the incident is terminated.

FIGURE 4-3 Implement the action. Once you are set up and ready to go, begin your operation.

FIGURE 4-4 Evaluate the effectiveness of your operation. How are your personnel, equipment, and other resources performing?

Terminate the Incident

The completion of a rescue or recovery is not the end of the operation. Depending on the circumstances of the emergency, many steps may be required to bring true closure to the incident. At a minimum, all entrants must be removed, accounted for, and decontaminated; the scene must be secured, and control of it must be returned to the AHJ; in some cases, OSHA notifications must be completed; equipment must be inspected and decontaminated; required maintenance must be performed; damaged equipment must be taken out of service; and required inspection, maintenance, and service information must be recorded.

Conduct the Debrief

That takes care of the equipment, but what about the people? All involved personnel must be debriefed to establish accurate information about the incident for analysis and critique. This analysis should prove the effectiveness of operations (or identify weaknesses), identify needed improvements for future incidents (as shown in **FIGURE 4-5**), and provide a factual basis for any investigation of the incident that may follow. If a serious injury or death occurs during an incident, an investigation will follow—and lawsuits may ensue. Recording all information at the time of the incident is much better than trying to remember it months or years later.

Incident debriefing has great value beyond simply listing lessons learned. It also provides an opportunity to present information to emergency responders

FIGURE 4-5 Debriefing rescue workers after an incident can provide valuable information for future use. These students are being debriefed after completing a training exercise.

about extraordinary circumstances or conditions they may encounter in their work. Suppose that hazardous materials are present at an emergency. In that case, personnel must be aware of the signs and symptoms of exposure to those hazardous materials—including those that may be delayed—as well as where and how to report such symptoms and where to seek treatment for them if they occur.

Whenever possible, get updates on the condition of victims and share that information with the rescue teams before they leave the scene. The news may be good, or it may be bad, but rescuers are entitled to have accurate information as it emerges and to hear it from a single, well-informed source.

The death or serious injury of a fellow emergency worker or a prolonged operation where there is little chance of rescuing a victim alive are examples of situations that can lead to **critical incident stress**, a strong emotional or physical reaction in response to a traumatic event or incident that can affect a person's ability to function at work, at home, or in other areas of life. Finding out that a victim has died or that injuries are more serious than first thought can evoke strong emotions, so a debriefing must consider the need to address critical incident stress. Emergency response personnel are highly trained, but they are also people with emotions and feelings. An emergency response incident is an abnormal situation, but responders' reactions to the situation, no matter how they manifest, are normal and must be respected.

Just as you take care to maintain the equipment you use, provide maintenance for human resources. As part of your team training, consider including recognition and awareness of critical incident stress. Resources such as critical incident stress teams may be available to you. Know how to activate them and how doing so will benefit your team.

Assess the Equipment

The type of equipment you will use for confined-space rescue can range from simple to sophisticated. The type of equipment you will need should be determined through preplanning. Simple equipment has the advantage of being easy to set up and use. It can require minimal training, and it can speed up a rescue. However, simpler is not always better. Whatever equipment you retain must be right for the job.

Think of the tools you use in everyday operations. To a firefighter, an axe isn't just a symbol of the fire service. It is a valuable tool used to open doors, break windows, and open walls, but it is seldom used as the primary tool to cut a ventilation hole in a roof. Instead, power saws are used to cut ventilation holes. Each tool has its place: You could use an axe to cut a vent hole, but would you want to? The axe requires less training, always starts, doesn't require fuel, and doesn't require much maintenance. But given the choice, which would you use?

The same rationale holds true during a confined-space rescue. Certain tools and equipment work better than others. Equipment requires maintenance for both its mechanical parts and its operator. Maintaining proficiency with equipment is as important as having the equipment.

The dangers presented by responders who don't know how to use equipment safely are self-evident. Select the tools you will need for the job, train people to use them correctly, and maintain the capabilities of both the equipment and the people who use it.

Gaining Information During a Size-Up

In Chapter 2, you learned about the incident size-up process. The information gathered during a size-up forms the foundation for an IAP and also informs subsequent operational decisions. Many of the steps in creating an IAP are performed simultaneously with conducting a size-up, so it is essential that one process does not distract from the other.

The size-up typically includes an assessment of the specific type of confined space involved (e.g., a tank, silo, vault, or tunnel) and the unique challenges and hazards associated with each. This assessment includes a record of the dimensions, configuration, and accessibility of the confined space, which is necessary to plan entry and rescue strategies. At this point, you must determine whether the mission of the response is rescue or recovery, as well as whether you will be performing an entry rescue or a nonentry rescue.

Background information on the incident itself is also part of the size-up. This includes the nature of the incident, a record of any previous rescue attempts or other interventions, and the ongoing operational procedures and controls that apply to the confined space.

During the size-up process, the location, status, number, and conditions of victims are recorded and evaluated. At this point, the specific needs of victims and any information regarding preexisting medical conditions must be gathered as well in order to ensure they are taken into account.

Witness interviews are a critical part of the size-up process during a confined-space incident. They provide firsthand information about the events leading up to the emergency, the location of victims, and any potential hazards. Interviewing individuals who were onsite when the incident occurred, such as co-workers or supervisors, can help rescuers understand the specifics of the confined space, including its configuration, atmosphere, and access points. Witnesses may also provide insights into what tasks were being performed, any safety protocols in place, and any known hazards such as gas leaks or equipment malfunctions. This information is invaluable in shaping the rescue strategy, ensuring that responders are prepared to handle the environment's risks effectively while prioritizing victim safety. Conducting these interviews swiftly and accurately allows for a more informed and targeted approach to the rescue.

The size-up also includes a thorough assessment of the potential hazards within the confined space, including atmospheric hazards (oxygen deficiency, toxic gases such as hydrogen sulfide or carbon monoxide, and flammable vapors or dusts); physical hazards (moving machinery, exposed wires, unstable structures, or design features of the space that pose risks); and environmental hazards, such as temperature extremes, poor ventilation, or hazardous materials. Accurate information on air condition gathered during the size-up provides guidance for setting up air-monitoring equipment and maintaining the safety of responders and victims.

An evaluation of the entry and rescue points of a confined space is also a critical element of the size-up. All potential entry points should be identified and assessed for their accessibility and safety. Any specialized equipment that will be necessary for the response effort, such as rescue kits or victim retrieval systems, will need to be weighed against these entry and rescue point evaluations.

Incorporating Information from a Size-Up into an IAP

If scene size-up information is not incorporated into an IAP for a confined-space incident, responders are unlikely to have a comprehensive understanding of the situation before initiating rescue operations. Scene size-up involves assessing the confined-space environment, identifying hazards, evaluating risks, and gathering critical information to inform decision-making and operational strategies. The information gathered during the size-up informs incident objectives, operational strategies, resource allocation, rescue tactics, and safety procedures—the fundamentals of any IAP. All responders onsite must be well trained in the art of the size-up, and regularly scheduled drills and exercises to practice these skills will reinforce their understanding. See **FIGURE 4-6** for a sample of the incorporation of a size-up into an IAP.

Sample Incorporation of a Size-Up into an IAP

Objective: Conduct a comprehensive scene size-up to assess confined-space conditions, identify hazards, evaluate victim status, and inform operational strategies.

Steps:

1. Hazard Identification:
   - Assess atmospheric hazards, physical obstacles, and other environmental risks.
   - Use air-monitoring equipment to measure oxygen levels and detect toxic gases and flammable or combustible products.
2. Victim Assessment:
   - Determine the number and location of victims inside the confined space.
   - Evaluate their condition and any immediate medical needs.
3. Resource Evaluation:
   - Identify available personnel, equipment, and specialized teams required for the rescue operation.
   - Assess the need for additional resources based on size-up findings.

Documentation: Record size-up findings in the IAP, including hazard assessments, victim status, and resource requirements.

Integration into Decision-Making: Use size-up results to establish incident objectives, develop operational strategies, and allocate resources effectively.

Communication and Reporting: Share size-up information with IC, rescue teams, and supporting agencies through established communication channels. Implement a CAN (Conditions – Actions – Needs) report and provide a needs assessment.

FIGURE 4-6 Sample incorporation of a size-up into an IAP.

The first step is to establish the objectives of the scene size-up. Specific objectives for the scene size-up phase include assessing the type of confined space; identifying potential atmospheric, mechanical, and environmental hazards that are present; and determining the condition and location of victims.

Next, take a systematic approach to conducting the scene size-up, with an emphasis on thoroughness and attention to detail. Outline the key steps in the size-up process, including observation, assessment, and information gathering. The components of the size-up should be specified, including an assessment of all atmospheric and physical hazards and other environmental factors (temperature, ventilation concerns); a full assessment of the number and condition of victims, as well as their locations within the space; and an evaluation of all available resources, including personnel, equipment, and specialized teams needed for the rescue operation. All findings of the size-up must be well documented, including observations, hazard assessments, victim status, revision of the IAP based on size-up results, and resource requirements.

An experienced IC knows that size-up is not a one-time process. The size-up continues throughout the incident response as conditions change and risks shift. Integrating the size-up approach throughout the incident response allows ICs to adapt their strategies accordingly as the situation changes.

Determining Operational Protocols

Determining operational protocols at the scene of a confined-space incident involves coordination and cooperation between the IC of the rescue team and the AHJ. Examples of AHJs include local emergency response agencies, regulatory authorities, and facility operators. All parties play critical roles in establishing and implementing operational protocols to ensure a safe and effective confined-space rescue.

IC Responsibilities

As the leader of the rescue team, the IC is responsible for managing all aspects of the confined-space incident response. The IC's role includes the following:

- **Assessment and size-up:** Conducting an initial assessment of the incident scene to gather critical information, including the type of confined space, potential hazards, and victim status, and performing a size-up to evaluate the scope of the incident and identify operational needs.
- **Establishing incident objectives:** Defining clear and achievable incident objectives based on the assessment findings, focusing on responder safety, victim rescue, and hazard mitigation.
- **Developing operational strategies:** Formulating operational strategies and tactics to address identified hazards, execute rescue operations, and allocate resources effectively; establishing an incident command structure; and assigning roles and responsibilities to personnel.
- **Implementing safety measures:** Ensuring the implementation of safety measures, including the use of PPE, atmospheric monitoring, and hazard control procedures.
- **Communicating with the AHJ:** Establishing communication channels with the AHJ to coordinate response efforts, provide incident updates, and request additional resources or support as needed.
- **Documenting incident actions:** Documenting incident actions, decisions, and operational activities for situational awareness, accountability, and post-incident analysis.

AHJ Responsibilities

The AHJ is responsible for overseeing and regulating emergency response activities within its jurisdiction. In the context of a confined-space incident, the AHJ's responsibilities include the following:

- **Enforcing regulations and standards:** Applying local, state, and federal regulations related to confined-space entry, rescue operations, and occupational safety standards.
- **Providing regulatory guidance:** Providing guidance and assistance to the rescue team regarding compliance with safety regulations and operational protocols.
- **Resource coordination:** Coordinating resources and support from other agencies or organizations, such as fire departments, hazardous materials teams, or specialized rescue units.
- **Approving operational plans:** Reviewing and approving the IC's operational plans, including incident objectives, strategies, and resource allocations.
- **Monitoring incident progress:** Monitoring the progress of the incident response and providing oversight to ensure compliance with regulatory requirements and safety standards.

- **Communicating with stakeholders:** Communicating with stakeholders, including the public, media, and elected officials, regarding incident status and response efforts.
- **Standard operating procedures (SOPs)/standard operating guidelines (SOGs):** All local SOPs and SOGs established by the AHJ must be adhered to, even by any outside or incoming rescue teams. Any differences must be communicated by the AHJ to the rescue team officer.

Shared Responsibilities of and Collaboration Between the IC and AHJ

Effective collaboration between the IC and AHJ is essential for determining operational protocols at the scene of a confined-space incident. This collaboration involves the following aspects:

- **Regular communication:** Maintaining open and clear communication channels to exchange information, coordinate actions, and address operational challenges.
- **Mutual support and cooperation:** Supporting each other's efforts and responsibilities to enhance incident management and achieve common objectives.
- **Shared decision-making:** Engaging in shared decision-making processes to ensure alignment with regulatory requirements, operational priorities, and safety considerations.
- **Continuous improvement:** Conducting postincident debriefings and evaluations to identify lessons learned and areas for improvement in operational protocols and response procedures.

By working together collaboratively, an IC and AHJ can establish effective operational protocols at the scene of a confined-space incident, promoting safety, efficiency, and successful outcomes in rescue operations. Regular training, exercises, and mutual understanding of roles and responsibilities contribute to effective incident management and enhanced emergency response capabilities.

Documenting Hazards Within the AHJ

The Federal Emergency Management Agency (FEMA) ICS provides forms that serve as essential tools for managing emergency responses efficiently. FEMA's standardized forms facilitate effective communication, documentation, and resource management during incidents. Key FEMA ICS forms commonly used in incident management include ICS 201, ICS 202, ICS 203, ICS 204, ICS 205, ICS 206, ICS 207, ICS 208, ICS 211, ICS 215, ICS 215A, and ICS 220. Each form is described in **TABLE 4-2**.

These forms provide a standardized approach to documenting, communicating, and managing resources and information, which is crucial for the success of any emergency response operation. See **FIGURE 4-7** for an example of one form—the ICS 201.

TABLE 4-2 Federal Emergency Management Agency (FEMA) Incident Command System (ICS) Forms and Their Uses

ICS 201 **Incident Briefing**	▪ Provides a concise summary of the incident situation and the initial response actions ▪ Includes incident name, situation summary, current and planned actions, incident objectives, resource summary, and organization chart ▪ Used during the initial phase of an incident to brief incoming incident commanders (ICs) and other key personnel
ICS 202 **Incident Objectives**	▪ Documents incident objectives for the operational period ▪ Lists specific, measurable, achievable, relevant, and time-bound (SMART) objectives, along with the general safety message ▪ Guides the development of an incident action plan (IAP) by outlining the primary goals for the response

Continues.

TABLE 4-2 Federal Emergency Management Agency (FEMA) Incident Command System (ICS) Forms and Their Uses CONTINUED

ICS 203 **Organization Assignment List**	▪ Identifies the ICS organization structure for the operational period ▪ Lists the names and contact information of personnel assigned to various ICS positions, including command and general staff ▪ Used to ensure all personnel are aware of their roles and responsibilities within the ICS structure
ICS 204 **Assignment List**	▪ Specifies assignments for each operational period ▪ Details work assignments, geographic areas, resources assigned, and contact information for team leaders ▪ Used to provide clear instructions to operational personnel on their tasks and locations
ICS 205 **Incident Radio Communications Plan**	▪ Outlines the radio communications plan for the incident ▪ Provides information on radio frequencies, channels, and assignments for each operational period ▪ Used to maintain effective communication among responders by providing a clear communication plan
ICS 206 **Medical Plan**	▪ Documents the medical response plan for the incident ▪ Identifies medical aid stations, hospitals, ambulance services, and procedures for medical emergencies ▪ Used to capture essential medical information to ensure responder safety and effective medical response
ICS 207 **Incident Organization Chart**	▪ Provides visual representation of the ICS organization structure ▪ Displays the command and general staff positions, along with subordinates ▪ Helps personnel understand overall organization and their place within it
ICS 208 **Safety Message/Plan**	▪ Communicates safety messages and plans for the operational period ▪ Includes specific safety hazards, precautions, and protective measures ▪ Ensures all personnel are aware of potential hazards and the measures in place to mitigate them
ICS 211 **Check-In List**	▪ Records arrival of personnel and resources at the incident ▪ Includes name, position, and contact information of each person or resource ▪ Maintains accountability for all personnel and resources on the scene
ICS 215 **Operational Planning Worksheet**	▪ Assists in planning resource needs and assignments for the operational period ▪ Itemizes resources by type and quantity, along with the specific assignments and locations ▪ Used during planning meetings to allocate resources effectively
ICS 215A **Incident Action Plan Safety Analysis**	▪ Identifies and mitigates potential safety hazards related to the operational assignments ▪ Lists potential hazards, risks, and mitigation strategies for each task ▪ Used to ensure safety considerations are integrated into the IAP
ICS 220 **Air Operations Summary**	▪ Summarizes the air operations plan for the incident ▪ Includes information on air resources, assignments, and contact information ▪ Used to ensure coordination and effective use of air resources during the incident

Incident Briefing (ICS 201)

Incident Name: Confined-Space Rescue - Industrial Plant

Date Initiated: June 29, 2025

Time Initiated: 0700 hours

Map/Sketch: *[Attach a detailed map or sketch of the incident location, including key landmarks, entry/exit points, and confined-space area.]*

Incident Objectives:

- Ensure the safety of all personnel involved in the rescue operation.
- Safely extricate the trapped worker from the confined space.
- Monitor and manage environmental hazards within the confined space.
- Provide medical assistance to the trapped worker and rescuers as needed.
- Document all actions and findings for postincident review.

Current Situation: A worker is trapped in a confined space (underground storage tank) at the ABC Industrial Plant, located at 456 Industrial Blvd. The worker was performing maintenance when they became unresponsive. Initial air monitoring indicates the presence of hazardous gases.

Resource Summary:

Resource	Resource Identifier	ETA	On Scene	Location/Assignment
Confined-space rescue team	R1, R2	5 min	Yes	Entry and extrication operations
HazMat team	H1	10 min	Yes	Air monitoring and hazard control
Ambulance	A1, A2	5 min	Yes	Medical triage area
Law enforcement	P1, P2	5 min	Yes	Perimeter security and traffic control
Safety officer	SO1	0 min	Yes	Overseeing safety protocols

Summary of Current Actions:

- Confined-space rescue team is preparing for entry and extrication operations.
- HazMat team is conducting air monitoring and setting up ventilation equipment.
- Safety officer is briefing all personnel on safety protocols and PPE usage.
- Law enforcement is securing the perimeter and managing traffic control.
- Ambulances are on standby at the medical triage area, ready to assist the trapped worker and any rescuers in need of medical attention.

Anticipated Actions:

- Rescue team will enter the confined space and attempt to safely extricate the worker.
- HazMat team will continue to monitor air quality and manage any detected hazards.
- Medical personnel will provide immediate care to the extricated worker.
- Safety protocols will be continuously enforced and monitored.
- Document all actions and environmental readings for postincident analysis.

Prepared By:

Name: John Doe

Position/Title: Incident Commander

Signature: *John Doe*

Attachments: Incident map/sketch; resource check-in list; safety message/plan

FIGURE 4-7 A sample FEMA ICS 201 form.

CHAPTER FEATURE: AFTER-ACTION

IN SUMMARY

- The criteria for conducting basic search operations in confined spaces include air-monitoring requirements, specialized equipment needs, and training in the use of ropes for rescue purposes.
- Determining the need for additional resources is a dynamic process that requires careful consideration of various factors, thorough assessment methodologies, and informed decision-making.
- The elements of an IAP are the incident overview, the incident objectives, the organizational structure and responsibilities, the operational strategies to be used, the incident tactics, continuity and follow-up, and incident review and debrief.
- The information gathered during an incident size-up process forms the foundation for an IAP and also informs subsequent operational decisions. Many of the steps of creating an IAP are performed simultaneously with conducting a size-up.
- Scene size-up involves assessing the confined-space environment, identifying hazards, evaluating risks, and gathering critical information to inform decision-making and operational strategies. The information gathered during the size-up informs incident objectives, operational strategies, resource allocation, rescue tactics, and safety procedures—the fundamentals of any IAP.
- The AHJ plays a critical role in defining the operational protocols that guide response efforts and establishes these protocols based on its legal authority, SOPs, best practices, and available resources.

KEY TERM

critical incident stress A strong emotional or physical reaction a person may have in response to a traumatic event or incident that can affect the ability to function at work, at home, or in other areas of life.

REFERENCE

National Fire Protection Association. 2021. *NFPA 1006: Standard on Technical Rescue Personnel Professional Qualifications, 2021 Edition.* Quincy, MA: National Fire Protection Association.

CHAPTER QUESTIONS

1. Which of the following can awareness level responders do?
   - a. conduct rescue operations
   - b. limit access to a confined space
   - c. start ventilation procedures
   - d. preplan a confined-space incident
2. Which of the following has to happen before a search can begin in a confined-space event?
   - a. A permit from the company must be secured.
   - b. A thorough risk assessment must take place.
   - c. An ambulance must be on the scene for transport.
   - d. Hazards must be documented with the AHJ.
3. A comprehensive risk assessment permits the IC to
   - a. eliminate potential hazards.
   - b. determine the severity of hazards.
   - c. document the event.
   - d. delegate work to others.
4. The incident objectives section of an IAP clearly states the
   - a. primary goal of the incident response.
   - b. incident tactics.
   - c. operational strategies of the incident response.
   - d. incident overview.

CHAPTER FEATURE: ON SCENE

A FIRE EMERGENCY IN A CONFINED SPACE

This case study examines a real-life incident where a fire broke out inside a confined space, endangering trapped workers. Rescue teams faced the challenge of coordinating fire-suppression efforts while conducting simultaneous rescue operations to ensure the safety and evacuation of the trapped workers.

This incident took place at a manufacturing facility that specialized in chemical processing. Inside one of the processing vessels, a chemical reaction went awry, leading to a sudden and intense fire. The vessel was located in a confined space, making access and rescue operations challenging.

The Emergency Response

Several workers were trapped inside the confined-space area, unable to escape on their own because of the fire and limited exit routes. The facility's emergency response team was immediately activated, along with local fire departments and specialized rescue teams trained in confined-space operations.

Upon arrival, the rescue teams faced a dual challenge of suppressing the fire to prevent further escalation while initiating rescue operations to evacuate the trapped workers. ICs established a unified command structure to coordinate fire-suppression efforts and rescue operations efficiently. Communication protocols were established to maintain constant updates and ensure safety.

The firefighters deployed specialized equipment and firefighting agents suitable for chemical fires to contain and extinguish the blaze. The confined-space environment added complexity to the firefighters' tactics, requiring careful consideration of ventilation and hazardous material control.

The Rescue Operations

Simultaneously, rescue teams equipped with confined-space rescue gear and tools entered the affected area to locate and evacuate the trapped workers. They followed established protocols for patient stabilization and extrication.

The Extrication and Treatment

Trapped workers were assessed for injuries and provided immediate medical care as needed before being safely extracted from the confined space.

The Outcome

The fire was successfully contained and extinguished within a few hours of intense effort by the combined response teams. All trapped workers were rescued and transferred to medical facilities for further evaluation and treatment.

Lessons learned from the incident highlighted the importance of preplanning, training, and interagency cooperation in confined-space emergencies involving fire. This incident underscores the critical nature of emergency response operations in confined spaces, particularly during fire incidents. Effective coordination, swift action, and adherence to safety protocols are paramount to ensuring successful outcomes in such challenging environments. Regular training and preparedness are essential for responders to mitigate risks and handle complex rescue scenarios effectively.

1. What caused the fire in the confined space in this incident?
   - **a.** electrical failure in the confined space
   - **b.** a chemical reaction inside a processing vessel
   - **c.** an explosion caused by gas leakage
   - **d.** overheating of the facility's machinery
2. What was the dual challenge faced by the emergency response teams upon arrival at the scene?
   - **a.** managing crowd control and the media
   - **b.** suppressing the fire while evacuating the trapped workers
   - **c.** containing chemical leaks and clearing debris
   - **d.** coordinating with facility managers while notifying families

A FIRE EMERGENCY IN A CONFINED SPACE CONTINUED

3. What specialized firefighting tactic was required because of the confined-space environment?
   - **a.** use of water cannons for immediate suppression
   - **b.** careful ventilation and hazardous material control
   - **c.** protocols for patient stabilization and extrication
   - **d.** use of explosives to create a ventilation pathway

4. Why was establishing a unified command structure critical in this incident?
   - **a.** It ensured efficient coordination of fire-suppression and rescue operations.
   - **b.** It helped reduce the number of emergency responders required.
   - **c.** It allowed the facility's management team to control the situation.
   - **d.** It provided a way to train new recruits during the incident.

SECTION

2

Operations

You Are the Rescuer

Along with your team of fellow responders, you have arrived on the scene of a terrible accident at a wastewater treatment facility. A worker there explains the situation to you: A public works foreman became trapped in a narrow access tunnel containing hazardous gases and limited ventilation. The local fire department responded to the call and attempted a rescue. Now, a seasoned member of the firefighting team with extensive training in confined-space rescue is also trapped in the tunnel alongside the foreman. No one has heard from either victim for some time now.

You immediately approach the incident commander (IC) to discuss the situation. She informs you that the team of firefighters donned basic personal protective equipment (PPE) and self-contained breathing apparatuses (SCBAs) and entered the confined space to assess the condition of the trapped foreman and provide medical assistance. Despite the hazardous conditions, they remained determined to perform a rescue. The team deployed mechanical advantage systems and manual hauling techniques to attempt to move the trapped foreman along the horizontal plane of the tunnel.

However, the confined space's constraints and hazardous atmosphere presented formidable challenges. The rescue operation encountered such obstacles as tight corners and debris within the tunnel, which required careful coordination and adjustments in rigging to safely navigate the confined space. Despite their best efforts, the firefighters were unable to safely extricate the trapped foreman from the confined space. At some point, the firefighters lost track of one of their colleagues, and now he, too, is a victim.

You Are the Rescuer Questions

1. Imagine that you are placed in charge of the investigation and analysis of this event. As the IC, what are the first three things you must address?
2. How do you think the situation may have changed while the responders were in the confined space, resulting in the firefighter getting separated from the others? What could the firefighter have done differently to survive the incident?

CHAPTER

5

Operations

Types of Confined Spaces, Operational Protocols, and Risk and Resource Management

KNOWLEDGE OBJECTIVES

After studying this chapter, you should be able to:

- List the multiple types of confined spaces. (**NFPA 1006 7.2.17**, pp. 115–117)
- List the functions established by the Incident Command System (ICS). (**NFPA 1006 7.2.18**, p. 118)
- Describe the operational protocols related to safety, communication, medical, and operations. (**NFPA 1006 7.2.10** [horizontal spaces] and **7.2.14** [vertical spaces], pp. 119–124)
- Discuss having operational protocols related to medical, emergency medical services (EMS) providers, and decontamination. (**NFPA 1006 7.2.17**, pp. 121–124)
- Discuss the importance of constant communication with outside rescuers. (**NFPA 1006 7.2.7**, pp. 120–121)
- Discuss the process for ensuring adequate resources. (**NFPA 1006 7.2.2**, p. 133)
- Describe the processes used for hazard identification, risk management, logistics, and resource management. (**NFPA 1006 7.2.18**, pp. 126–134)
- Discuss the need for personnel and equipment resource lists. (**NFPA 1006 7.2.17**, p. 134)
- Discuss the need for personnel accountability systems. (**NFPA 1006 7.2.18**, p. 115; pp. 137–139)
- Identify the process for ensuring rescuers have been cleared for confined-space operations. (**NFPA 1006 7.2.9** [horizontal spaces] and **7.2.13** [vertical spaces], pp. 124–125)
- Discuss procedures and protocols involving personnel rehabilitation. (**NFPA 1006 7.2.18**, p. 115; p. 126)
- Describe the risk–benefit analysis process. (**NFPA 1006 7.2.4**, pp. 135–136)

SKILL OBJECTIVES

- Ensure that operational protocols are instituted. (**NFPA 1006 7.2.10** [horizontal spaces] and **7.2.14** [vertical spaces], pp. 123–124)
- Ensure that operational protocols are instituted and reassessed as needed. (**NFPA 1006 7.2.14** [vertical spaces], pp. 123–124)
- Demonstrate pre-entry personnel evaluations. (**NFPA 1006 7.2.9** [horizontal spaces] and **7.2.13** [vertical spaces], pp. 124–125)
- Deploy scene control measures. (**NFPA 1006 7.2.5**, pp. 118–119)
- Demonstrate the use of the accountability system. (**NFPA 1006 7.2.18**, pp.138–139)
- Identify communication needs. (**NFPA 1006 7.2.9** [horizontal spaces] and **7.2.13** [vertical spaces] , pp. 119–121)
- Develop communication methods so that information is adequately sent and received. (**NFPA 1006 7.2.6**, pp. 119–121)

- Demonstrate communication methods. (**NFPA 1006 7.2.6**, pp. 119–121)
- During a size-up, gather all pertinent information and relay it to the incident commander (IC). (**NFPA 1006 7.2.2**, pp. 134–135)
- Interpret size-up information and perform a risk–benefit analysis. (**NFPA 1006 7.2.4**, pp. 135–136)
- Manage the termination of a confined-space incident. (**NFPA 1006 7.2.18**, p. 115; pp. 136–137)

Introduction

Effective incident command and management are critical to ensure the safety of both rescuers and victims. The Incident Command System (ICS) provides a structured framework for managing emergency response operations. Functions of the ICS include establishing a clear chain of command and organizational structure. This allows for command and control of incident operations. Coordination of resources and personnel is essential for a successful operation and can only be accomplished with proper **communications management**, which is the systematic planning, implementation, monitoring, and control of all channels of communication to ensure that information is efficiently and effectively exchanged. Equally important for a successful outcome is having proper incident documentation and a reporting process.

ICS plays a crucial role in confined-space rescue operations by doing the following:

- Establishing a clear chain of command
- Coordinating resources and personnel
- Managing communications
- Ensuring proper documentation and reporting

ICS is crucial for the effective management of confined-space rescue operations. In addition to establishing a clear chain of command and organizational structure, the ICS for confined-space rescues should include the following:

- Designation of a safety officer specifically trained in confined-space hazards
- Establishment of an entry team supervisor to oversee rescuer entry and exit
- Integration of technical specialists (e.g., atmospheric monitoring experts) into the command structure
- Creation of a medical group to manage victim care and rescuer health monitoring

These ICS components help ensure that all aspects of the confined-space rescue are coordinated and managed effectively, enhancing both rescuer safety and operational efficiency.

Safety management is the systematic process of developing, implementing, and maintaining policies, procedures, and practices that ensure the safety and well-being of individuals within an organization or environment; it involves identifying potential hazards, assessing risks, and taking proactive measures to prevent accidents, injuries, and illnesses. **Risk assessment** is the systematic process of identifying, evaluating, and analyzing potential risks that could negatively affect individuals, assets, operations, or the environment; the goal is to determine the likelihood of these risks occurring and the potential consequences, enabling organizations to implement appropriate measures to mitigate or eliminate them. These both work hand in hand with **hazard identification**—the process of recognizing and documenting potential sources of harm or adverse effects in the workplace, environment, or any operational setting—and **risk management**, which is the process of identifying, assessing, prioritizing, and mitigating risks to minimize their impact on an organization, project, or individual. It is critical to conduct thorough assessments of confined-space hazards, including atmospheric, physical, and biological hazards. Regardless of rank or position, implementation of control measures to mitigate identified risks depends on each and every person involved in a rescue. It is the responsibility of confined-space rescuers to continuously monitor and reassess hazards throughout the duration of an operation.

Risk–benefit analyses are essential to ensure the safety of responders and victims while maximizing the effectiveness of the rescue effort. This process involves evaluating the potential risks associated with various rescue strategies against the benefits of successful victim extraction.

Logistics and **resource management** processes involve procuring and allocating necessary resources, including personnel, equipment, and supplies. A chief logistics officer is responsible for establishing staging areas and incident command posts and managing communications and coordination between different response teams and agencies. The logistics officer is also responsible for tracing resources and requesting additional resources as needed.

Personnel accountability systems are an essential part of emergency incident responses. These include systems to track the location and status of all personnel involved in the operation, often using personnel tags (**FIGURE 5-1**), check-in/checkout procedures, and accountability boards (**FIGURE 5-2**). Without them, it is difficult to impossible to monitor personnel as they enter and exit a confined space.

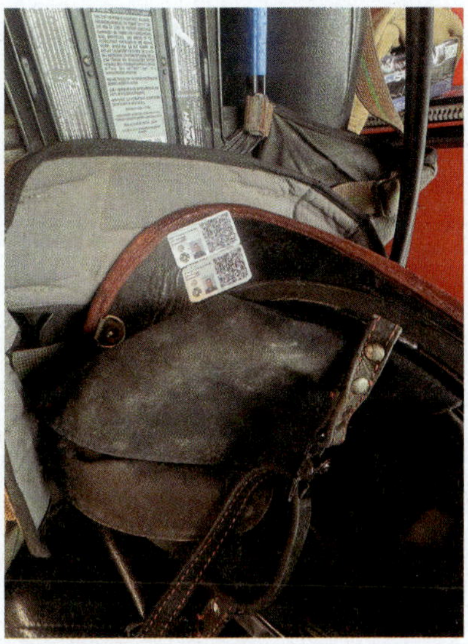

FIGURE 5-1 Accountability tags are used to track personnel during emergency operations, ensuring firefighter safety and proper team management in the field.

Courtesy of Eric Scruggs.

FIGURE 5-2 An accountability board is used during emergency operations to monitor and track the location and assignment of all personnel on the scene, enhancing safety and coordination.

Used with permission from American Trade Mark Co.

> **TIP**
>
> All personnel are responsible for safety. Any individual can call for a stop operations when a hazard is recognized.

Personnel rehabilitation requires the establishment of designated areas for personnel to rest. These areas require provisions for medical monitoring and treatment as needed, and personnel must be rotated in shifts to prevent fatigue and ensure operational readiness.

The first step in the process of terminating an incident is to assess when a confined-space incident has been successfully resolved and the scene is safe for secure demobilization. This is followed by debriefs and postincident evaluations designed to identify lessons learned and areas for improvement. Ensuring proper documentation and reporting of the incident for regulatory compliance and future reference is also key.

Types of Confined Spaces and Their Potential Hazards

Just as there are varied industrial, commercial, and residential settings, there are varied types of confined spaces. Rescuers have the potential to encounter a diverse range of hazardous materials and substances depending on the nature of a confined space, whether it is a tank; a railcar; a silo; or any number of spaces within, say, a shipyard or a commercial plant. These environments present unique challenges because of the variety of products stored or processed within them. Types of confined spaces and their associated potential hazards are shown in **TABLE 5-1**.

> **TIP**
>
> In incidents like a trench collapse or a cave-in at a mine or excavation, the associated hazards to watch for are oxygen deficiency or enrichment in the atmosphere, toxic gases from soil or underground air pockets, and entrapment by heavy machinery or falling debris. Whether or not incidents like these would be considered confined spaces is up to the authority having jurisdiction (AHJ).

TABLE 5-1 Types of Confined Spaces and Associated Hazards

Type of Confined Space	Potential Hazards
Tanks, bins, or vats	▪ Engulfment in collapsing material ▪ Oxygen deficiency or enrichment ▪ Toxic atmospheres from stored substances ▪ Flammable or explosive atmospheres from chemical reactions or residual gases
Silos or grain bins	▪ Engulfment in grain or other bulk materials ▪ Oxygen deficiency or enrichment ▪ Grain dust explosions ▪ Toxic gases produced by decaying grain
Sewers, manholes, or storm drains	▪ Oxygen deficiency or enrichment ▪ Toxic atmospheres from sewer gases ▪ Engulfment in flowing liquids or solids ▪ Physical hazards from sharp objects or moving sewage
Trenches, excavations, or underground mines	▪ Trench collapse or cave-in ▪ Oxygen deficiency or enrichment ▪ Toxic gases from soil or underground air pockets ▪ Entrapment by heavy machinery or falling debris
Crawl spaces or ductwork	▪ Limited space for movement and egress ▪ Oxygen deficiency or enrichment ▪ Exposure to electrical hazards ▪ Entrapment by structural elements or equipment
Wells, cisterns, or water tanks	▪ Oxygen deficiency or enrichment ▪ Drowning or suffocation in water ▪ Exposure to toxic gases from stagnant water ▪ Entrapment by collapsing structures or debris
Confined spaces within vehicles	▪ Oxygen deficiency or enrichment ▪ Exposure to hazardous materials transported in tanks or containers ▪ Mechanical hazards from moving parts or equipment ▪ Limited egress in case of accident or rollover

For example, tank cars and other railcars are used to transport a wide range of liquids, solids, and gases. These may include petroleum products, chemicals, liquefied gases, granular products (grains, plastics, etc.), and other types of hazardous materials (acids, caustics, flammable liquids, etc.). Responders to scenes involving these types of confined spaces may encounter spills, leaks, or releases of hazardous substances that may pose the risk of fire, explosion, chemical exposure, or asphyxiation. Confined spaces within tank cars and railcars also often have limited ventilation and may contain residual vapors or gases, which increase the risk of atmospheric hazards. The consolidation of the hazardous materials standard in 2022 incorporated the content of National Fire Protection Association (NFPA) standards 472, 473, and 1072 into a single document, NFPA 470 (NFPA 2022b), which provides guidelines for responders to hazardous materials incidents, including those involving tank cars and railcars. *NFPA 704: Standard System for the Identification of the Hazards of Materials for Emergency Response and the Hazardous Materials Identification System (HMIS)* (NFPA 2022c) provides a system for identifying the hazards associated with chemicals stored in these containers.

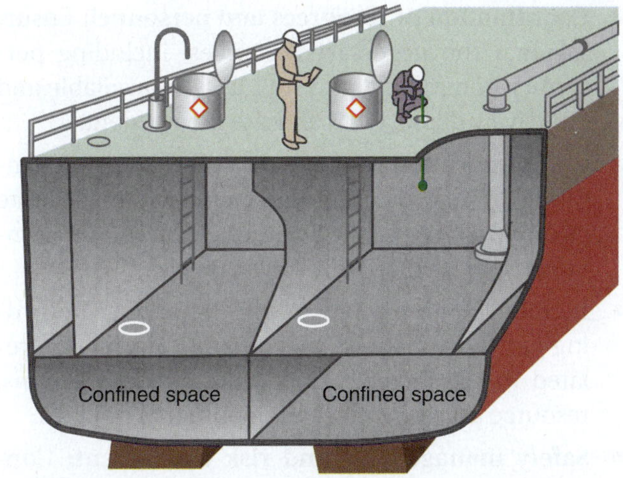

FIGURE 5-3 This cross-section illustration of a ship shows where the confined spaces are located.

Courtesy of OSHA/U.S. Department of Labor.

> **TIP**
>
> Determining the contents of a confined space involves first identifying all stored materials, chemicals, or chemical by-products present within the space. By understanding the contents, responders can predict potential risks associated with these substances and choose appropriate monitoring methods to ensure safety during confined-space entry and rescue operations.

An incident at a shipyard may involve a confined space in a ship compartment, tank, bilge, or engine room (see **FIGURE 5-3**). Any of these areas may contain various hazardous materials and products used during ship construction, maintenance, and operation, such as fuels, lubricants, paints, solvents, welding gases, and cleaning chemicals. Older vessels sometimes contain asbestos or materials derived from asbestos, which pose exposure risks as well.

Storage tanks, silos, and process vessels may contain various chemicals and hazardous substances used in industrial processes, such as acids, bases, solvents, fuels, lubricants, and toxic gases. Chemical exposures can lead to burns to the skin, respiratory maladies, poisonings, or other acute or chronic health effects. Certain chemicals may also be flammable, explosive, or reactive (or all or any combination of these).

Any of these spaces may contain mechanical or other physical hazards, such as mechanical equipment, machinery, piping systems, and structural components that pose physical hazards to rescuers and victims. These hazards can include moving parts, pinch points, entanglement hazards, sharp edges, and falling objects, and their consequences can be severe. Mechanical hazards can cause crushing injuries, lacerations, punctures, amputations, and other traumatic injuries to personnel working in confined spaces. Confined spaces are also vulnerable to structural collapse or failure.

Some confined spaces, such as sewers, manholes, and underground vaults, may harbor biological hazards like mold, bacteria, viruses, or human or animal waste. These hazards can pose health risks to responders, especially if they come into contact with contaminated surfaces or materials. Exposure to biological hazards can result in infections, respiratory illnesses, skin irritation, and other health effects. Certain pathogens may also cause serious or life-threatening diseases if proper precautions are not taken.

In certain industrial or research facilities, confined spaces may contain radiological or nuclear materials used in medical imaging, research, or industrial processes. These materials can emit ionizing radiation, such as gamma rays or neutron radiation. Exposure to ionizing radiation can cause tissue damage, radiation sickness, cancer, and other adverse health effects. Rescuers must take appropriate precautions and use specialized equipment to minimize radiation exposure during rescue operations.

In any confined space, a dangerous atmosphere may be present, whether it is oxygen deficient (levels below 19.5 percent); oxygen enriched (levels above 23.5 percent); or contaminated with toxic gases, vapors, or particulate matter. Oxygen deficiency can cause asphyxiation, whereas toxic gases and vapors can lead to respiratory distress, chemical burns, and systemic poisoning. Toxic atmospheres (those including carbon monoxide, hydrogen sulfide, and other specific gases) and flammable/explosive atmospheres (those including methane, propane, and other combustible gases) pose fire and explosion hazards, especially in confined spaces with limited ventilation.

> **TIP**
>
> Typically, trenches deeper than four feet (1.2 m) may be considered confined spaces because of the limited natural ventilation and the potential for hazardous atmosphere accumulation. Narrow trenches that restrict movement and limit escape routes are more likely to be classified as confined spaces. The presence of hazardous atmospheres, engulfment hazards, or other serious safety risks can qualify a trench as a confined space.

ICS Functions

The ICS establishes several key functions for effective management of confined-space incident response. (See Figure 2-1 in Chapter 2 for an incident command chart and positions; meanwhile, **TABLE 5-2** shows the ICS sections and their purposes.) These functions work together to provide a structured framework for managing confined-space incidents, promoting a coordinated and efficient response while prioritizing responder safety and the successful resolution of the incident. They are as follows:

- **Establishing a clear chain of command:** Designating the incident commander (IC) and establishing a clear organizational structure to oversee the response operation.
- **Unified command:** Plays a role in unifying command in situations involving multiple agencies or jurisdictions, which is common in complex confined-space rescues.
- **Command and control of incident operations:** Directing and coordinating all response activities, including rescue efforts, hazard mitigation, and victim care.
- **Coordination of resources and personnel:** Ensuring that the necessary resources, including personnel, equipment, and supplies, are available and effectively deployed to address the incident.
- **Communication management:** Establishing communication protocols and channels to facilitate effective communication among responders, command staff, and other involved parties.
- **Incident documentation and reporting:** Recording and documenting all relevant information related to the incident, including incident actions, resource assignments, and operational updates.
- **Safety management and risk assessment:** Conducting ongoing risk assessments to identify hazards and implement measures to mitigate risks to responders and victims.

TABLE 5-2 Incident Command System (ICS) Sections

Section	Purpose
Operations section	Plays a crucial role in directly managing the hands-on rescue efforts in a confined-space incident, ensuring that all tactical operations are carried out safely and effectively in line with the overall incident objectives
Planning section	Develops incident action plans (IAPs), gathers and analyzes incident information, and tracks resources
Logistics section	Provides support, resources, and services to meet the incident needs, including specialized equipment for confined-space rescue
Finance/ administration section	Tracks costs and manages financial aspects of the incident, especially for prolonged or complex rescues

Scene Control

Scene control measures for confined-space rescue operations must be implemented in accordance with NFPA 1006 standards (NFPA 2021a). The key components of scene control are establishing command, securing the scene, and controlling access to the confined space, but much more is necessary to truly control the safety and security of the scene. Additional elements of scene control include the following:

- Assessment of the specific hazards and risks present at a confined-space rescue scene and development of appropriate control measures to mitigate those hazards
- Identification and deployment of the tools and equipment necessary to control the scene, including barricades, signage, communication devices, and appropriate personal protective equipment (PPE)
- Establishment of incident command and implementation of a unified command structure to coordinate rescue efforts with multiple agencies and stakeholders
- Establishment of clear communication among team members, ICs, and other emergency responders to ensure clear and concise information exchange and coordination
- Preservation of situational awareness among all responders
- Continuous reassessment of the scene for changing conditions or hazards
- Management and control of bystanders and onlookers to prevent them from interfering with rescue operations and ensure their safety

- Documentation of all scene control measures, including hazard assessments, control measures implemented, and communication protocols, for postincident analysis and debriefing

In a typical hazardous materials response, work areas are divided into three zones: hot, warm, and cold. These zones can also be utilized for a confined-space event. The hot zone would include the immediate area around the confined space, the ingress and egress areas of the space, and where the anchor systems are located. The warm zone would be the area where equipment drops can be established and backup teams can be staged—it should include anything that may be needed urgently but is located outside the hot zone. The cold zone would include the areas for equipment and other resources, incident command, and the rehabilitation area.

Operational Protocols

Operational protocols play a crucial role in ensuring the safety, effectiveness, and coordination of confined-space emergency response efforts. The following operational protocols related to safety, communication, medical procedures, and operational tasks during confined-space emergencies align with the guidelines provided in NFPA 1006 (NFPA 2021a).

Safety Protocols

The primary safety protocols to follow are risk assessment, use of PPE, atmospheric monitoring, and proper entry and exit procedures.

- **Risk assessment:** Conduct thorough and continuous risk assessments before and during entry; identify such hazards as toxic atmospheres, mechanical equipment, and structural instability; and implement control measures to mitigate risks to rescuers and victims.
- **Use of PPE:** Ensure all personnel wear appropriate PPE, including self-contained breathing apparatus (SCBA), full-body harnesses for fall protection, and protective clothing resistant to chemicals and physical hazards, as required by NFPA 350 (NFPA 2022a) and NFPA 1006 (NFPA 2021a).
- **Atmospheric monitoring:** Continuously monitor atmospheric conditions inside the confined space for oxygen levels, toxic gases, and combustible gases, adhering to NFPA 1006 (NFPA 2021a) standards for atmospheric testing and monitoring.
- **Proper entry and exit procedures:** Establish clear entry and exit procedures, including access control, entry permits, and retrieval systems, to facilitate safe ingress and egress for rescuers and victims.

Communication Protocols

The primary communication protocols to follow are the establishment of clear communication channels, two-way communication, emergency signals, and hand signals.

- **Clear communication channels:** Establish clear communication channels between rescuers, incident command, safety officers, and medical personnel. Use multiple methods such as radios, intercom systems, and hand signals to relay critical information and coordinate rescue operations effectively.
- **Face-to-face communication:** This is the best and most preferred method of communication, but it is not always possible.
- **Two-way communication:** Maintain two-way communication between rescuers inside the confined space and personnel stationed outside, using reliable communication devices such as radios, intercom systems, or hand signals, as specified in NFPA 1006 (NFPA 2021a).
- **Emergency signals:** Implement and practice standardized emergency signals for distress situations, evacuation orders, and other critical messages to ensure rapid response and coordination.
- **Hand signals:** Implement emergency hand signals as a backup measure if communication via radio is lost. However, these hand signals must be well established and understood before entering the space. This works for any time the entrants and outside personnel are visible.
- **OATH and SUDRH methods:** The OATH method involves tugging on a rope (O for okay = one tug; A for advance = two tugs; T for take up slack = three tugs; H for help = four tugs), and the SUDRH method uses air horn blasts (S for stop = one short blast; U for up = two short blasts; D for down = three short blasts; R for rope free = four short blasts; and H for help = one long blast). These are covered in Chapter 2.

Identifying Communication Needs and Methods

Communication plays a critical role in ensuring the safety and effectiveness of confined-space rescue operations. A confined-space emergency requires multiple methods, including verbal, nonverbal, visual, and radio communication.

Communication strategies and protocols must be established before entering a confined space. An assessment of the potential barriers to communication in that particular space, such as noise, limited visibility, and distance, helps narrow down what strategies may work best. An assessment of the team dynamics will yield specific communication needs based on team members' roles, responsibilities, and experience levels.

Reliance on verbal commands, signals, and cues for conveying information or nonverbal methods, such as body language, gestures, and facial expressions, within a confined space requires appropriate visibility and sound conditions—relative quiet, adequate scene lighting, and clear lines of sight, to name only a few. Lights and markers (**FIGURE 5-4**) can help with visual modes of communication, but they are not always available or adequate to meet the needs of responders. Radio communication, whether it is through two-way radios or other types of wireless communication devices, is perhaps the most reliable method of real-time team communication.

The best strategy for ensuring the effectiveness of communication strategies is to conduct hands-on exercises that simulate various rescue scenarios to evaluate which strategies prove most effective, given the circumstances of each simulation. Practicing scheduled and timed check-ins can help reinforce periodic communications and maintain constant communication. These types of exercises can reinforce team bonds and put problem-solving skills to the test without the typical risks to health and safety that real emergencies entail.

FIGURE 5-4 Glow sticks are used to safely illuminate confined spaces during rescue operations, providing a reliable, nonelectric light source.

In these exercises, participants should play different roles within the team and strive to determine potential communication challenges. Featuring radio communication drills within these exercises also gives the team an opportunity to demonstrate their efficacy with clear and concise communication, proper channel usage, and understanding of emergency procedures.

The Importance of Maintaining Constant Communication

Confined spaces are characterized by their limited openings for entry and exit, poor ventilation, and the potential for hazardous atmospheres. Rescuers entering these spaces face numerous risks, including oxygen deficiency, toxic gases, engulfment, and physical hazards. The complexity and danger of these environments make effective communication a cornerstone of successful rescue efforts. *NFPA 1006: Standard on Technical Rescue Personnel Professional Qualifications* (NFPA 2021a) outlines the necessary skills and knowledge required for technical rescuers; specifically, section 7.2.7 of NFPA 1006 highlights the critical importance of constant communication with outside rescuers during a confined-space emergency.

Physical barriers pose some of the most severe communication challenges. Obstructions, such as structural elements and equipment within confined spaces, can impede sound transmission and line-of-sight communication. The acoustic properties of the space can create echoes and muffle sounds, which hinders verbal communication.

Hazardous atmospheres also cause communication hurdles. Rescuers often wear atmosphere-supplying respirators, such as a supplied-air respirator (SAR) or SCBA, which can muffle voices and make communication challenging. Noise levels from machinery, ventilation systems, and other operational equipment can drown out verbal communication. Nonessential auditory noise should be silenced during operations.

For the most part, confined-space responses are complex operations that require coordination. Rescues frequently involve multiple teams performing different tasks simultaneously, necessitating clear and constant communication to coordinate efforts effectively. The physical safety of rescuers and victims is always the first priority, so as rescuers, we must have a high degree of hazard awareness. Continuous updates on environmental conditions, potential hazards, and changing situations are crucial to maintaining situational awareness. We must also maintain a high state of readiness and be cognizant of any emergency alerts. Immediate communication of emergencies, such as sudden gas leaks or structural collapses, enables a faster response and evacuation.

Clear communication ensures that each team member understands their specific roles and responsibilities. It is also critical for resource allocation: Efficiently managing resources, such as personnel, equipment, and medical supplies, requires constant communication between inside and outside rescuers.

ICs are tasked with decision-making responsibilities that require real-time information. All emergency incidents have a three-part structure:

1. **Past:** What happened prior to the rescue team's arrival
2. **Present:** The conditions found upon arrival
3. **Future:** The potential conditions or events that may occur as the incident progresses or is controlled

Ongoing communication provides decision-makers with real-time information, allowing for informed and timely decisions. Flexibility is also key because it allows us to develop adaptive strategies. As conditions change, continuous communication allows teams to adapt their strategies and tactics accordingly.

Communication Tools and Techniques

Handheld two-way radios, which are both portable and reliable (for the most part), are essential tools for communication in confined spaces. Effective channel management requires dedicating certain channels to specific teams or tasks to help minimize crosstalk and ensure clear communication. However, in certain types of situations (e.g., on ships or in large steel structures), portable radios can suffer performance issues.

Another communication device used for confined-space rescue is a hard-wired communication system. A tethered system's wired connections can be more reliable in environments where radio signals may be obstructed. Some hands-free devices are available, including a headset-based device that allows rescuers to communicate freely without hindering physical tasks.

Sometimes, verbal communication is not possible in a confined-space incident. In such instances, visual signals, including hand and light signals, can be a better alternative. Both hand and light signals, which involve flashlights or other light sources, can be used to signal messages or alerts in low-visibility environments, but they are dependent on training: These signals will not be effective unless every individual involved in the -confined-space response knows and understands each of the signals.

Communication Training and Preparation

Adherence to the NFPA 1006 7.2.7 (NFPA 2021a) requirement for constant communication with outside rescuers during a confined-space emergency is vital for ensuring the safety and efficiency of rescue operations. Effective communication mitigates risks, enhances coordination, supports informed decision-making, and increases the likelihood of success of a rescue mission.

Communication methods for confined-space emergencies include verbal communication (both face to face and over communication systems, such as radios or headsets) and visual communication (hand signals [**FIGURE 5-5**] or the OATH and SUDRH methods, which are covered in Chapter 2).

Setting up standard operating procedures (SOPs)—written guidelines for handling a specific situation—for communication during confined-space emergencies helps standardize responses and ensures consistency. Integrating communication protocols within the ICS framework lends a structured and hierarchical approach to incident management.

Technology integration with advanced communication systems provides additional benefits. Investing in and integrating advanced communication technologies, such as digital radios and wireless headsets, can enhance the reliability and clarity of communication. Communication equipment must be regularly tested and maintained to prevent failures during actual emergencies.

Adherence to the NFPA 1006 7.2.7 requirement (NFPA 2021a) for constant communication with outside rescuers during a confined-space emergency is vital for ensuring the safety and efficiency of rescue operations. Effective communication mitigates risks, enhances coordination, supports informed decision-making, and increases the likelihood of success of a rescue mission.

Medical and EMS Protocols

During confined-space emergencies, the coordination of medical and emergency medical services (EMS) providers is crucial for ensuring the safety and well-being of both victims and rescuers. Upon arrival at the scene, medical personnel assess the condition of victims and prioritize treatment based on the severity of injuries or medical conditions. When conditions permit, or after any incident hazards have been addressed or controlled, they provide necessary medical interventions to stabilize victims' conditions before extrication or evacuation. These may include airway

A.

B.

C.

FIGURE 5-5 (**A**) A hand on top of the head making a tapping motion indicates "I'm okay." (**B**) Hands across the throat indicate "air problems." (**C**) Hands over the head in a waving motion or both hands tapping the head is a signal for "trouble." This could indicate a suit problem or that something is amiss with the task or situation on the part of the responder or responders.

management, hemorrhage control, or spinal immobilization, among other treatments. Occupational Safety and Health Administration (OSHA) standard 1910.146(k)(2) requires that at least one team member must be trained in cardiopulmonary resuscitation (CPR) and first aid ("Permit-Required Confined Spaces" 1993).

Throughout the response, effective communication must be maintained between EMS providers and rescuers, including the smooth handover of victims from rescue personnel to EMS and the provision of all necessary equipment to EMS providers, including medical supplies, stretchers, and ambulances, to support patient care and transport operations. See **TABLE 5-3** for a sample medical protocol; this one is for confined-space rescue airway management.

The primary medical protocols to follow are medical assessment, spinal injury precautions, and coordination with EMS.

- **Medical assessment:** Conduct initial and ongoing medical assessments of victims to evaluate their condition and provide appropriate medical interventions, including first aid, CPR, and advanced life support, if necessary.
- **Spinal injury precautions:** Implement spinal injury precautions during victim extrication and movement, including proper spinal immobilization and minimization of the risk of exacerbating spinal injuries.
- **Coordination with EMS:** Coordinate with EMS providers for timely transportation of injured victims to medical facilities, following established protocols for patient handover and transfer of care.

TIP

Providing advanced life support inside a confined space is typically not practical because confined spaces rarely have enough room to manage such care. The better objective is to remove the victim from the space as quickly as possible to get them the definitive treatment needed.

Operational Task Protocols

Operational protocols ensure the safety, communication, medical care, and overall effectiveness of response efforts during confined-space emergencies. They play a critical role in the effective management of confined-space operations and the mitigation of potential risks.

TABLE 5-3 Medical Protocol for Confined-Space Rescue Airway Management

Initial assessment	▪ Upon arrival at the confined-space rescue site, the medical responder assesses the victim's airway status as part of the primary survey. ▪ Evaluate the victim's level of consciousness, breathing effort, and **airway patency** (the ability of the victim's airway to remain open and unobstructed for sufficient airflow).

Airway-opening maneuvers	▪ If the victim is conscious and maintaining their own airway, monitor closely for any signs of deterioration.
	▪ If the victim is unconscious or exhibiting signs of airway obstruction, initiate airway-opening maneuvers:
	  ▪ **Head-tilt chin-lift maneuver:** Gently tilt the victim's head backward while lifting their chin forward to open the airway.
	  ▪ **Jaw-thrust maneuver:** Place fingers behind the angles of the victim's jaw and lift forward to move the jaw without tilting the head.

Airway adjuncts	▪ If the victim's airway remains obstructed despite manual maneuvers, consider the use of airway adjuncts:
	  ▪ **Oropharyngeal airway (OPA):** Insert an appropriately sized OPA to maintain airway patency in an unconscious victim without a gag reflex.
	  ▪ **Nasopharyngeal airway (NPA):** Insert an NPA into the nostril of a conscious or semiconscious victim to maintain airway patency.

Airway suctioning	▪ If the victim has secretions or foreign material obstructing the airway, use a portable suction device to clear the airway as needed. ▪ Ensure proper suction catheter size and suction pressure to avoid causing trauma to the airway tissues.
Continuous monitoring	▪ Continuously monitor the victim's airway status, breathing pattern, and oxygen saturation throughout the rescue operation. ▪ Reassess the effectiveness of airway management interventions and adjust as necessary based on the victim's condition.

Advanced airway management	▪ If the victim's airway cannot be adequately managed with basic maneuvers and adjuncts, consider advanced airway interventions:
	  ▪ **Endotracheal intubation:** Establish a definitive airway by inserting an endotracheal tube into the trachea to provide mechanical ventilation.
	  ▪ **Supraglottic airway device:** Use a supraglottic airway device as an alternative airway management option in cases where intubation is not feasible.
	Note that these interventions require advanced-level training.

Documentation	▪ Document all airway management interventions, including the techniques used, airway adjuncts inserted, and the victim's response to treatment, in the patient care record.

Operational protocols for confined-space emergencies include communication procedures, incident command structure, hazard mitigation strategies, and rescue procedures, all developed in accordance with *NFPA 1006: Standard on Technical Rescuer Professional Qualifications* (NFPA 2021a) and other relevant NFPA standards related to confined-space rescue operations. Also essential are the designation of key roles and responsibilities within the rescue team; establishment of communication channels; and definition of protocols for incident assessment, victim assessment, and rescue operations. Because a rescue team must always be ready, the time for this preparation is before an incident occurs. Established operational protocols allow a team to spring into action immediately.

Once operational task protocols have been established, they must be communicated to all team members. As team members receive assignment of their roles and responsibilities, their actions must also be coordinated to address identified hazards and execute rescue operations safely and efficiently.

Continuous reassessment of the established operational protocols must be conducted throughout the course of a confined-space emergency to adapt to changing conditions and emerging risks. Close monitoring of environmental conditions, victim status, rescuer safety, and operational effectiveness allows the team to make adjustments to protocols as needed to ensure the safety and well-being of all personnel involved.

The primary operational task protocols to follow are victim assessment, rescue techniques, and proper equipment use.

- **Victim assessment:** Perform rapid victim assessments to identify injuries, prioritize treatment needs, and determine the appropriate rescue techniques and equipment required.
- **Rescue techniques:** Employ safe and effective rescue techniques, such as vertical or horizontal extrication, patient packaging, and rope rescue systems, in accordance with NFPA 1006 standards (NFPA 2021a) and best practices.
- **Proper equipment use:** Ensure proper use, maintenance, and inspection of rescue equipment, including ropes, harnesses, litters, and mechanical advantage systems, following manufacturer guidelines and NFPA 1006 requirements (NFPA 2021a).

Decontamination Protocols

The decontamination protocols to follow are taking precautionary measures, using appropriate PPE, providing decontamination stations, and conducting medical monitoring.

- **Precautionary measures:** Implement decontamination protocols to minimize the risk of exposure to hazardous substances present in confined spaces.
- **Appropriate PPE:** Require all personnel involved in rescue and medical operations to wear appropriate PPE, including gloves, eye protection, and respiratory protection, to prevent contamination.
- **Decontamination stations:** Establish designated decontamination stations equipped with showers, cleansing agents, and protective clothing for personnel exposed to hazardous substances (**FIGURE 5-6**). Ensure that these stations are easily accessible and properly equipped.
- **Medical monitoring:** Closely monitor the health status of rescue and medical personnel after decontamination to detect any signs of contamination-related illnesses or injuries.

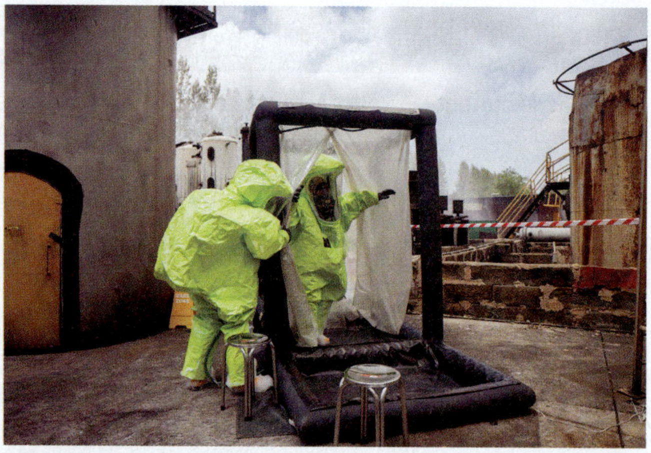

FIGURE 5-6 A decontamination station is set up to safely clean and remove hazardous materials from personnel and equipment after an emergency operation.

TABLE 5-4 illustrates a decontamination procedure for confined-space rescue equipment.

Pre-Entry Personnel Evaluations and Rescuer Clearance

Before anyone enters a confined space for an operation, a pre-entry personnel evaluation is necessary to ensure that all personnel involved in the entry are physically and mentally fit to perform their assigned tasks safely. To properly conduct a pre-entry evaluation, you must first define criteria for assessing personnel fitness, including physical health, mental alertness, training, and PPE readiness.

A pre-entry personnel evaluation consists of the following:

- **Medical assessments:** Assess team members' medical fitness, focusing on their physical health and any preexisting conditions that could hinder safe performance in confined spaces; ensure that personnel are free from illnesses or injuries that could compromise their safety or the safety of the team.
- **Fitness for duty:** Evaluate for fatigue, stress, intoxication, or other impairments; ensure that all personnel are mentally alert and physically capable of performing their tasks.
- **Training and competency:** Verify that personnel have completed all required training and possess the necessary skills and knowledge for

TABLE 5-4 Equipment Decontamination Procedure for Confined-Space Rescue

Precleaning	▪ Inspect the equipment for visible contamination, debris, or residues. ▪ Remove any large debris or foreign materials using gloved hands or appropriate tools. ▪ Brush or wipe off loose contaminants using disposable towels or brushes. ▪ If necessary, rinse the equipment with water to remove surface dirt or grime.
Washing	▪ Prepare a cleaning solution according to manufacturer's instructions or use a suitable disinfectant solution. ▪ Immerse the equipment in the cleaning solution, ensuring all surfaces are submerged. ▪ Use a soft brush or cloth to scrub the equipment, paying attention to crevices, seams, and hard-to-reach areas. ▪ Agitate the solution to facilitate the removal of contaminants from the equipment surfaces. ▪ Rinse the equipment thoroughly with clean water to remove any residual cleaning solution.
Disinfection	▪ Prepare a disinfectant solution appropriate for the type of contaminants and the material of the equipment. ▪ Immerse the cleaned equipment in the disinfectant solution, ensuring all surfaces are covered. ▪ Allow the equipment to soak in the disinfectant for the recommended contact time specified by the manufacturer. ▪ If required, agitate the solution periodically to ensure uniform disinfection of all surfaces. ▪ After the contact time has elapsed, remove the equipment from the disinfectant solution.
Drying	▪ Drain excess water from the equipment by shaking or using a clean, absorbent cloth. ▪ Place the equipment in a well-ventilated area or use drying racks to allow air circulation. ▪ Ensure the equipment is completely dry before storage or reuse to prevent the growth of mold or bacteria. ▪ If necessary, use clean, dry towels to wipe down the equipment and remove any remaining moisture. ▪ Store the dried equipment in a clean, dry environment away from potential sources of contamination.
Documentation	▪ Maintain records of equipment decontamination procedures, including the date, type of equipment, cleaning and disinfection agents used, and any observations or notes. ▪ Document any instances of equipment damage, malfunction, or contamination that may require further action or disposal.

confined-space entry and rescue operations; confirm familiarity with confined-space protocols and emergency procedures.

- **PPE readiness:** Ensure that all personnel are equipped with the necessary PPE and are trained in its proper use, check that all PPE is in good working condition and appropriately fitted to each team member, and ensure proper fit test certification for all respiratory protection being used.

Pre-entry evaluation documentation needs to include the recording of personnel assessments, training records, and any relevant medical history or clearance.

TIP

To train teams for pre-entry personnel evaluations, consider role-playing scenarios. In these hands-on practice sessions, participants can take on the roles of evaluators, entry supervisors, and entrants to simulate pre-entry evaluation processes. Sample evaluation forms or checklists can also be used to help guide participants through the assessment process and ensure consistency in evaluating personnel fitness.

Personnel Rehabilitation

Personnel rehabilitation plays a crucial role in maintaining the health and well-being of responders. Confined-space operations can be physically and mentally demanding, exposing responders to such hazards as heat stress, exhaustion, and psychological stress. Personnel rehabilitation aims to mitigate the impact of these hazards by providing responders with rest, rehydration (note that responders should always be well hydrated, but certainly the stresses of an incident demand significant rehydration), medical assessment, and emotional support. By prioritizing the welfare of the responders, rehabilitation efforts contribute to maintaining operational effectiveness and ensuring safety throughout the incident.

The rehabilitation area should be designated near the incident command post or staging area and located away from immediate hazards of the confined space but within close proximity for easy access. It should be equipped with the following to accommodate the needs of responders:

- **Shelter:** Tents or canopies provide protection from the elements and privacy for responders during rehabilitation.
- **Seating:** Folding chairs or benches offer comfortable seating for rest and evaluation.
- **Hydration stations:** Coolers or hydration packs with water, electrolyte drinks, and snacks ensure responders remain hydrated and nourished.
- **Medical supplies:** First-aid kits, vital sign monitors, thermometers, and cooling/heating devices are essential for medical assessment and treatment.
- **Psychological support resources:** Counseling materials, stress relief tools, and trained personnel facilitate psychological support for responders.
- **Communication tools:** Radios, cell phones, or communication systems enable coordination between the rehabilitation area and incident command.
- **Privacy partitions:** A temporary setup of partitions (such as portable screens, tarps, or modular panels) is used to ensure privacy and comfort for personnel during rehabilitation efforts.

Procedures and protocols for setup and maintenance of a rehabilitation area include the following:

- **Initial assessment:** Upon arrival at the rehabilitation area, responders undergo a brief initial assessment by medical personnel to evaluate their physical and mental condition.
- **Rest and hydration:** Responders are encouraged to rest and hydrate to replenish fluids lost during the operation. Hydration stations with water, electrolyte drinks, and snacks should be readily available.
- **Medical evaluation:** Medical personnel conduct periodic medical evaluations to monitor responders for signs of heat stress, fatigue, injuries, or other health concerns. Any issues identified are addressed promptly.
- **Psychological support:** Trained mental health professionals or peer support teams provide psychological support and debriefing sessions to help responders cope with the emotional stress of the incident.
- **Rotation and relief:** Implement rotation schedules to ensure that responders are not overexerted and have adequate time for rest and recovery. Relief personnel are available to replace responders who require extended rehabilitation.
- **Documentation:** Maintain accurate records of responder rehabilitation activities, including initial/baseline assessments, hydration status, medical interventions, and rotation schedules, for accountability and future reference.

Risk and Resource Assessment and Management

Risk and resource management are cornerstones of effective confined-space rescue operations. Before risk can be assessed and managed in a confined-space rescue response, however, potential hazards—whether they are atmospheric, physical, or biological in nature—must be identified, measured, and documented.

Risk Assessment

Risk assessment methodologies play a crucial role in identifying, evaluating, and mitigating hazards. These methodologies provide guidelines and frameworks for systematically analyzing risks and making informed decisions to enhance safety and operational effectiveness. A risk assessment is far more valuable when it is collaborative and inclusive, engaging and soliciting the input of all stakeholders (e.g., entry supervisors, safety officers, technical experts) and leveraging their individual expertise and knowledge.

The first step in the risk assessment process—the process of identifying hazards and the risks they represent—begins with a cautious and thorough inspection of the confined space and a review of the incident history, as well as the history of the confined space, including previous incidents or near misses. This should include interviewing the entry supervisor or responsible party for the confined space and a review of the entry permit, if available. The risk assessment process feeds into the development and ongoing modification of the incident action plan (IAP), and risk assessment findings influence tactical decisions and resource allocation within the IAP.

Risk assessment tools, whether they are qualitative or quantitative, identify the likelihood of the presence and severity of the identified hazards. Examples include risk matrices, hazard and operability studies (HAZOPs), failure mode and effects analyses (FMEAs), and job safety analyses (JSAs). These can be used to evaluate the potential consequences of identified hazards, including injury or illness to personnel, environmental impact, property damage, and operational disruptions.

A risk assessment necessarily includes a careful review of the efficacy of the control measures (e.g., engineering or administrative controls, PPE in use) already in place. During this review, gaps or deficiencies in these control measures will surface, and the assessment can prioritize actions to address them. A crucial part of risk assessment is to evaluate how well existing control measures are working, identify any shortcomings, and prioritize improvements to them.

Risk Mitigation and the Hierarchy of Controls

An effective risk assessment includes well-developed risk mitigation strategies, and chief among these is the Hierarchy of Controls. OSHA (2021) established a system known as the Hierarchy of Controls to determine which measures are most appropriate for reducing or removing hazards, as you learned in Chapter 1. Confined spaces are inherently hazardous environments, but the Hierarchy of Controls provides a systematic approach to dealing with and mitigating hazards within them. Through understanding and implementing these control measures, employers and employees can create safe systems of work to mitigate the risks associated with these environments, in turn ensuring the safety of everyone involved.

OSHA's hierarchy is structured as an inverted pyramid, with the most effective measures at the top and the least effective at the bottom, as shown in **FIGURE 5-7**.

- **Elimination:** The most preferable control measure, elimination, involves removing the hazard entirely, rendering exposures impossible.
- **Substitution:** This option, which is used when elimination is not reasonable or practical, involves minimizing the risks by replacing the task itself or changing the work process.
- **Engineering controls:** Engineering controls change the hazard at its source and involve either modifying

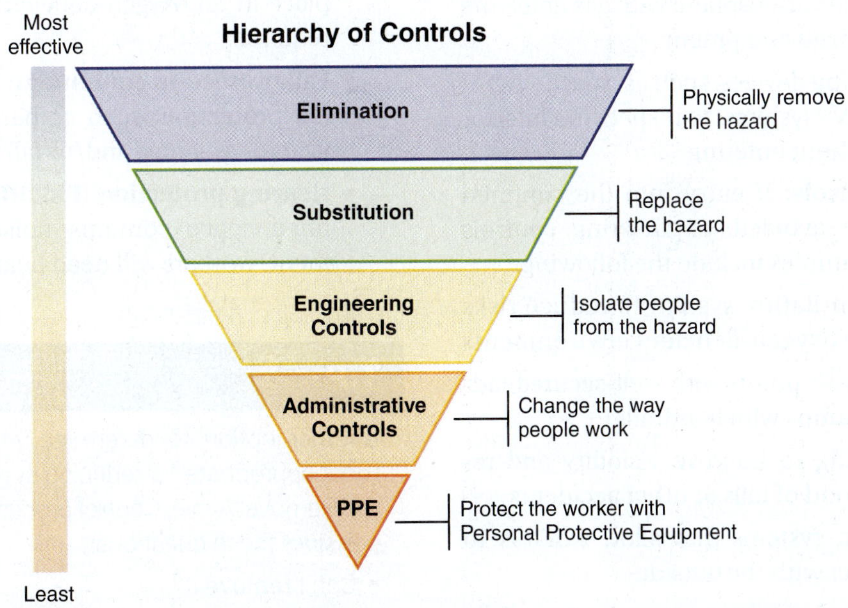

FIGURE 5-7 OSHA's Hierarchy of Controls.

Courtesy of NIOSH/CDC.

the work activities or the workspace itself to reduce exposure or protect workers from exposure.

- **Administrative controls:** Development of safe work procedures is the hallmark of administrative controls, which organize work in a way that minimizes exposure potential.
- **PPE:** When any or all four measures have been attempted and are determined to be insufficient, PPE is used as the last line of defense.

Applying the Hierarchy of Controls to Confined-Space Work

Following are some concrete applications of OSHA's (2021) Hierarchy of Controls for confined-space work:

- **Elimination:** No person shall enter a confined space to carry out work for any purpose unless it is not reasonably practicable to achieve that purpose without such entry.
- **Substitution:** Examples of modifying work practices to avoid the need for entry into a confined space include the following:
  - Cleaning/removing residues from the space from outside by using high-pressure water sprays or other means
  - Performing a remote visual inspection (perhaps with a drone) to examine the space
  - Clearing obstructions via mechanical equipment or vibration devices
  - Testing the space's atmosphere with gas detectors or other specialized equipment
  - Deploying existing devices, such as closed-circuit television (CCTV) systems, to inspect the interior of the space without entering
- **Engineering controls:** If entry into the confined space cannot be avoided, engineering controls must be used. Examples include the following:
  - An effective ventilation system to reduce risks associated with oxygen-deficient environments
  - Safe entry and exit points with well-secured ladders, stairs, or ramps with handrails
  - Sufficient lighting to improve visibility and reduce the likelihood of falls or other accidents
  - Communication systems that allow workers to maintain contact with the outside
  - Guardrails, barriers, or other systems to prevent falls
  - Appropriate rescue equipment, such as hoists, winches, safety harnesses, and retrieval systems
- **Administrative controls:** Although they are not as effective as engineering controls or substitution, administrative controls play an important role in increasing overall safety, and they complement other control measures. Examples include the following:
  - Permit-to-work systems that ensure only trained and competent workers may enter
  - Completion of all relevant work permits, such as a hot-work permit if welding will be performed
  - Sufficient training to ensure workers are aware of hazards, emergency procedures, communication protocols, and safe work practices
  - Verification that the supervisor is trained to respond to emergencies
  - Requirement that the supervisor always remains outside the confined space
  - Pre-entry checks, such as checking the space's atmospheric conditions before entry and at established periodic intervals
  - Emergency procedures and ensuring that all workers are aware of the emergency drills
- **PPE:** Specific requirements are dependent on the nature of the confined space, the work activities being carried out there, and the specific hazards of those activities. Common PPE used in confined spaces includes but is not limited to the following:
  - **Respiratory protective equipment** (RPE) (**FIGURE 5-8A**): In spaces with a risk of exposure to hazardous substances or if work is taking place in an oxygen-deficient atmosphere, RPE is required.
  - **Fall protection equipment:** Workers may require fall protection, such as harnesses, retrieval devices or systems, and/or fall arrest equipment.
  - **Hearing protection** (**FIGURE 5-8B**): If the confined space contains noise-generating equipment, workers will need hearing protection.

TIP

Another option to determine which measures are most appropriate for reducing or removing hazards is the Remove, Avoid, Control process. For each hazard, consider these questions:

- Can I **remove** it?
- Can I **avoid** it?
- Can I **control** it?

- **Head protection** (**FIGURE 5-8C**): If the space contains a risk of falling objects or other overhead hazards, hard hats or other types of helmets are required.

Documenting and Communicating Risk Assessment Findings

The last steps in any risk assessment are to (1) document all findings of the assessment, including identified hazards, risk levels, control measures, and recommended actions, and (2) subsequently communicate these findings to all stakeholders through clear and concise reports, briefings, or safety meetings. See **TABLE 5-5** for a template that provides a structured approach to assessing and managing risks associated with various tasks or activities.

Risk Management

In the context of confined-space rescue operations, the risk management process consists of four primary steps:

- **Step 1:** Assess identified hazards to determine their severity and likelihood of occurrence.
- **Step 2:** Develop risk management plans that outline strategies for mitigating identified hazards and reducing risks to an acceptable level.
- **Step 3:** Implement control measures to eliminate or minimize hazards, including engineering controls, administrative controls, and the use of PPE.
- **Step 4:** Continuously monitor and reassess risks throughout the operation to identify any changes or new hazards that may arise.

Logistics

During a confined-space rescue, logistics support systems are necessary to ensure the effective coordination and deployment of resources, including personnel, equipment, and supplies. Careful planning of the logistics operation addresses the transportation, storage, and distribution of resources and coordination with external agencies, such as emergency services and specialized rescue teams, to access additional resources as needed.

Resource Management

As you learned earlier in this chapter, resource management is the process of determining the proper allocation of resources based on operational needs and priorities, taking into account the nature and severity of identified

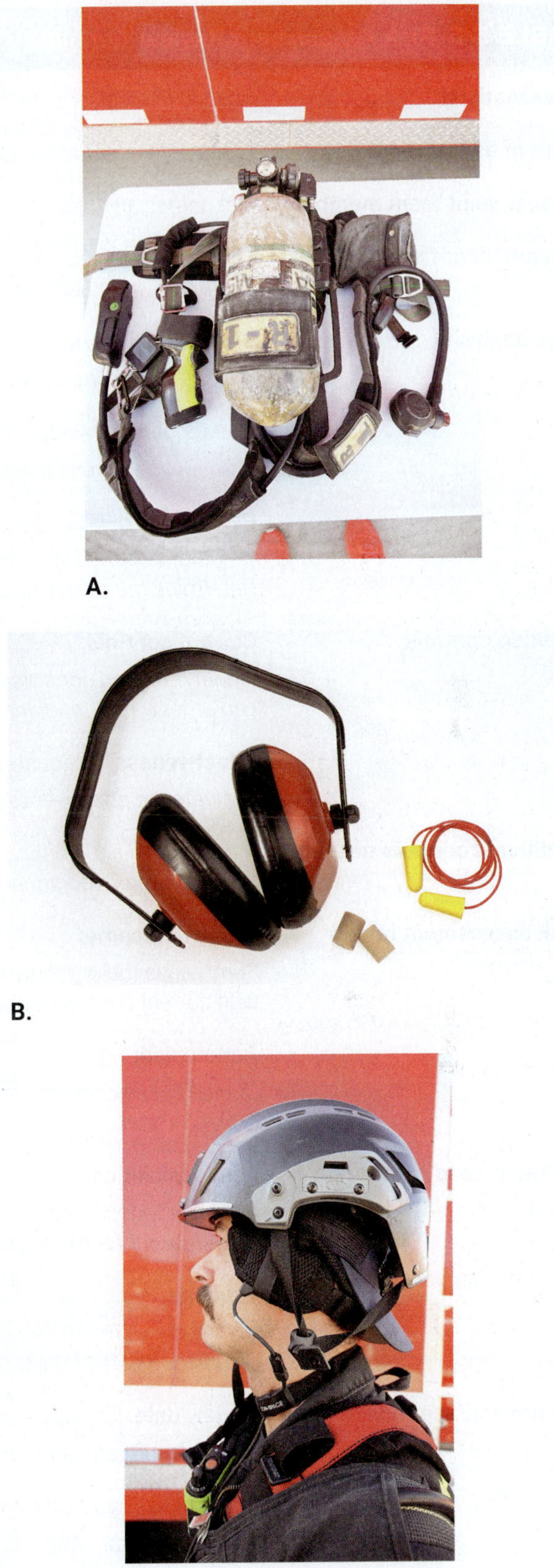

A. B. C.

FIGURE 5-8 PPE commonly used in confined spaces includes (**A**) RPE; (**B**) various types of hearing protection; and (**C**) head protection, such as a hard hat.

B: © TonyLomas/iStock/Getty Images Plus/Getty Images

TABLE 5-5 Risk Assessment Review Template

Task/Activity	*Specify the task or activity being assessed.*
Date of assessment	*List the date that the risk assessment is conducted.*
Assessment team members	*List names and roles of individuals involved in the assessment.*
Hazard identification	*List all potential hazards associated with the task/activity (e.g., exposure to hazardous chemicals, confined-space entry, working at heights).*
Risk analysis	**Likelihood:** Low/Medium/High/Very High *Assess the likelihood of each identified hazard occurring.*
	Consequence: Minor/Moderate/Major/Critical *Evaluate the potential consequences if the hazard were to occur.*
	Risk level: Low Risk (Acceptable)/Medium Risk (Manageable with Controls)/High (Requires Immediate Action or Mitigation) *Determine the overall risk level based on likelihood and consequence.*
Existing controls	**Current controls:** *Identify existing controls in place to mitigate identified risks (e.g., personal protective equipment [PPE], engineering controls, administrative controls).*
	Effectiveness: Adequate/Inadequate/Not Applicable *Evaluate the effectiveness of existing controls in managing risks.*
Additional controls needed	*Recommend additional controls or measures to further mitigate risks (e.g., enhanced training, use of additional PPE, implementation of new engineering controls).*
Risk assessment review	**Review outcome:** *Summarize the outcome of the risk assessment review (e.g., identified risks are acceptable with current controls; additional measures are required to mitigate risks).*
	Action plan: *Outline specific actions to be taken based on the assessment findings (e.g., implement recommended controls, conduct further training, revise procedures).*
Follow-up and monitoring	**Responsibilities:** *Assign responsibilities for implementing actions and monitoring progress (specifying the individuals or teams responsible for each action item).*
	Timeline: *Set deadlines for implementing actions and conducting follow-up reviews (specifying target completion dates for action items).*
Review and approval	**Review date:** [Specify review date] *Schedule a future date for reviewing the effectiveness of implemented controls.*
	Approvals: [Signatures or approvals from relevant stakeholders] *Obtain necessary approvals from management or safety personnel.*

hazards. The process involves maintaining an inventory of available resources and tracking their deployment and utilization throughout the operation; ensuring that personnel are adequately trained and equipped to perform their assigned tasks safely and effectively; and implementing tracking systems for resource usage, including equipment maintenance and replenishment, to ensure readiness for future operations.

TABLE 5-6 Assessment of Resource Needs for Confined-Space Rescue

Personnel requirements	▪ Determine the number and qualifications of personnel needed for the rescue operation based on the complexity of the confined space and the potential hazards involved. ▪ Assess the need for specialized rescue teams, including confined-space rescue technicians, medical personnel, and incident commanders (ICs) with appropriate training and certifications as per National Fire Protection Association (NFPA) 1006 (NFPA 2021a).
Equipment and tools	▪ Identify and procure specialized equipment required for confined-space entry, rescue, and patient care, including the following: ▪ Confined-space entry and retrieval systems (e.g., tripod, harnesses, lifelines). ▪ Atmospheric monitoring and gas-detection equipment to assess air quality and detect hazards. ▪ Lighting equipment (e.g., portable lights, headlamps) for visibility inside the confined space. ▪ Communication devices (e.g., radios, intercom systems) to maintain constant contact between rescuers. ▪ Medical supplies and first-aid kits for treating injured individuals.
Technical expertise and support	▪ Evaluate the need for technical experts or consultants, such as structural engineers, hazardous materials specialists, or industrial hygienists, to assess specific hazards and provide guidance during the rescue operation.
Safety and personal protective equipment (PPE)	▪ Ensure all personnel involved in the rescue operation have appropriate PPE based on identified hazards, including the following: ▪ Respiratory protection (e.g., self-contained breathing apparatuses [SCBAs]) for atmospheric hazards. ▪ Protective clothing (e.g., coveralls, gloves, helmets) to prevent exposure to chemicals or physical hazards. ▪ Fall protection equipment (e.g., harnesses, lanyards) for working at heights or during vertical rescues.
Logistical support	▪ Plan for logistical support, including transportation of personnel and equipment to the incident site, establishment of incident command post, and coordination with external agencies if additional resources are needed.
Training and readiness	▪ Ensure all responders are adequately trained and prepared for confined-space rescue operations, including administration of regular drills and exercises to maintain proficiency. ▪ Provide continuous education and training on the latest rescue techniques, equipment operation, and safety protocols.
Risk assessment and contingency planning	▪ Conduct a thorough risk assessment to identify potential hazards and develop contingency plans for unforeseen circumstances or emergencies during the rescue operation. ▪ Prepare for confined-space rescue emergencies, optimize response capabilities, and enhance the safety of personnel involved in the rescue operation. ▪ Perform regular reviews and updates to resource assessments to ensure readiness and adaptability for evolving rescue scenarios.

TABLE 5-6 describes the requirements for an assessment of resource needs for a confined-space rescue.

During an ongoing confined-space incident, tracking and managing resources is crucial for ensuring safety and efficiency. The process typically involves logging the arrival, deployment, and status of personnel, equipment, and supplies. This includes maintaining a detailed record of who enters the confined space, what equipment is being used, and the condition of all resources. Accurate tracking helps prevent overcrowding, ensures that the right tools are available, and allows for quick response to emergencies.

Resource staging is a critical aspect of this process, involving the strategic placement of resources in a designated area close to the incident. Staging ensures that necessary resources are readily available for immediate deployment, reducing response times and minimizing the risk of delays. Proper staging also enhances safety by preventing unnecessary personnel and equipment from entering the hazardous zone until they are needed, thereby reducing congestion and potential hazards.

Mutual aid agreements play a vital role in resource management, especially for large-scale or prolonged incidents. These agreements allow for the sharing of resources, personnel, and expertise between different agencies or jurisdictions. When an incident exceeds the capacity of local resources, mutual aid can provide additional support, ensuring that the incident is managed effectively and safely. These agreements ensure that resources can be quickly mobilized from neighboring areas, helping to sustain operations over extended periods and improving overall incident response.

Integration with the Incident Management System (IMS)

An **Incident Management System (IMS)** is a series of protocols that are put in place during an incident. It provides a way to organize, structure, and coordinate personnel, equipment, facilities, and communications that will allow the incident response to proceed more effectively. Integration with IMS is a critical aspect of size-up during any emergency response, including confined-space rescue operations, because size-up provides essential situational awareness that informs and supports the overall incident management approach within the IMS framework.

As you have learned, in the context of confined-space rescue, size-up data are utilized to establish clear roles, responsibilities, and operational protocols for responders, ICs, and support personnel within the IMS. This integration ensures efficient coordination and communication among all stakeholders involved in the rescue operation. Key elements of integration with the IMS during size-up include the following:

- **Establishment of clear roles and responsibilities:** Size-up information helps define specific roles and responsibilities for responders based on their expertise and operational requirements. ICs allocate tasks and resources strategically to address identified needs and priorities.
- **Implementation of operational protocols:** Size-up data inform the development and implementation of operational protocols within the IMS. This includes communication protocols, incident reporting procedures, resource allocation guidelines, and safety protocols tailored to the confined-space rescue scenario.
- **Using size-up data to determine their relevancy to SOPs**: When using data to determine their relevancy to SOPs for confined-space operational protocols, it is essential to understand both the nature of the data and how the data align with safety and operational requirements. By systematically analyzing size-up data in the context of SOPs for confined-space operations, you can ensure that your protocols are both relevant and effective, enhancing the safety and efficiency of your confined-space operations.
- **Coordination and communication:** The IMS facilitates seamless coordination and communication among responders, the IC, and support personnel. Clear lines of communication are established to ensure timely information sharing, standardized language and terminology, status updates, and decision-making throughout the rescue operation.
- **Resource management:** Size-up data assist in resource management within the IMS framework. These data help prioritize resource allocation based on identified risks, rescue priorities, and operational requirements to optimize efficiency and effectiveness during the response.
- **Adaptation and continual improvement:** Integration with the IMS allows for adaptation and continual improvement based on evolving conditions and feedback during the rescue operation. Size-up information is continuously evaluated and used to refine strategies, scale the incident (expand or shrink it), adjust tactics, and enhance overall incident management capabilities.

Risk Mitigation

Risk mitigation, a subset of risk management, specifically focuses on the actions taken to reduce the likelihood or impact of risks identified during the risk management process. In the context of confined-space incident response, risk mitigation includes the following:

- Conducting a comprehensive risk assessment
- Implementing engineering controls
- Establishing administrative controls
- Providing proper training and education
- Using PPE based on guidelines set forth by the AHJ
- Monitoring and testing atmospheric conditions throughout the response

- Implementing emergency response and rescue plans
- Conducting regular inspections and routine maintenance
- Fostering a culture that encourages the reporting of hazards, near misses, and unsafe conditions
- Complying with all regulatory requirements

Roles of NFPA Standards and OSHA Regulations in Ensuring Adequate Resources

Confined-space incidents requiring technical rescue operations demand careful planning and allocation of resources to ensure responder safety and effective victim extrication. Compliance with NFPA and OSHA standards is essential to uphold safety protocols and operational guidelines.

During the risk assessment and resource identification process, NFPA 1006 (NFPA 2021a) provides clear guidelines on risk assessment methodologies specific to technical rescue operations, and OSHA 29 C.F.R. 1910.146 ("Permit-Required Confined Spaces" 1993) contains specifics for confined-space entry, rescue, and emergency response requirements.

Regarding resource allocation and deployment, NFPA 350 (NFPA 2022a) and NFPA 1006 (NFPA 2021a) provide guidance on determining the type and level of technical rescue response that is required (based on the findings of the risk assessment), as well as information on resource capabilities, equipment requirements, and personnel qualifications and recommendations for allocation of appropriate personnel and equipment to support rescue operations.

NFPA 1006 (NFPA 2021a) also outlines training, qualification, and certification requirements and qualifications for technical rescue team members. NFPA 2500 (NFPA 2022d) lists standards for technical rescue training and operational proficiency.

When procuring and maintaining specialized rescue equipment suitable for confined-space operations, including PPE, communication devices, ventilation systems, and retrieval systems, review OSHA 29 C.F.R. 1910.134 ("Respiratory Protection" 1993), which also provides requirements for respiratory protection and atmospheric monitoring equipment.

As previously discussed, confined-space rescue operations must be integrated with the IMS, and NFPA 1561 (NFPA 2020) sets standards for this integration, including clear incident command roles, responsibilities, and communication protocols that streamline resource coordination and deployment.

Continuous evaluation and improvement can be achieved through NFPA 1006 (NFPA 2021a) compliance audits. These identify areas for improvement in both resource management and operational readiness. The findings from compliance audits can be used to implement corrective measures and training enhancements.

Assessing Resource Needs Through a SWOT Analysis

Evaluation of the personnel, equipment, and tools required to safely conduct the rescue operation must take place prior to any emergency response operation. Command and organizational leadership must identify and perform a **strengths, weaknesses, opportunities, and threats (SWOT) analysis** of their response teams and response organizations and determine if they meet or exceed the requirements outlined in *NFPA 1006: Standard on Technical Rescue Personnel Professional Qualifications* (NFPA 2021a). A SWOT analysis is a strategic planning tool widely used in businesses, but the foundational strategy is applicable to confined-space emergency responses as well.

The first step of a SWOT analysis is to determine what internal factors affect the organization's (in this case, the response team's) performance, including expertise, resources, and operational efficiencies or inefficiencies—in effect, its strengths as well as its weaknesses. Next, an evaluation of the external forces that influence the organization's environment (in this case, the confined space and the emergency within it) will reveal the opportunities and threats it faces. In business terms, these external forces would include an assessment of market trends, regulatory changes, technological advancements, and competition, but in the context of confined-space emergency response, the opportunities and threats are quite different. External factors could create opportunities for growth or enhancement of rescue capabilities or threats that pose risks or challenges to your team's effectiveness and readiness (**TABLE 5-7**).

Once all information has been gathered, the analysis can take place. Assessment of the relationships that exist among the team's internal strengths and weaknesses and the external opportunities and threats presented during the emergency will yield valuable insights. Team members can formulate recommendations to capitalize on their strengths and address any identified weaknesses, then determine ways to maximize and exploit potential opportunities and devise plans to mitigate threats.

TABLE 5-7 A Strengths, Weaknesses, Opportunities, and Threats (SWOT) Analysis in the Context of Confined-Space Technical Rescue

Aspect	Description
Strengths (what we do or did well)	Internal attributes and resources that contribute positively to achieving rescue objectives
Weaknesses (what we do or did incorrectly, or what we struggled with)	Internal factors that hinder effective rescue operations or pose challenges
Opportunities (where we can improve)	External factors and situations that can be leveraged to enhance rescue capabilities
Threats (what the risks and threats are)	External challenges and obstacles that could affect rescue operations

TABLE 5-8 shows an example of a SWOT analysis for a confined-space technical rescue team, including recommendations and action plans.

Personnel and Equipment Resource Lists

Assembling a team with the requisite expertise and capabilities to address the unique challenges of confined-space environments can be challenging. *NFPA 1006: Standard on Technical Rescue Personnel Professional Qualifications* (NFPA 2021a) emphasizes the need for a well-trained and proficient rescue team comprising personnel with diverse skill sets. These may include rescuers certified in confined-space rescue, rope rescue technicians, medical responders, and communication specialists.

Equipment needs are also set forth in *NFPA 1006: Standard on Technical Rescue Personnel Professional Qualifications* (NFPA 2021a). These include but are not limited to PPE, communication devices, lighting, rescue tools, and rigging hardware. In addition to this standard equipment, confined-space rescue operations may necessitate specialized tools, such as tripod-based systems or high directionals, mechanical advantage systems, and devices for atmospheric monitoring and ventilation.

Effective confined-space rescue operations hinge on the availability of skilled personnel and appropriate equipment resources. By adhering to NFPA 1006 standards and guidelines (NFPA 2021a), rescue agencies can ensure the readiness and proficiency of their teams, thereby enhancing the safety and success of confined-space rescue response operations.

Post-Size-Up Information Relay

During a size-up of a confined-space incident, all pertinent information is gathered and relayed to the IC in compliance with NFPA standards and OSHA regulations. So what is meant by "pertinent information"?

For starters, key information about the confined space, including its size, configuration, entry points, and potential hazards (e.g., atmospheric, physical, biological), is essential to make good decisions regarding the tactical response. Also important, of course, is information on the nature of the incident, such as the reported problem or emergency situation. Witnesses, reporting parties, and onsite personnel can provide firsthand information regarding the incident.

Once this information has been gathered, it must be relayed to the IC. It should be organized and prioritized based on its relevance and urgency. The most critical details include the space's characteristics and conditions, the identified hazards, the risks associated

TIP

The data collected through size-up information-gathering techniques are systematically analyzed and utilized to develop a comprehensive understanding of the rescue scenario. These data inform decision-making related to rescue priorities, resource allocation, safety measures, and operational strategies. Key size-up information-gathering techniques include the following:

- **Visual assessments:** Visually inspecting the confined-space environment to identify potential hazards, access points, and conditions that may affect rescue efforts.
- **Interviews:** Gathering firsthand accounts from witnesses, reporting parties, or individuals familiar with the incident provides valuable insights into the circumstances leading to the rescue situation.
- **Atmospheric monitoring:** Conducting air quality assessments using specialized equipment to detect and analyze atmospheric hazards such as low oxygen levels, toxic gases, or combustible vapors.
- **Hazard evaluations:** Systematically assessing potential hazards within the confined space, including physical, chemical, biological, and environmental risks.

TABLE 5-8 Example of a Confined-Space Technical Rescue Team Strengths, Weaknesses, Opportunities, and Threats (SWOT) Analysis and Results

Aspect	Description
Strengths	▪ Highly trained and certified rescue team members ▪ Specialized equipment for confined-space rescue operations that is current and up to date, with a detailed maintenance and inspection plan ▪ Established standard operating procedures (SOPs) for safe and effective rescue that are current and approved by the authority having jurisdiction (AHJ) ▪ Strong collaboration and teamwork among team members
Weaknesses	▪ Limited budget for acquiring advanced rescue equipment ▪ Occasional gaps in training due to scheduling constraints ▪ Communication challenges during rescue operations ▪ Lack of experience in handling specific types of confined-space incidents
Opportunities	▪ Increased demand for confined-space rescue services in industrial sectors ▪ Partnerships with other emergency response agencies to enhance capabilities ▪ Adoption of new technologies for enhanced communication and monitoring ▪ Expansion of training programs to address evolving rescue techniques
Threats	▪ Regulatory changes affecting rescue protocols and safety standards ▪ Budget cuts affecting funding for training and equipment ▪ Competitive pressures from other rescue teams and service providers ▪ Evolving safety risks associated with new industrial processes and technologies
Potential recommendations	▪ Secure additional funding to invest in advanced rescue equipment and technology. ▪ Enhance communication protocols and conduct regular team training sessions to improve coordination and efficiency. ▪ Strengthen partnerships with industry stakeholders and emergency response agencies to broaden capabilities and resources. ▪ Stay updated on regulatory requirements and industry trends to ensure compliance and readiness for emerging challenges.
Action plans	▪ Develop a strategic plan to secure funding through grants, sponsorships, or budget allocations for equipment upgrades. ▪ Implement enhanced communication tools and protocols, including training on effective radio communication and Incident Command Systems (ICSs). ▪ Foster collaborations with local fire departments, emergency medical services (EMS), and industrial partners to establish mutual aid agreements and joint training exercises. ▪ Establish a continuous improvement program to monitor and address evolving threats and opportunities in confined-space rescue operations.

with the incident, the last reported location of victims (if known), and any observed changes or developments.

The Risk–Benefit Analysis

A comprehensive risk–benefit analysis is essential to ensure the safety of responders and victims while maximizing the effectiveness of the rescue effort. This process involves identifying and evaluating the potential risks associated with various rescue strategies against the benefits of successful victim extraction. You must be capable of determining whether the risks of a rescue will outweigh the benefit of performing it, or vice versa.

When performing a risk–benefit analysis, segregate the risks into three groups: atmospheric hazards, physical hazards, and operational risks. For atmospheric hazard assessment, evaluate the presence of toxic gases, deficient or enriched oxygen levels (in accordance with 19 C.F.R. 1910.146 ["Permit-Required Confined Spaces" 1993]), or combustible atmospheres within the confined space and consider the potential for exposure to hazardous substances and airborne contaminants. Assess the risks for physical hazards by considering the space's configuration (e.g., confined movement, potential entanglement or engulfment hazards, or collapse) and also any sharp edges, protruding objects, or unstable surfaces that may be present. Operational risks may include limited access, communication challenges, or potential equipment failures; also possible are secondary incidents, such as rescuer injuries or entrapment of additional victims.

The benefit analysis should include consideration of the likelihood of survival for the victims, the minimization of harm, and the potential legal and ethical obligations the rescue will face. Consider how likely survival will be if rescue efforts are successful, as well as such factors as the conditions of the victims, their medical needs, and the time sensitivity of the rescue. Assess the impact of a successful rescue on the minimization of harm to victims, including further injury or deterioration of their conditions. Remember that compliance with legal and regulatory requirements for rescue operations is paramount, so the ethical imperative to intervene and mitigate suffering whenever possible must be considered. Other ethical considerations include the duty to rescue, moral responsibility, and public perception.

Performing a Risk–Benefit Analysis

Size-up information is important at all stages of a response, but it is particularly important in a risk–benefit analysis. In the size-up, key elements of a confined-space environment, including entry and exit points, the viability of victims (rescue vs. recovery), hazards, and potential obstacles, are identified. Environmental factors such as ventilation, temperature, and the presence of hazardous substances gathered during the size-up must be considered in order to fully assess the risks and challenges of the rescue. Size-up information regarding the condition of the confined space and its potential hazards is also necessary.

A risk–benefit analysis (**FIGURE 5-9**) has three primary phases: a probability assessment, a consequence analysis, and a benefit analysis. First, estimate the likelihood of various risks occurring during rescue operations based on available information and your own past experience, coupled with such factors as environmental conditions, rescuer expertise, and equipment reliability.

Next, assess the potential consequences of the identified risks on the safety of responders and victims, taking into consideration the severity of victims' injuries, whether fatalities have occurred, the environmental impact of performing the rescue, and the financial implications of the rescue.

In the last phase, the consideration of benefits, quantify the potential benefits of a successful rescue in terms of lives saved, injuries prevented, and societal impact. Be sure to consider qualitative factors as well, such as the emotional well-being of rescuers, the importance of the public's trust in the rescue services, and the organization's reputation.

Terminating a Confined-Space Incident

Effective management of the termination phase of a confined-space incident, ensuring that all necessary actions are taken to conclude the operation safely and efficiently, involves six key steps:

1. Incident assessment
2. Coordination and communication
3. Personnel accountability
4. Equipment and resource management
5. Documentation and reporting
6. Demobilization

During the incident assessment, you determine the current status of the confined-space incident, including the completion of rescue and recovery operations, hazard mitigation efforts, and the condition of personnel involved, and evaluate any remaining risks or hazards within the confined space and its surrounding environment.

Coordination involves working with incident command and other key personnel to confirm readiness for terminating the incident. The next step is to communicate plans for termination with all involved parties, including responders, support personnel, and affected stakeholders, regarding the planned termination process and any remaining tasks.

Verifying the accountability of all personnel involved in the confined-space operation requires making sure that everyone is present, safe, and accounted for. At this point, you should also collect and document any equipment or resources used during the incident, ensuring that nothing is left behind or misplaced. Following established decontamination and storage procedures, properly clean, inspect, and maintain all equipment and resources used during the response. Once all items are accounted for and cleaned, inventory and restock consumable items, such

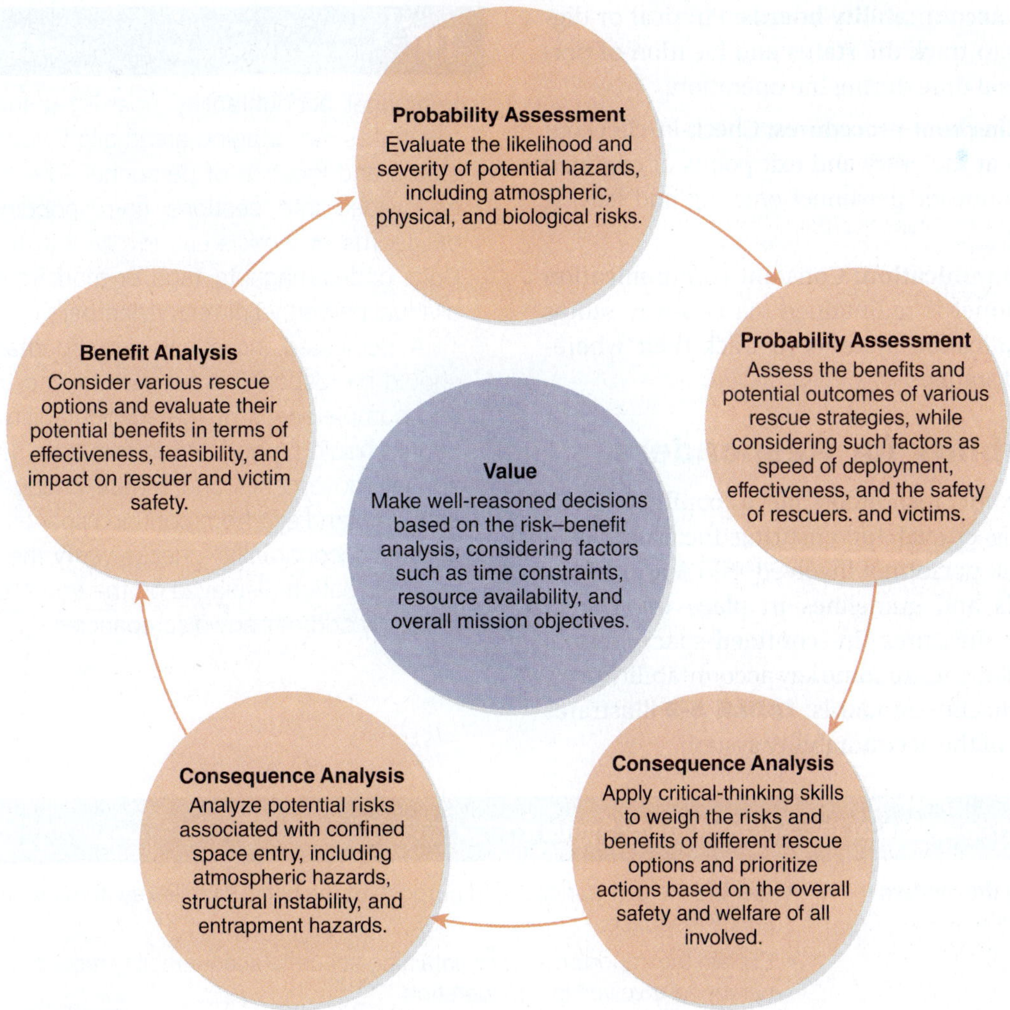

FIGURE 5-9 The process and value of a risk–benefit analysis.

as medical supplies, PPE, and rescue equipment, to ensure their readiness for future incidents.

All necessary documentation must be completed and retained, including incident reports, log entries, and any required regulatory reporting. This documentation must include all relevant information, observations, and lessons learned for future reference and analysis. After-action debriefings, where any clarification of data can occur, are encouraged and should be included in the final documentation.

The last stage, demobilization, involves the removal of all response resources and personnel from the confined-space incident site in an organized and systematic manner. Any temporary barriers, signage, or hazard controls established during the incident response must be removed or secured, returning the area to its preincident condition. Demobilization may involve a "hotwash," which is a debrief with all participants that immediately follows the incident.

Personnel Accountability Systems

Personnel accountability systems help track the location and status of all individuals involved in the operation, enabling rapid deployment of resources when emergencies occur. Accountability systems also facilitate effective communication and coordination among team members and incident command personnel.

Commonly used personnel accountability systems include the following:

- **Name tags or identification cards:** Personnel wear visible name tags or identification cards that include essential information such as name, team assignment, and contact details.
- **Personnel rosters:** Up-to-date rosters list all personnel involved in the operation, along with their assigned roles and responsibilities.

- **Personnel accountability boards:** Physical or digital boards to track the status and location of personnel in real time during the operation.
- **Check-in/checkout procedures:** Check-in/checkout procedures at the entry and exit points of confined spaces account for personnel entering and leaving the area.
- **Radio communication:** Constant communication with personnel is maintained via two-way radios or communication systems to track their whereabouts and status.

Accountability for Responders

Accountability for responders at a confined-space rescue scene is crucial for ensuring the safety and well-being of all personnel involved. NFPA and OSHA have standards and guidelines in place to address accountability measures in confined-space rescue operations. Following are some key accountability principles and applicable standards. **TABLE 5-9** illustrates the proper use of the accountability system.

TIP

Personnel accountability boards at incident command posts or staging areas allow you to track the status and location of personnel. The board should be divided into sections corresponding to different teams or work areas involved in the operation. Color-coded magnets, tags, or markers represent individual personnel or team members on the board.

A dedicated individual or accountability officer should be responsible for maintaining the accuracy and completeness of the accountability board. The board should be updated in real time based on personnel movements and status changes, including entry into and exit from confined spaces. Regular roll calls or accountability checks verify the accuracy of the information displayed on the board and help you promptly address any discrepancies.

TABLE 5-9 Proper Use of the Accountability System

Understanding the system	▪ Identify the components and purpose of the accountability system used in confined-space response. ▪ Explain the importance of maintaining accurate accountability records for all personnel involved in the operation.
Check-in procedures	▪ Demonstrate the correct procedure for personnel check-in upon arrival at the incident scene or staging area. ▪ Follow established protocols for recording personnel information, including name, team assignment, and time of arrival.
Personnel tracking	▪ Utilize tracking tools, such as accountability boards or electronic systems, to monitor the location and status of personnel throughout the operation. ▪ Update the accountability system in real time to reflect changes in personnel assignments, rotations, or movements within the confined space.
Implementing accountability measures	▪ Ensure that all personnel wear or carry accountability tags, badges, or identifiers that are easily visible and accessible. ▪ Conduct periodic roll calls or checks to verify the presence and status of personnel at designated intervals.
Responding to challenges	▪ Address any discrepancies or issues identified in the accountability system promptly and effectively. ▪ Follow established protocols for resolving missing or unaccounted-for personnel, including search and rescue procedures if necessary.
Communication and coordination	▪ Communicate effectively with team members and incident command regarding accountability updates, changes, or concerns. ▪ Collaborate with accountability personnel and team leaders to maintain situational awareness and ensure accurate record-keeping.

The Buddy System

The buddy system, a fundamental accountability measure in confined-space rescue operations, pairs each responder with a designated buddy who monitors their safety and provides assistance if needed. *NFPA 2500: Standard for Operations and Training for Technical Search and Rescue Incidents and Life Safety Rope and Equipment for Emergency Services* (NFPA 2022d) outlines the use of the buddy system for accountability during technical rescue operations, including confined-space rescues.

Check-In/Checkout Procedures

Check-in/checkout procedures require responders to formally sign in before entering the confined-space rescue area and sign out upon exiting. This helps track the whereabouts of personnel and ensures that no one is left unaccounted for. OSHA's "Permit-Required Confined Spaces" standard (1993) requires employers to implement procedures for the safe entry and exit of confined spaces, which may include check-in/checkout protocols for personnel.

Personnel Tracking Systems

Personnel tracking systems use electronic or manual methods to monitor the location and status of responders during rescue operations. These systems may include visual tracking boards, radio-frequency identification (RFID) tags (**FIGURE 5-10**), or Global Positioning System (GPS) technology. *NFPA 1500: Standard on Fire Department Occupational Safety, Health, and Wellness Program* (NFPA 2021b) mandates the use of personnel accountability systems, including tracking methods, to monitor personnel during emergency operations, including confined-space rescues.

FIGURE 5-10 An RFID tag is used for tracking equipment and personnel during emergency operations, enhancing accountability and safety.

The Incident Command System (ICS)

The ICS provides a structured framework for managing emergency incidents, including confined-space rescues. It designates specific roles and responsibilities for ICs, safety officers, and other personnel involved in the rescue operation. *NFPA 1561: Standard on Emergency Services Incident Management System and Command Safety* (NFPA 2020) mandates the use of ICS for managing emergency incidents, ensuring clear lines of communication, and maintaining accountability for responders.

Briefings and Debriefings

Pre-entry briefings and post-rescue debriefings are essential for communicating safety protocols, assigning roles, and reviewing incident objectives. A thorough postincident analysis will identify lessons learned and improve future risk management strategies, as well as provide guidance on incorporating findings into training and procedural updates. These sessions provide an opportunity to address any concerns, reinforce accountability measures, and identify lessons learned for future operations. *NFPA 1006: Standard on Technical Rescue Personnel Professional Qualifications* (NFPA 2021a) outlines the requirements for pre- and post-operation briefings and debriefings to ensure effective communication and accountability among rescue personnel.

CHAPTER FEATURE: AFTER-ACTION

IN SUMMARY

- Types of confined spaces include tanks, bins, vats, silos, grain bins, sewers, manholes, storm drains, excavations, underground mines, crawl spaces, ductwork, wells, cisterns, water tanks, and confined spaces within vehicles.
- The ICS provides a structured framework for managing emergency response operations.
- ICS functions include establishing a clear chain of unified command, taking command and control of incident operations, coordination of resources and

IN SUMMARY CONTINUED

personnel, communication management, incident documentation and reporting, and safety management and risk assessment.

- Safety management and risk assessment work hand in hand with hazard identification and risk management.
- Personnel rehabilitation aims to mitigate the impact of hazards by providing responders with rest, rehydration, medical assessment, and emotional support.
- Personnel accountability systems are used to track the location and status of all personnel involved in the operation, often by the use of personnel tags, check-in/checkout procedures, and accountability boards.
- Risk assessment methodologies provide guidelines and frameworks for systematically analyzing risks and making informed decisions to enhance safety and operational effectiveness.
- Logistics and resource management processes involve procuring and allocating necessary resources, including personnel, equipment, and supplies.
- Risk–benefit analyses involve evaluating the potential risks associated with various rescue strategies against the benefits of successful victim extraction.
- During a size-up of a confined-space incident, all pertinent information is gathered and relayed to the IC in compliance with NFPA standards and OSHA regulations.
- Terminating an incident involves assessing whether a confined-space incident has been successfully resolved and the scene is safe for secure demobilization, as well as debriefing and conducting post-incident evaluations designed to identify lessons learned and areas for improvement.

KEY TERMS

airway patency The ability of a victim's airway to remain open and unobstructed for sufficient airflow.

communications management The systematic planning, implementation, monitoring, and control of all channels of communication within a project, organization, or team to ensure that information is efficiently and effectively exchanged.

hazard identification The process of recognizing and documenting potential sources of harm or adverse effects in the workplace, environment, or any operational setting. It is a critical step in risk management because it allows organizations to anticipate and mitigate risks before they result in accidents, injuries, or damage.

Incident Management System (IMS) A recognized system for providing management of personnel, resources, and activities during emergency operations by defining roles and responsibilities and standard operating procedures.

logistics The process of planning, implementing, and controlling the efficient and effective flow of goods, services, and information from the point of origin to the point of consumption; it encompasses all the activities required to move and store products, raw materials, or information, ensuring that they are delivered to the right place, at the right time, in the right condition, and at the right cost.

personnel accountability systems Systems used to track the location and status of all personnel involved in the operation, often by the use of personnel tags, check-in/checkout procedures, and accountability boards.

personnel rehabilitation The process of restoring and supporting the physical, mental, and emotional well-being of individuals who have experienced stress, injury, trauma, or other adverse conditions, particularly in demanding environments such as military operations, emergency services, or high-intensity work environments; this process aims to ensure that individuals can return to their duties or normal life in a healthy, effective, and sustainable manner.

resource management The process of determining the proper allocation of resources based on operational needs and priorities, taking into account the nature and severity of identified hazards.

risk assessment The systematic process of identifying, evaluating, and analyzing potential risks that could negatively affect individuals, assets, operations, or the environment; the goal is to determine the likelihood of these risks occurring and the potential consequences, enabling organizations to implement appropriate measures to mitigate or eliminate them.

risk–benefit analysis A process involving identifying and evaluating the potential risks associated with various rescue strategies against the benefits of successful victim extraction.

risk management The process of identifying, assessing, prioritizing, and mitigating risks to minimize their impact on an organization, project, or individual; it involves a series of steps designed to manage uncertainty by taking proactive measures to avoid, reduce, or transfer risks and prepare for their potential consequences.

safety management The systematic process of developing, implementing, and maintaining policies, procedures, and practices that ensure the safety and well-being of individuals within an organization or environment; it involves identifying potential hazards, assessing risks, and taking proactive measures to prevent accidents, injuries, and illnesses.

strengths, weaknesses, opportunities, and threats (SWOT) analysis A strategic planning tool used to identify and evaluate the strengths, weaknesses, opportunities, and threats related to a project, business venture, or organization.

REFERENCES

National Fire Protection Association. 2020. *NFPA 1561: Standard on Emergency Services Incident Management System and Command Safety, 2020 Edition.* Quincy, MA: National Fire Protection Association.

National Fire Protection Association. 2021a. *NFPA 1006: Standard on Technical Rescue Personnel Professional Qualifications, 2021 Edition.* Quincy, MA: National Fire Protection Association.

National Fire Protection Association. 2021b. *NFPA 1500: Standard on Fire Department Occupational Safety, Health, and Wellness Program, 2021 Edition.* Quincy, MA: National Fire Protection Association.

National Fire Protection Association. 2022a. *NFPA 350: Guide for Safe Confined Space Entry and Work, 2022 Edition.* Quincy, MA: National Fire Protection Association.

National Fire Protection Association. 2022b. *NFPA 470: Standard for Competence of Responders to Hazardous Materials/Weapons of Mass Destruction Incidents, 2022 Edition.* Quincy, MA: National Fire Protection Association.

National Fire Protection Association. 2022c. *NFPA 704: Standard System for the Identification of the Hazards of Materials for Emergency Response, 2022 Edition.* Quincy, MA: National Fire Protection Association.

National Fire Protection Association. 2022d. *NFPA 2500: Standard for Operations and Training for Technical Search and Rescue Incidents and Life Safety Rope and Equipment for Emergency Services, 2022 Edition.* Quincy, MA: National Fire Protection Association.

Occupational Safety and Health Administration. 2021. "Identifying Hazard Control Options: The Hierarchy of Controls." https://www.osha.gov/sites/default/files/Hierarchy_of_Controls_02.01.23_form_508_2.pdf.

"Permit-Required Confined Spaces." 1993. 29 C.F.R. 1910.146. https://www.ecfr.gov/current/title-29/part-1910/section-1910.146.

"Respiratory Protection." 1993. 29 C.F.R. 1910.134. https://www.ecfr.gov/current/title-29/subtitle-B/chapter-XVII/part-1910/subpart-I/section-1910.134.

CHAPTER QUESTIONS

1. A toxic atmosphere from stored substances with oxygen deficiency or enrichment is considered a
   a. non-permit-required space.
   b. purge-required space.
   c. nonflammable atmosphere.
   d. potential hazard.
2. If a victim is unconscious or exhibiting signs of airway obstruction, which of the following is an airway-opening technique?
   a. airway-opening maneuver
   b. jaw-thrust maneuver
   c. administering three palm thrusts to the back
   d. placing a non-rebreather over the mouth and nose
3. Which of the following is a decontamination protocol?
   a. exiting the confined space
   b. transporting entry personnel for full medical evaluations
   c. establishing designated decontamination stations
   d. setting up a rehabilitation area
4. What does the acronym *SWOT* stand for?
   a. strengths, weaknesses, opportunities, and technical specifications
   b. strengths, weaknesses, opportunities, and threats
   c. strengths, weaknesses, opportunities, and time required
   d. strengths, weaknesses, opportunities, and training

CHAPTER FEATURE: ON SCENE

A CHEMICAL FIRE EMERGENCY

At a manufacturing facility that specialized in chemical processing, a chemical reaction went awry inside one of the processing vessels, leading to a sudden and intense fire. The vessel was in a confined-space area, making access and rescue operations challenging. Rescue teams also faced the challenge of coordinating fire-suppression efforts while conducting simultaneous rescue operations to ensure the safety and evacuation of the trapped workers.

The Emergency Response

The facility's emergency response team was immediately activated, along with local fire departments and specialized rescue teams trained in confined-space operations. Upon arrival, the rescue teams faced the dual challenge of suppressing the fire to prevent further escalation while initiating rescue operations to evacuate the trapped workers. The IC established a unified command structure to coordinate fire-suppression efforts and rescue operations efficiently. Communication protocols were established to maintain constant updates and ensure safety.

The Rescue Operations

Firefighters deployed specialized equipment and firefighting agents suitable for chemical fires to contain and extinguish the blaze. The confined-space environment added complexity to their firefighting tactics, requiring them to carefully consider ventilation needs and hazardous material control. The situation presented quite a few challenges:

- Limited access and egress points because of the confined-space environment
- Heightened risk of exposure to hazardous materials and toxic fumes
- Coordination of multiple agencies and teams involved in fire-suppression and rescue operations
- Maintaining situational awareness and ensuring responder safety throughout the incident

The Extrication and Treatment

Rescue teams equipped with confined-space rescue gear and tools simultaneously entered the affected area to locate and evacuate the trapped workers. They followed established protocols for patient stabilization and extrication. The trapped workers were assessed for injuries and provided immediate medical care as needed before being safely extracted from the confined space.

The Outcome

The fire was successfully contained and extinguished within a few hours of intense effort by the combined response teams. All trapped workers were rescued and transferred to medical facilities for further evaluation and treatment. Lessons learned from the incident highlighted the importance of preplanning, training, and interagency cooperation in confined-space emergencies involving fire.

1. What was the main cause of the chemical fire in the facility?
   - **a.** a toxic gas leak
   - **b.** an electrical malfunction inside a processing vessel
   - **c.** a chemical reaction inside a processing vessel
   - **d.** a human error during handling
2. What was the major challenge faced by rescue teams during the fire?
   - **a.** lack of proper equipment
   - **b.** limited access and egress points in the confined space
   - **c.** poor communication between team members
   - **d.** unfamiliarity with chemical fires
3. What was the role of the unified command structure during the emergency response?
   - **a.** to ensure that only one agency was involved in the rescue operation
   - **b.** to coordinate fire-suppression efforts and rescue operations efficiently
   - **c.** to control public access to the facility
   - **d.** to assess the financial impact of the fire
4. Which of the following was NOT a challenge the rescue teams faced during the operation?
   - **a.** risk of exposure to hazardous materials and toxic fumes
   - **b.** coordination of multiple agencies and teams
   - **c.** ensuring responder safety during the incident
   - **d.** lack of specialized firefighting equipment for chemical fires

CHAPTER

6

Operations

Search, Rescue, and Victim Retrieval

KNOWLEDGE OBJECTIVES

After studying this chapter, you should be able to:

- Describe the process for establishing search parameters. (**NFPA 1006 7.2.1**, p. 144; pp. 145–146)
- Indicate the process for interviewing witnesses. (**NFPA 1006 7.2.1**, p. 144; pp. 145–147)
- Determine the type of rescue. (**NFPA 1006 7.2.2**, p. 144; pp. 147–148)
- Describe the process for establishing a rescue area. (**NFPA 1006 7.2.5**, pp. 148–149)
- Identify a means of retrieval for rescuers. (**NFPA 1006 7.2.4**, pp. 162–166)
- Describe how to build victim removal systems. (**NFPA 1006 7.2.4**, pp. 152–153; p. 156; pp. 160–161)
- Explain the importance of spinal immobilization. (**NFPA 1006 7.2.11** [horizontal spaces] and **7.2.15** [vertical spaces], pp. 152–161)
- Discuss the hazards of removal of victims on immobilization devices. (**NFPA 1006 7.2.8**, pp. 169–174)
- Discuss the types of rescue and retrieval systems and the equipment needed to build them. (**NFPA 1006 7.2.17**, pp. 157–161)
- Explain the process for ensuring a clear and unobstructed space for rescuer retrieval systems. (**NFPA 1006 7.2.6**, pp. 163–166)
- Describe the process for immobilizing and securing victims to retrieval devices. (**NFPA 1006 7.2.8**, p. 158; pp. 166–170)

SKILL OBJECTIVES

- Ensure victim and rescuer communications are established. (**NFPA 1006 7.2.13** [vertical spaces], p. 152)
- Ensure rescue systems are operational. (**NFPA 1006 7.2.4**, p. 161)
- Demonstrate evacuation methods. (**NFPA 1006 7.2.9** [horizontal spaces] and **7.2.13** [vertical spaces], pp. 165–166)
- Using two rescuers, demonstrate the process for immobilizing and securing a victim to a retrieval device. (**NFPA 1006 7.2.8**, pp. 167–168)
- Establish rescue systems and perform a rapid victim extraction. (**NFPA 1006 7.2.16**, pp. 153–155)
- Utilizing appropriate personal protective equipment (PPE) and retrieval systems, remove all victims from a confined-space incident. (**NFPA 1006 7.2.17**, pp. 172–174)

Introduction

In confined-space rescue operations, the initiation of a search within a confined space is a critical and challenging phase that demands careful planning, specialized skills, and strict adherence to safety protocols. The primary objective of this chapter is to provide a comprehensive guide for initiating search operations within areas immediately visible from the confined-space entry portal. By following these essential steps and considerations, rescue teams can enhance the effectiveness and safety of their search efforts while minimizing risks to both victims and rescuers.

Search teams must establish effective search parameters that align with National Fire Protection Association (NFPA) 1006 (NFPA 2021) operations level training standards via hazard identification, risk mitigation, effective communication, and continuous assessment. Regular training and adherence to established protocols are essential for maintaining operational readiness and response capabilities in confined-space rescue scenarios.

Interviewing witnesses to a confined-space emergency incident is a necessary part of any size-up; after all, the best way to gain a comprehensive understanding of a situation is to elicit accurate information from the witnesses who saw it happen.

Determining the type of rescue for a confined-space incident involves a systematic approach that considers the unique challenges and requirements of each rescue scenario. Thorough assessments and risk analyses help responders effectively plan and execute rescue operations to achieve successful outcomes while prioritizing safety and operational efficiency.

Identifying space-specific hazards within confined spaces is another critical aspect of ensuring the safety and well-being of workers and responders. Systematic approaches to hazard identification, appropriate methodologies, and a culture of continuous monitoring and review can effectively mitigate risks posed by hazards and promote a safe working environment within confined spaces.

Immediate and ongoing monitoring within confined spaces, as outlined in NFPA 1006 7.2.4 (NFPA 2021), is essential for safeguarding the safety of personnel and effectively managing confined-space incidents. By prioritizing atmospheric monitoring, visual inspections, and victim assessment, responders can mitigate hazards, maintain situational awareness, and ensure timely and appropriate response efforts.

Building victim removal systems for confined-space rescue operations that meet or exceed NFPA 1006 and Occupational Safety and Health Administration (OSHA) standards requires careful planning, design, construction, and verification. By following established guidelines, employing proper techniques, and prioritizing safety, responders can develop robust and reliable systems capable of safely extracting victims from confined spaces while minimizing risks to all involved parties.

Rescue and retrieval systems play a critical role in confined-space operations, enabling rescuers to safely extract victims from hazardous environments (**FIGURE 6-1**). By understanding the various types of systems available and the equipment needed to build them, rescue teams can effectively plan and execute confined-space rescue operations, ensuring the safety and well-being of all involved parties.

Spinal immobilization of victims inside a confined space is essential to rescue and retrieval processes (**FIGURE 6-2**). The process presents unique challenges that require careful planning, coordination, and adherence to established protocols. In a confined-space incident requiring spinal precautions, the safety and well-being of the victim are paramount. Whether the victim has sustained a suspected spinal injury or there is a risk of exacerbating existing spinal conditions, it is crucial to ensure that proper spinal immobilization protocols are followed to minimize the risk of further injury during rescue operations.

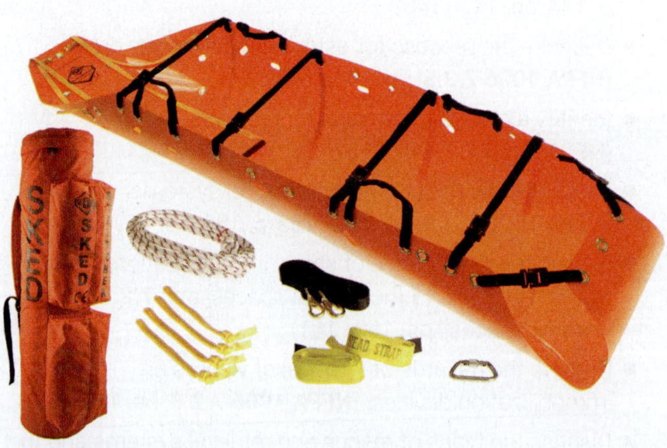

FIGURE 6-1 A rescue/retrieval system.

Used with permission from SKEDCO.

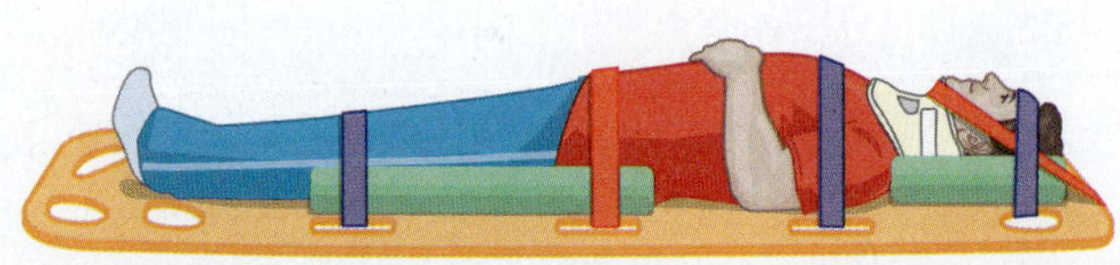

FIGURE 6-2 Spinal immobilization of a victim.

Establishing Search Parameters

Establishing search parameters in accordance with NFPA 1006 7.2.1 (NFPA 2021) for operations level training involves a systematic, six-step approach to defining the scope and boundaries of a search operation within a confined space. The process is critical to ensure the safety and effectiveness of search teams. The steps involved in establishing search parameters are as follows:

1. Assess the confined space.
2. Define the search areas.
3. Mitigate risks and establish safety measures.
4. Communicate the search parameters.
5. Continuously assess and adjust the search parameters.
6. Document and report.

To properly assess a confined space for the purposes of establishing search parameters, you must evaluate the space's layout, identify the hazards present, and determine the considerations that must be taken for the weather and temperature conditions. The space evaluation should include consideration of the space's size, configuration, entry points, access routes, and any areas of concern. Potential hazards could include atmospheric problems (e.g., oxygen deficient [levels below 19.5 percent], oxygen enriched [levels above 23.5 percent], or contaminated with toxic gases, vapors, or particulate matter), physical dangers like machinery or limitations on the access to space, or biological hazards (e.g., mold, bacteria, or viruses). When assessing weather conditions or temperature extremes, consider whether they will affect the victims or rescuers operating within the impact zone and how they may be controlled.

The definition of the search areas includes the establishment of the primary entry point or portal, the division of the space into manageable search zones, and consideration of victim profiles. The zones of search should be selected according to their accessibility and visibility from the space's entry point, as well as the overall size and layout of the space. The victim profiles provide the information currently available on victims' last known locations and conditions to help you prioritize the search.

Risk mitigation requires the implementation of security measures, necessary controls, and safety procedures. Security measures include ensuring the entry point is secured, preventing unauthorized access, and maintaining accountability among the search team members. Other controls include atmospheric monitoring, ventilation of the space (**FIGURE 6-3**), and designation and use of appropriate personal protective equipment (PPE).

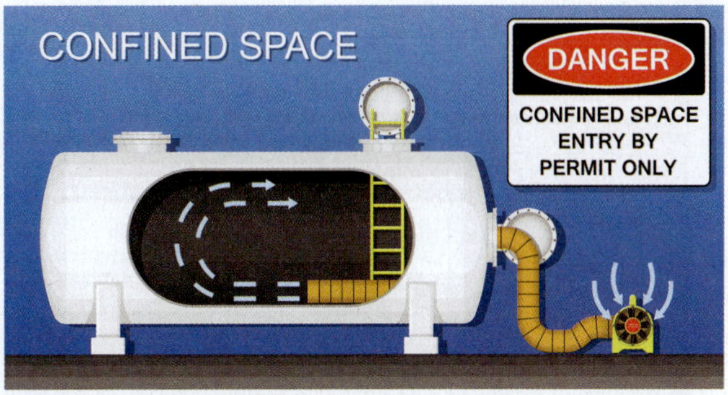

FIGURE 6-3 Ventilating a confined space ensures proper airflow and removes hazardous gases, improving safety for rescue operations.

Effective communication involves a detailed briefing to the entire search team that outlines the established search parameters, including the hazards that have been identified, the designated search zones, and all safety protocols. The incident commander (IC) must also be informed of the search parameters and consistently updated on the progress of the search to ensure alignment with the incident's overall objectives and strategies.

As with any process in a confined-space rescue, there must be room for continuous assessment and adjustment to the search parameters. Readings from atmospheric monitoring systems, assessments of the structural stability of the space, and changes in hazard levels must be accounted for as a search progresses. These ongoing assessments may require the parameters of the search to shift to ensure the safety and efficacy of the search for all.

Documenting and maintaining detailed records of the established search parameters (including any shifts that took place as the search progressed), findings, actions taken, operational activities, and outcomes will allow for review and analysis following completion of the rescue operation.

FIGURE 6-4 is a checklist for Incident Command System (ICS) search parameters. Use this sheet during operations to ensure that critical steps are completed systematically and documented accurately. Feel free to customize the sheet based on specific incident needs and organizational protocols.

Interviewing Witnesses

Interviewing witnesses is a critical step in gathering information to aid rescue efforts and ensure the safety of all involved. Your interview process should comply

Step	Task	Start Time*	End Time*	Completed**
1. Assess the Confined Space	Evaluate the layout • Conduct thorough assessment of layout • Identify size, configuration, entry points, access routes, areas of concern			☐
	Identify hazards • Determine what specific atmospheric, physical, and other hazards are present • Document hazards (e.g., low oxygen, toxic gases, confined-space configuration)			☐
	Weather/temperature considerations • Assess weather and temperature impacts on victims and rescuers • Implement control measures to manage weather-related risks			☐
2. Define Search Areas	Establish entry point • Identify primary entry point or portal			☐
	Divide into zones • Divide space into manageable search zones based on accessibility and visibility • Prioritize search based on victim profiles and incident circumstances			☐
	Consider victim profiles • Use available victim information to prioritize search areas • Focus on likely victim locations based on last known location or circumstances			☐
3. Mitigate Risks and Establish Safety Measures	Implement controls • Deploy necessary controls to mitigate identified hazards • Conduct atmospheric monitoring & ventilation, use appropriate PPE			☐
	Secure entry point • Ensure entry point is secured and controlled; no unauthorized access • Maintain accountability of search team members			☐
4. Communicate Search Parameters	Brief search team • Provide detailed briefing on search parameters, hazards, safety protocols			☐
	Update IC • Communicate search parameters and progress to incident command • Align with overall incident objectives and strategies			☐
5. Continuously Assess and Adjust	Monitor conditions • Continuously monitor confined-space conditions • Assess atmospheric conditions, structural stability, hazard levels			☐
	Adapt search strategy • Adjust search parameters based on ongoing assessments • Optimize search efforts for safety and effectiveness			☐
6. Document and Report	Document findings • Maintain detailed records of search parameters, findings, and actions			☐
	Report to authorities • Prepare comprehensive reports documenting search parameters and outcomes			☐

* Record the start and end times for each task.
** Check the box once the task is completed.

FIGURE 6-4 A confined-space rescue ICS search parameter checklist.

with *NFPA 1006: Standard on Technical Rescue Personnel Professional Qualifications* (NFPA 2021) and OSHA regulations related to confined-space operations and incident response. Quickly identifying key witnesses and prioritizing interviews based on their potential knowledge of the incident is essential to get the most updated and critical information as soon as possible.

First, ensure the immediate safety of the interview area and of all witnesses before initiating interviews by addressing any ongoing hazards or risks. Choose a suitable location for conducting interviews that is private and away from distractions and noise.

Next, prepare for the interview by gathering the necessary tools: interview forms, notepads, pens, and any other essentials. Familiarize yourself with available information about the incident, including the nature of the confined space, all victim profiles, and initial observations.

As you begin your interviews, be sure to clearly identify yourself, your role, and the purpose of the interview. Be sensitive to cultural differences and communication styles when conducting your interviews, and be sure to maintain your professionalism and exhibit empathy throughout the process. Reassure witnesses that you will respect their confidentiality and privacy rights, both during and after their interviews. Build rapport by establishing a respectful and supportive atmosphere that encourages them to feel comfortable and share information.

Ensure that the questions you ask are open-ended and not easily answered with a yes or no. For example, you could ask questions like these:

- "Can you describe what you observed when the incident occurred?"
- "What tasks were being performed before the emergency?"
- "Did the tasks you were performing require or involve any equipment or products?"

Questions about safety measures followed will help you identify any gaps or deficiencies; this can also be a good opportunity to encourage witnesses to suggest potential improvements to safety protocols and rescue procedures. As witnesses provide details about what they have seen, note any comments that are unclear or ambiguous and ask for clarification.

Documentation of your interviews should include timelines, all actions taken, and each observation they share with you. If conditions are appropriate and witnesses provide consent, record their statements using audio- or video-recording devices.

TIP

One of the primary objectives of a confined-space rescue size-up is swift identification of the number and last reported locations of victims who may be trapped inside. This critical information is required to efficiently plan and execute the rescue operation.

Interviewing witnesses and reporting parties plays a crucial role in obtaining firsthand information about the incident. By gathering accounts from those present during the event, rescuers can gain valuable insights into the circumstances leading to the incident. These interviews provide key details that help rescuers understand how the situation unfolded and refine search parameters accordingly.

If possible, rescue entrants should be present for interviews so that they may hear the firsthand account and, if needed, ask for clarifying information. By leveraging this firsthand information, rescuers can enhance their situational awareness and make better-informed decisions.

Determining the Type of Rescue

Determining the appropriate type of rescue is dependent on various factors, including the nature of the confined space, the condition of the victims, and the specific hazards present. NFPA 1006 outlines different levels of response and rescue techniques that may be appropriate, depending on operational requirements and resources available (NFPA 2021). Types of technical rescues that may be applicable to confined-space incidents include the following:

- **Nonentry rescue:** Rescuing victims without the need for responders to physically enter the confined space. This may involve using equipment such as retrieval systems or mechanical devices.
- **Entry rescue:** Sending in trained personnel to perform the rescue. Entry rescues require proper safety protocols, including atmospheric monitoring, ventilation, and PPE.
- **Complex rescue:** Challenging scenarios, such as high-angle, trench, or structural collapse rescues within confined spaces, requiring specialized techniques and equipment.
- **Rescue versus recovery:** In some instances, victim rescue is not possible, and the mission becomes one of victim recovery instead.

Operational and technical level responders may be involved in either the entry or nonentry types of rescues. Complex rescues could be either entry or nonentry, depending on the situation. See **TABLE 6-1** for key differences among awareness, operations, and technician level responders.

TABLE 6-1 Key Differences Among Awareness, Operations, and Technician Level Responders

	Training and Skill Level	Scope of Practice	Decision-Making Authority
Awareness	▪ Basic knowledge and awareness of rescue incidents and procedures	▪ Focused on recognizing and reporting rescue incidents	▪ Limited decision-making authority; primarily focus on safety and notification
Operations	▪ Intermediate skills and training to assist with rescue operations under supervision	▪ Active participants in rescue operations under supervision	▪ Work under the direction of technician level responders; follow established operational guidelines
Technician	▪ Highly trained and experienced in conducting complex technical rescues independently	▪ Lead and execute technical rescues, including planning, coordination, and implementation	▪ Have the authority to make critical decisions and lead rescue operations based on their expertise and training

To determine which type of rescue is appropriate, conduct a thorough size-up and assessment of the confined space that identifies the space type (e.g., vertical, horizontal, or inclined) and assesses potential hazards (e.g., atmospheric, physical, or biological). Evaluate the condition and number of victims trapped inside the confined space, considering such factors as injuries, accessibility, and available resources.

A risk analysis will determine the rescue's level of complexity and potential dangers. Once this is performed, you can assess the need for specialized equipment, technical skills, and operational procedures based on identified risks and hazards.

The operational capabilities and limitations of the rescue team must be considered when determining the appropriate type of response. If trained personnel, appropriate equipment, and support resources required for a particular type of rescue are not available, your options are more limited.

Once the size-up, risk analysis, and resource assessment are complete, you will be ready to develop an incident action plan (IAP) that specifies the type of rescue operation required; outlines your objectives, strategies, and tactics; and allocates resources within the framework of NFPA 1006 guidelines. (See Figure 4-2 in Chapter 4 for an example IAP.)

Establishing a Rescue Area

Establishing the rescue area involves creating a designated zone where rescue operations will be conducted, controlling access to the incident scene, and implementing measures to protect personnel from hazards present in the confined-space environment.

Before establishing the rescue area, a thorough risk assessment of the confined space and surrounding area should be conducted. This assessment helps identify potential hazards, such as atmospheric dangers, physical obstacles, and other risks that could pose a threat to personnel during rescue operations.

A suitable location for the rescue area should be selected based on the results of the risk assessment. The area should be as close to the confined space's entry point as possible while ensuring the safety of personnel. It should provide adequate space for staging rescue equipment, medical treatment, and coordination of rescue efforts.

Once the rescue area is identified, it must be clearly marked and secured (**FIGURE 6-5**) to control access and prevent unauthorized entry. This may involve setting up physical barriers such as cones, tape, or barricades around the perimeter of the rescue area. Signage (e.g., signs stating "Restricted Area" or "Confined-Space Entry in Progress") should be prominently displayed to warn others of potential hazards.

Communication protocols for personnel involved in the rescue operation should be owned by a communication coordinator. This individual is responsible for relaying information between the rescue team, IC, and other stakeholders. Two-way radios or other communication devices should be used to maintain constant communication and coordination throughout the rescue operation.

Entry and exit points for rescuers must be clearly identified and marked. These points should be easily accessible and provided with proper lighting and signage guiding personnel safely in and out of the rescue area. If possible, provide a secondary means of egress in case the primary entry/exit is compromised.

Setting up a buddy system will help ensure that no rescuer works alone inside the confined space. In a buddy system, each rescuer has a designated partner responsible for monitoring their safety and providing them with assistance, if needed. Accountability measures such as sign-in/sign-out procedures and personnel tracking will help you track the whereabouts and status of all personnel involved in the rescue operation.

The rescue area should be fully equipped with all necessary safety equipment and resources, including first-aid kits, fire extinguishers, lighting, and emergency response tools. Specific hazards present in the confined-space environment will require appropriate PPE for rescuers, such as respiratory protection, harnesses, helmets, and gloves.

The next step should be to conduct a pre-entry briefing with all personnel involved in the rescue operation. At the briefing, review safety procedures, hazard controls, and assigned roles and responsibilities. Coordination with the IC and other response agencies will ensure a unified approach to the rescue operation and facilitate seamless communication and coordination among all stakeholders.

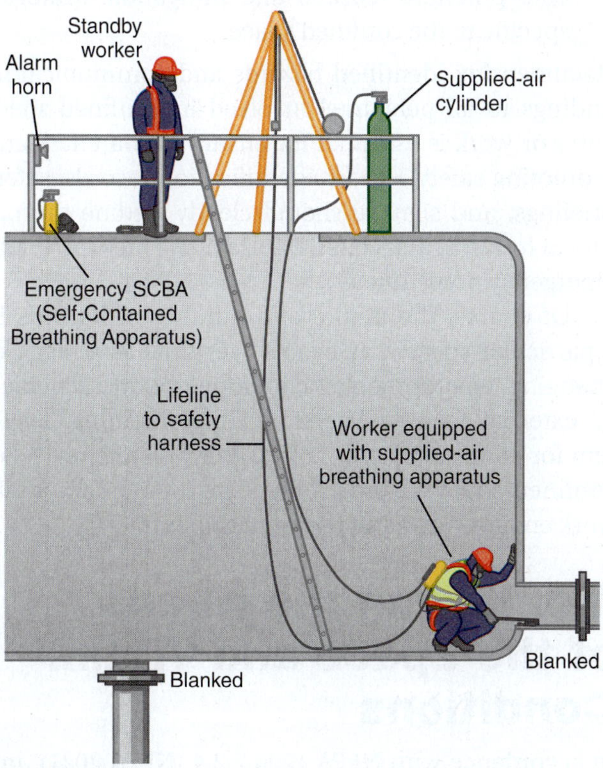

FIGURE 6-5 A properly set up rescue area.

Operations Level Size-Up Information Gathering and Interview Techniques

Size-up information and effective interview techniques are critical for assessing the situation and developing an informed rescue plan. Operations level rescuers must adhere to NFPA 1006 standards (NFPA 2021), and they must possess the knowledge and skills to gather pertinent information and conduct interviews to ensure the safety and success of rescue operations.

As discussed previously, size-up is the systematic assessment of a confined-space environment. Its purpose is to gather critical details that will inform rescue strategies. Operations level responders must be proficient in collecting the following size-up information:

- **Space configuration:** Assess the dimensions, layout, and structural features of the confined space, including entry points, internal obstacles, and potential hazards.
- **Atmospheric conditions:** Use monitoring equipment to evaluate air quality, identifying oxygen levels and the presence of toxic gases or combustible vapors that may pose risks to rescuers and victims.
- **Hazards:** Recognize common confined-space hazards, such as engulfment (**FIGURE 6-6**), entrapment, falls, electrical hazards, and potential atmospheric contaminants.

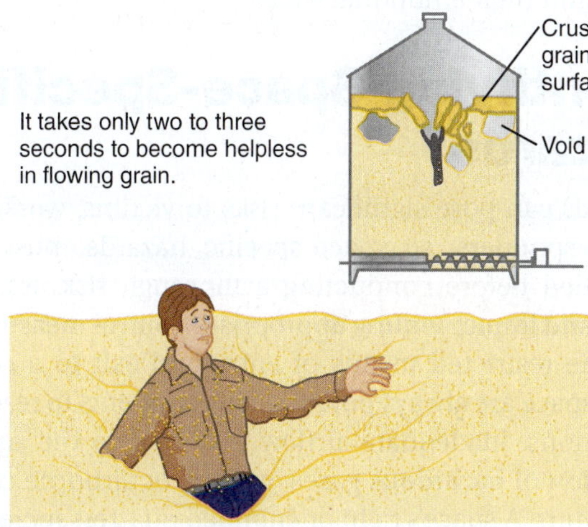

FIGURE 6-6 Engulfment inside a grain bin.

Courtesy of OSHA/U.S. Department of Labor.

- **Victim assessment:** Gather information about the number and condition of victims inside the confined space, including their last known location and activity, as well as any known medical conditions.
- **Entry and egress points:** Determine safe entry and exit points, ensuring clear pathways for rescue operations. In many cases, the entry and egress points are in the same location; often, confined spaces have only one means of access.

Effective interview techniques are necessary to gather essential information from witnesses, bystanders, and individuals familiar with an incident. Operations level responders must be skilled in conducting interviews to obtain accurate and actionable details. The following protocols are recommended for operations level responders:

- **Establish communication:** Approach witnesses and individuals calmly and respectfully, establishing rapport to encourage open communication.
- **Ask open-ended questions:** Pose questions that elicit detailed responses, allowing witnesses to provide comprehensive accounts of the incident.
- **Actively listen:** Listen attentively to witnesses, focusing on key details and clarifying information as needed.
- **Document:** Record interview findings systematically, documenting witness accounts, last known locations of victims, and other pertinent details.
- **Consider psychological factors:** Be mindful of emotional reactions from witnesses and victims, maintaining empathy and professionalism throughout the interview process. Assess the validity of witnesses' statements while taking into account their emotional states.

Identifying Space-Specific Hazards

Hazards can pose significant risks to victims, workers, and responders, so space-specific hazards must be identified before conducting a thorough risk assessment and implementing appropriate control measures.

The restricted means of entry and exit to a confined space are always potential impediments to rescue operations. Inadequate airflow may lead to the accumulation of hazardous gases, vapors, or airborne contaminants. A space's tight or cramped quarters increase the likelihood of physical dangers such as trip hazards, confined movement, and entanglement risks. These spaces are often isolated from external assistance or communication, which exacerbates the challenges of an emergency response.

Space-specific hazards may include atmospheric issues (oxygen enrichment or deficiency, toxic vapors, airborne contaminants); physical objects (uneven surfaces, sharp edges, falling objects); or chemical (corrosive substances, irritants, carcinogens), biological (mold, bacteria, viruses), or radiological (ionizing radiation) contaminants. The following methodologies can be used to identify space-specific hazards:

- **Visual inspection:** A systematic visual examination will reveal visible hazards, such as protruding objects, machinery, electrical components, or structural deficiencies.
- **Atmospheric monitoring:** Use gas detectors and atmospheric monitors (see Figure 2-7b in Chapter 2) to continuously measure the levels of hazardous gases and oxygen. Ensure the proper calibration of these devices before use.
- **Review of safety data sheets (SDSs):** SDSs document substances stored or used in the confined space and detail their hazardous properties and associated risks.
- **Consultation with subject-matter experts:** Experts such as industrial hygienists, safety engineers, or environmental consultants can be engaged to evaluate potential hazards and mitigation strategies specific to the confined space.

Documenting identified hazards and communicating findings to all personnel involved in confined-space entry or work is essential for ensuring awareness and promoting safety. Hazard identification records, safety briefings, and signage should clearly outline the nature of hazards, associated risks, control measures, and emergency procedures.

Of course, the nature and number of hazards in a particular space may evolve over time as a result of changing environmental conditions, work activities, or external factors. Therefore, implementing a system for continuous monitoring and periodic review of confined-space hazards is essential to maintain a safe work environment and prevent accidents.

Monitoring the Needs of the Space and Victim Conditions

In accordance with NFPA 1006 7.2.4 (NFPA 2021), immediate and ongoing monitoring within confined spaces is required to mitigate hazards and facilitate

Victim Information	
	Details
Name:	
Age:	
Physical Description:	
Medical Conditions/History:	
Allergies:	
Special Considerations:	
Confined-Space Information	
	Details
Location:	
Entry Points:	
Dimensions:	
Hazardous Atmospheres:	
Timeline of Events	
Time of Incident:	
Tasks Being Performed:	
Witness Statements:	
Initial Signs/ Symptoms:	
Changes in Conditions:	
Additional Notes:	

FIGURE 6-7 A sample victim information form.

prompt response efforts. Both immediate and ongoing monitoring needs consist of continuous atmospheric monitoring, repeated periodic visual inspections of the space, and regular victim assessments. Victim assessments should include information regarding each victim's level of consciousness, vital signs, and potential injuries.

FIGURE 6-7 is an example of a victim information form for confined-space rescue. This tool can capture essential details required for developing comprehensive victim profiles during confined-space rescue operations, and it can be adapted and customized based on specific organizational needs and incident response protocols. The form should be continuously updated with any new information or developments during the rescue operation.

Victim and Rescuer Communications

NFPA and OSHA standards and regulations provide guidance on effective communication between victims and rescuers during a confined-space emergency. Clear communication channels between victims and rescuers during confined-space emergencies ensure coordination, safety, and efficient rescue operations. Communication methods commonly used in confined-space environments include verbal communication, hand signals, electronic communication devices, rope signals, and the confined-space hard-wired system (see Chapter 11).

Communication requirements must be determined based on the size and complexity of the confined space, potential hazards, and operational conditions. Depending on the specific confined-space environment, appropriate methods of communication may be two-way radios, intercom systems, signal devices, or **safety/communication ropes**, which are high-performance life safety ropes that contain built-in communication wires (**FIGURE 6-8**).

Effective verbal communication requires the use of clear, simple, and concise language during communication to avoid misunderstandings and ensure accurate information exchange. Responders must also rely on active listening skills to attentively receive and understand messages from victims and fellow rescuers and solicit feedback and clarification when necessary. Backup emergency communication measures include the OATH (O for okay = one tug; A for advance = two tugs; T for take up slack = three tugs; H for help = four tugs) and SUDRH (S for stop = one short blast; U for up = two short blasts; D for down = three short blasts; R for rope free = four short blasts; and H for help = one long blast) methods of nonverbal communication, which are covered in Chapter 2.

FIGURE 6-8 The Rescom® Safety/Communication Rope™, with built-in communication wires, is used in confined-space operations for clear communication between team members.

Used with permission from Rescom Sales Inc.

Verbal communication is not always possible, and other means of communication can fail. For this reason, contingency communication plans are needed to continue the flow of information despite equipment failure, loss of signal, or other communication disruptions. Changing conditions, such as noise, darkness, or distance, often require the communication strategies and techniques deployed during a rescue to adapt. Any communication issues, equipment malfunctions, or concerns must be promptly reported to the IC.

All communication exchanges, including instructions, acknowledgments, and incident reports, must be documented for the purposes of regulatory compliance, incident investigation, and future reference.

Victim Removal and Rescue Systems

Building Victim Removal/Retrieval Systems

NFPA 1006 (NFPA 2021) outlines standards for technical rescuer professional qualifications and provides guidance on rescue operations, including confined-space rescue. It emphasizes the need for safe and effective victim removal/retrieval techniques, equipment, and procedures to minimize risks to responders and victims. OSHA regulations—in particular, 29 *Code of Federal Regulations* (C.F.R.) 1910.146 ("Permit-Required Confined Spaces" 1993)—set forth requirements for confined-space entry and rescue procedures to protect workers from hazards associated with confined spaces. OSHA requires that a retrieval system must be used if the incline or decline of the space is five or more feet (1.5 or more meters).

> **TIP**
>
> Rescuers must take at least one rope systems class that can teach them the techniques described in this text and that provides them with practical experience in building rope rescue systems. It is important to acquire this basic knowledge prior to engaging in confined-space rescue training.

Victim removal/retrieval systems, which are customized to each confined-space response, must be designed and built in line with these standards. A compliant system is built from components that meet relevant safety standards and are compatible with each other. Components of these systems include anchor points, harnesses, ropes, **pulleys** (devices with free-turning, grooved metal wheels, known as **sheaves**, that are used to reduce rope friction; **FIGURE 6-9**), mechanical advantage systems, and rescue stretchers. To better understand how to perform a rapid victim extraction, follow the steps in **SKILL DRILL 6-1**.

One of the most critical considerations when designing a victim removal/retrieval system is the maximum anticipated load that the system must support, including the weight of victims, rescuers, and equipment. Each component within the system must have sufficient load capacity to safely accommodate all anticipated loads—without compromising the system's integrity.

FIGURE 6-9 A rescue pulley designed for technical rescue operations, providing smooth and efficient movement during confined-space extractions.

SKILL DRILL 6-1

Performing a Rapid Victim Extraction Using a 4:1 Raise and Lowering System with Tripod

Step 1: Set Up the Tripod over the Confined Space

Set up the tripod over the entry point of the confined space. Ensure the tripod is on stable ground and properly secured, with no visible defects.

Step 2: Attach Pulleys to the Tripod Apex

Attach pulleys to the apex of the tripod. The pulleys are essential for creating a mechanical advantage. Ensure proper alignment and that all hardware is rated for load.

Continues.

SKILL DRILL 6-1 CONTINUED

Performing a Rapid Victim Extraction Using a 4:1 Raise and Lowering System with Tripod

Step 3: Connect the Rescue Line to the Lowering/Raising System

Thread the rope through the pulleys and connect it to the mechanical device for lowering or raising. The system creates a 4:1 mechanical advantage, making lifting easier. Check rope integrity and ensure it is suitable for the load.

Step 4: Attach the Harness to the Rescue Line

Connect the confined-space entrant's full-body harness to the rescue line using a locking **carabiner**, a load-bearing metal connector with a self-closing gate used to link the elements of the system. Ensure all connections are secure and that the carabiner is rated for life safety.

Step 5: Lower the Entrant into the Confined Space

Begin lowering the entrant into the confined space using the 4:1 system. One team member should control the descent while others monitor for any signs of difficulty or hazards.

SKILL DRILL 6-1 CONTINUED

Performing a Rapid Victim Extraction Using a 4:1 Raise and Lowering System with Tripod

Step 6: Monitor the System During Descent

The team should monitor the system, checking for smooth operation and ensuring the rope does not tangle or kink. Communication between the entrant and the safety team is continuous.

Step 7: Raise the Entrant Out of the Confined Space

Raise the entrant back out of the confined space using the 4:1 mechanical advantage system. This effort is distributed among team members, allowing for a controlled ascent.

Step 8: Secure the Entrant After Extraction

After the entrant is safely extracted and removed from the system, perform a safety check on the individual and begin stowing the equipment. Proper debriefing and equipment inspection should follow.

Redundancy can minimize the risk of single points of failure within a system. Backup attachment points (**FIGURE 6-10**), redundant attachment systems, and secondary safety devices provide extra layers of safety in case of equipment malfunction.

A system is only as good as the rigging techniques deployed to build it. Proper rigging techniques must be used to secure anchor points, set up mechanical advantage systems, and connect components within the system. Rescuers should follow manufacturer recommendations and industry best practices for rigging configurations and load distribution.

The anchor points used to base the system must be structurally sound and capable of supporting the anticipated loads without compromising stability or integrity. Avoid using makeshift or untested anchor points; instead, use anchor systems that are unquestionably sound or bomb-proof for confined-space rescue operations. Every component of a victim removal/retrieval system must be inspected and tested before each use to ensure it is in good working condition and free from defects or damage. Be sure to check for signs of wear, corrosion, or deterioration, and replace any compromised components immediately. Rigorous testing of the system should be routinely performed using simulated rescue scenarios to validate its performance and identify potential issues. See **FIGURE 6-11** for an inspection checklist.

FIGURE 6-10 The dorsal attachment point on a full-body harness.

Mechanical Advantage in Confined-Space Rescue Operations

Mechanical advantage is the amplification of force achieved by using a mechanical system to manipulate loads. In confined-space rescue operations, where rescuers often face challenges such as limited space, heavy loads, and difficult access, mechanical advantage systems play a crucial role in overcoming these obstacles and safely extracting victims. Mechanical advantage allows rescuers to reduce the amount of physical effort required to move heavy loads, improve control over lifting and hoisting operations, and enhance overall safety during rescue operations.

Confined-space environments may involve moving heavy equipment, machinery, or debris, making it difficult for rescuers to lift or maneuver manually. Mechanical advantage systems enable rescuers to exert less force while achieving greater lifting or pulling power, making it easier to manage heavy loads effectively. Confined spaces also often have restricted access points, narrow passages, and obstacles that hinder manual movement and maneuverability, so mechanical advantage systems can be used to navigate these tight spaces more efficiently by providing alternative methods for lifting, pulling, and hoisting loads.

☐	**Ropes**	Check for frays, cuts, and wear
☐	**Harnesses**	Inspect buckles, stitching, and overall condition
☐	**Pulleys**	Ensure smooth operation and no signs of damage
☐	**Anchors**	Verify structural integrity and secure attachment
☐	**Carabiners**	Test locking mechanisms and check for deformation

FIGURE 6-11 A sample victim removal/retrieval system inspection checklist.

Types of Mechanical Advantage Systems

A **1:1 mechanical advantage system**, also known as a *direct pull system*, provides a mechanical advantage of 1:1, meaning that the force applied to the system equals the force exerted on the load. This system is simple and straightforward, involving a single rope or cable directly attached to the load, with no additional pulleys or mechanical devices to amplify force. Although a 1:1 system does not provide significant mechanical advantage, it is suitable for light loads or situations where minimal force is required.

A **2:1 mechanical advantage system** provides a mechanical advantage of 2:1, meaning that for every unit of force applied to the system, two units of force are exerted on the load. This system typically involves a pulley arrangement where the load is attached to one end of the rope, and the rope runs through a pulley that redirects the force. The other end of the rope is then pulled, requiring half the effort needed to lift the load directly. In this setup, the distance the load moves is half the distance the rope is pulled, effectively doubling the force applied to the load.

A **3:1 mechanical advantage system** provides a mechanical advantage of 3:1, meaning that for every unit of force applied to the system, three units of force are exerted on the load. This system typically involves multiple pulleys arranged in a block-and-tackle configuration, with the rope threaded through the pulleys to create a mechanical advantage. By using pulleys to redirect the direction of force, a 3:1 system allows rescuers to lift or pull heavier loads with less effort compared with a direct pull system.

A **4:1 mechanical advantage system** provides a mechanical advantage of 4:1, meaning that for every unit of force applied to the system, four units of force are exerted on the load. This system typically involves a combination of pulleys in a block-and-tackle arrangement. In this setup, multiple pulleys are used, with the load attached to the block of pulleys. The rope is threaded through these pulleys, creating a situation where the force applied to the rope is multiplied. The user pulls on the rope, and because of the multiple pulleys, the load is lifted with only one-fourth of the effort that would be needed to lift it directly. The trade-off is that the distance the rope must be pulled is four times the distance the load is lifted.

A **5:1 mechanical advantage system** provides a mechanical advantage of 5:1, offering even greater amplification of force compared to a 3:1 system. This system utilizes additional pulleys and ropes arranged in a compound pulley system, with multiple stages of mechanical advantage to achieve a 5:1 ratio. Although more complex than a 3:1 system, a 5:1 system allows rescuers to lift extremely heavy loads with minimal effort, making it well suited for demanding rescue scenarios.

Types of Rescue and Retrieval Systems

Effective rescue and retrieval systems ensure the safety of both rescuers and victims. Different types of rescue and retrieval systems are used in horizontal and vertical confined-space rescue operations.

Vertical Rescue Systems

Vertical rescue systems are divided into two categories: rope rescue systems and mechanical advantage systems. The following are types of rope rescue systems:

- Main line: The primary line.
- **Tag line**: A non-load-bearing line that connects to a litter to help it move from side to side in order to avoid obstacles, prevent swing-fall, or attain a more advantageous position.
- **Belay line** or safety line: A secondary (backup) line to protect against a fall or system failure.
- **Haul line**: A type of tag line that a haul team uses for pulling on a load; in most cases, the main line becomes the haul line.
- Twin-line system: A rope rescue system involving the use of two separate lines—a primary line and a secondary line—to ensure worker safety or facilitate rescue.

One of each type of rope rescue system is required for each rescuer. There must also be an additional belay line (safety line) set on hand for each backup responder (on call in case of an emergency during the rescue).

The equipment required for rope rescue systems includes **kernmantle rescue ropes** (a kernmantle is a rope design consisting of an interior core [the kern], which usually supports the major portion of the load on the rope, and an outer sheath [the mantle] that serves to protect the core but also may support a minor portion of the load; **FIGURE 6-12**); static or low-stretch rescue ropes; harnesses and hardware (e.g., carabiners) (**FIGURE 6-13**); mechanical **ascenders** (rope grab devices used to ascend a fixed rope) (**FIGURE 6-14**) and **descent control devices (DCDs)** (rappel devices that create friction by means of a rope running through them) (**FIGURE 6-15**); and anchor devices (e.g., bolts, slings, tripods).

FIGURE 6-12 A kernmantle rope.

Courtesy of Steve Hudson.

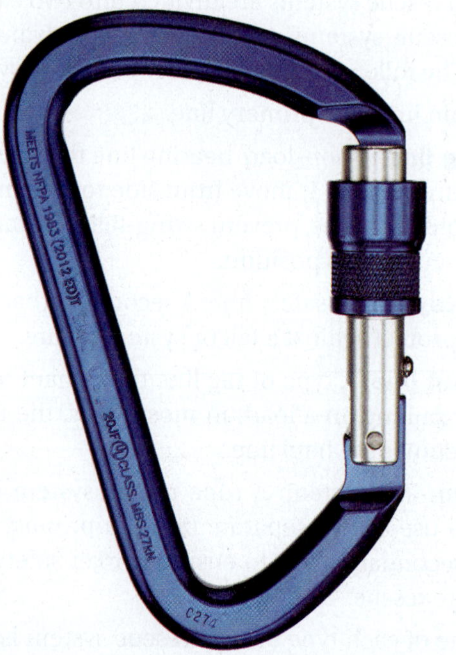

FIGURE 6-13 An NFPA-rated carabiner for rescue operations.

Mechanical advantage systems have the following types:

- 3:1 haul system
- 4:1 or 5:1 haul system (team/situation dependent)
- Piggyback or tandem systems
- Stretcher haul systems

A **haul system** is an arrangement of ropes, pulleys, and connectors that is arranged in a manner to provide mechanical advantage to raise a load under controlled circumstances. A **hitch** is a method of tying a rope such that when the object or rope is removed, the knot falls apart. The equipment required for mechanical advantage systems includes pulleys and rescue haul systems; mechanical ascenders or **Prusik loops**, a configuration of rope tied into a loop for use in a **Prusik hitch**, a type of friction hitch used in ascending and belaying (**FIGURE 6-16**); anchor devices (e.g., bolts, rigging plates); and load-releasing straps or systems.

FIGURE 6-14 A rescue climbing ascender, used for secure and controlled upward movement during technical rescue and confined-space operations.

© shoricelu/iStock/Getty Images Plus/Getty Images

TIP

Sked® is a brand of stretcher that is commonly used for technical rescue operations. It is a patient immobilization device made of hard plastic that is designed to envelope and package a victim. It can be used for both vertical lifting and horizontal movement within confined spaces. However, Sked® stretchers do not provide spinal immobilization; if a victim has neck or back injuries, they can be used in combination with a long **spine board** (or backboard), which is a rigid, flat surface made of plastic or wood to which the victim is secured using straps or a harness. Newer **Kendrick extrication devices (KEDs)**—multipart immobilization devices that provide full-body support and stabilization—have a vertical rating, and are small devices that offer spinal immobilization and vertical applications. Other devices, such as **Yates Spec Paks** (victim immobilization and evacuation systems that also function as a lifting harness), are better suited for use in vertical victim removal scenarios.

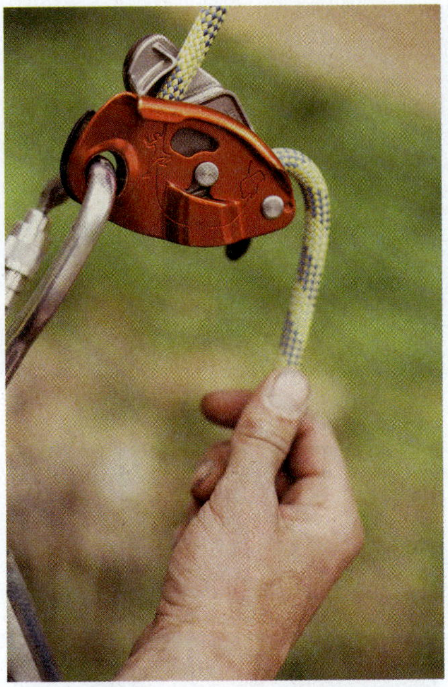

FIGURE 6-15 A DCD used for safe and controlled lowering in rescue and rope access operations.

FIGURE 6-16 A Prusik hitch, formed with a Prusik loop tied around a climbing rope, provides a secure grip.

TIP

A litter is a stretcher or basket designed for use in rescues where physical obstacles impair movement. Examples include confined spaces, slopes, uneven terrain, or densely forested areas. Litters are designed for both vertical and horizontal lifting operations. They must be used with a rated **lifting bridle**, which is a safety device used in rescue operations and lifting scenarios to securely attach a load or patient to a lifting mechanism.

A **Reeves™ Sleeve** is a medical device used in emergency situations, particularly in rescue and transport scenarios. It is a type of extrication device designed to immobilize and protect a patient's spine during transport, allowing first responders to safely move individuals who may have sustained spinal injuries. The sleeve is typically padded and can be adjusted to fit the patient securely while providing support.

The Yates Spec Pak Rescue Kit can be used in confined-space operations in a vertical environment. In addition, the Spec Pak system functions as a lifting harness, eliminating the need for a separate victim harness. Its self-adjusting (sliding) cervical restraint addresses the challenges associated with lifting patients vertically while they are secured to a rigid backboard.

The kit includes a storage bag equipped with a shoulder strap for easy carrying and a top connection point for lowering when not in use with a patient. It is important to note that the Spec Pak is intended as a limited spinal immobilization system and must be used in conjunction with proper spinal immobilization methods, such as a **cervical collar** (or C-collar)—a rigid or semirigid device placed around the victim's neck to restrict movement and maintain alignment of the cervical spine—where conditions and protocols dictate.

The Yates Spec Pak comes with two sets of forehead and chin restraint straps: One set is two inches wide with Velcro, and the other is made of vinyl-coated nylon and foam. The materials used in the Spec Pak allow for easy decontamination, making it suitable for use in diverse environments.

Horizontal Rescue Systems

There are three primary types of horizontal rescue systems: Sked® or stretcher haul systems, low-profile packaging systems, and **basket stretchers** (sturdy, elongated, and open stretchers covered by fabric mesh or canvas) or **litter baskets** (basket stretchers designed for horizontal retrieval that provide support and stability for victims; **FIGURE 6-17**). All three types share the same basic equipment needs: anchors and rigging hardware, rope or webbing for rigging, and mechanical advantage systems (for hauling). The only difference is the victim retrieval device used for each: a Sked® or stretcher in the first, a **vacuum mattress** (a mattress-like structure made of a durable, flexible material, typically reinforced with plastic or metal inserts to maintain rigidity, that

FIGURE 6-17 A litter basket being lifted and suspended for entry or retrieval from a confined space.

Courtesy of Illinois Fire Service Institute.

conforms to the shape of the victim's body when air is extracted) or a backboard for patient immobilization in the second, and a basket stretcher or litter basket in the third. See later in this chapter for details on these types of immobilization devices.

Building a 4:1 Haul Line with a Belay System

A 4:1 haul line with a belay system is often used in confined-space rescue operations to safely extract victims.

You will need the following equipment to build a 4:1 haul line belay system:

- **Static rope:** A high-strength static rope appropriate for the expected load and conditions.
- **Pulleys:** Pulleys must be rated for rescue operations with high efficiency.
- **Locking carabiners:** Only locking carabiners suitable for rescue applications are appropriate for this purpose.
- **Anchor point:** The anchor point must be secure and capable of withstanding the expected load.

The first step in building a 4:1 haul line with a belay system is to locate two anchor points—one for the haul line and one for the belay line (**FIGURE 6-18A**). Next, rig the haul line by tying a secure anchor knot (e.g., a figure-eight knot) to the haul line's anchor point (**FIGURE 6-18B**). Then, attach a pulley to the anchor knot using a locking carabiner and extend the haul line to the victim's location, making sure that the line remains free from knots or obstructions (**FIGURE 6-18C**).

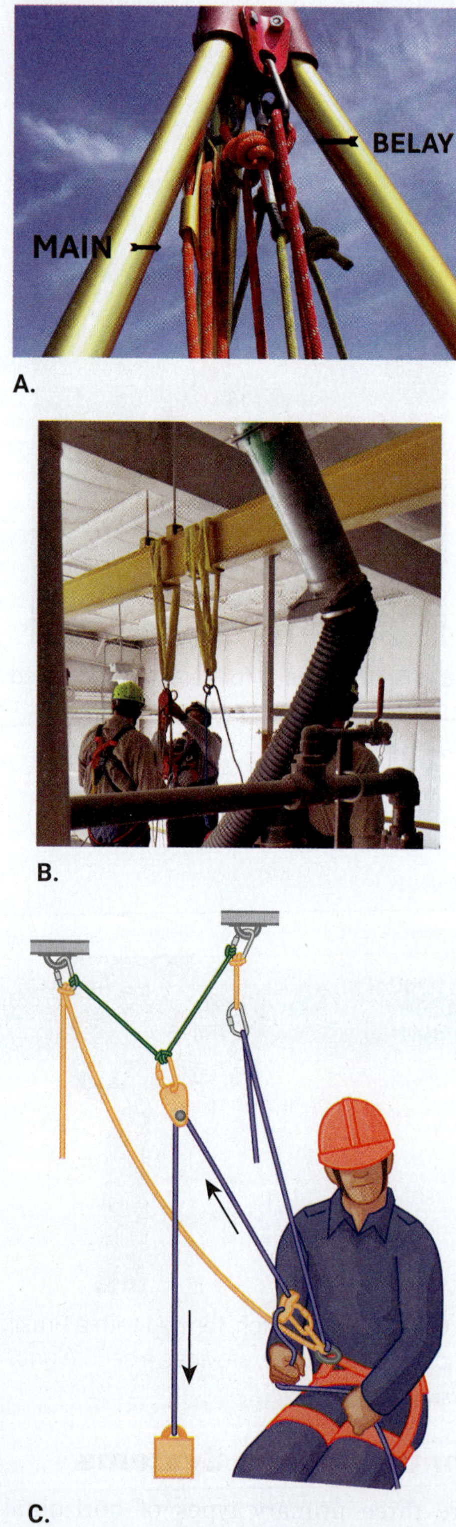

FIGURE 6-18 (**A**) A rope rescue system attached to multiple anchor points. The haul line and the belay line are attached to two different anchor points. (**B**) A rescuer rigging a haul line by tying a secure anchor knot to the haul line's anchor point. (**C**) A rescuer using a deactivated cam with line through a pulley, making sure that the line remains free from knots or obstructions.

A and **B:** Courtesy of Illinois Fire Service Institute.

To rig the belay line, repeat the same process: Tie a secure anchor knot to the belay line's anchor point and attach a pulley to the anchor knot using a locking carabiner. Extend the belay line parallel to the haul line, ensuring that it remains separate from and does not interfere with the haul line.

Rig a **belay device** into the belay line. A belay device is a braking mechanism through which the secondary line is rigged. The device must allow the free run of the belay rope through it when the system is operating correctly; it must exert a slowing or stopping action on the belay line if an uncontrolled descent or fall occurs on the mainline.

Next, attach a pulley to the victim's harness or litter using a suitable attachment method (e.g., carabiner). Ensure the pulley is properly aligned and oriented to reduce friction and maximize efficiency.

At this point, you are ready to rig the 4:1 mechanical advantage. Pass the haul line through the pulley attached to the anchor point. Route the haul line through the pulley attached to the victim's harness or litter. Pass the haul line through a second pulley attached to the anchor point and bring the haul line back to the rescuer's position, ensuring it remains parallel to the initial segment.

Securely attach locking carabiners to the haul line between the pulleys, creating a progress capture system. Use a Prusik or similar friction hitch attached to the haul line as a **progress capture device** (or PCD), which holds the rope in place to prevent the load from sliding back down when the pulling force is released (**FIGURE 6-19**).

FIGURE 6-19 A Prusik hitch attached to a haul line as a progress capture device (PCD).

Once the system has been built, it must be inspected and tested before it is used in a rescue. Conduct a thorough safety check of the entire system, ensuring all components are properly rigged and secure. Perform a function test by applying tension to the haul line and verifying that the system operates smoothly and efficiently.

Ensuring Rescue Systems Are Operational

Verifying that rescue systems are operational and ready for use in confined-space rescue operations requires a thorough understanding of the components of each type of rescue system and their functionality. These components include ropes, harnesses, pulleys, carabiners, anchors, and auxiliary equipment.

A comprehensive inspection of a rescue system requires three checkpoints—visual inspection, functional testing, and equipment checks—to ensure that they are in good working condition and compliant with relevant safety standards and regulations. The assembly, installation, and rigging of rescue systems, including correct attachment points, functional testing, load distribution, and redundancy measures, must be thoroughly checked to ensure no weak points or vulnerabilities are present.

The best way to ensure that rescue systems are fully operational is to conduct hands-on practice activities. In addition to being good practice for the rescue teams, these activities can identify potential issues or deficiencies in rescue systems, such as damaged ropes, malfunctioning pulleys, or improperly rigged anchors. See **TABLE 6-2** for best practices for operational readiness.

TABLE 6-2 Best Practices for Operational Readiness

Regular drills	Conduct quarterly drills simulating different rescue scenarios.
Documentation	Maintain detailed logs of equipment inspections and system tests.
Feedback loop	After each drill or real operation, conduct a debrief to identify lessons learned and areas for improvement.

Identifying a Means of Retrieval

Victim Retrieval Methods

Six primary methods are relied on to retrieve victims during confined-space emergencies: direct lifting, mechanical advantage systems, **self-retracting lifelines (SRLs**: fall arrest devices that contain a drum-wound line that pays out from and automatically retracts back onto an enclosed drum during use and that automatically locks at the onset of a fall) (**FIGURE 6-20**), confined-space entry and exit systems, basket stretchers or rescue slings, and rope access and rope rescue techniques.

Direct lifting involves manually lifting or pulling the victim out of the confined space using physical strength. Rescuers may use harnesses, straps, or clothing to secure the victim and hoist them to safety. Direct lifting is suitable for situations where the victim is conscious, able to cooperate, and not seriously injured. It requires sufficient physical strength and may be limited by the weight and condition of the victim.

Mechanical advantage systems, such as pulley systems or winches, provide leverage and mechanical advantage to lift or pull the victim out of the confined space. Rescuers rig the system to create a pulley configuration that amplifies force, allowing for easier extraction of the victim. These systems are effective for lifting heavy loads or overcoming obstacles within the confined space. They require proper setup, rigging, and operation by trained personnel to ensure safety and effectiveness.

SRLs with rescue capabilities feature built-in winches or retrieval systems that allow for self-rescue or assisted rescue of a fallen worker. The victim or rescuer activates the retrieval mechanism to raise or lower the victim to safety. SRLs are particularly useful for vertical or elevated confined spaces where traditional retrieval methods may be impractical. They provide controlled descent or ascent, minimizing risks associated with free fall or uncontrolled movement.

Ideally, a confined space will contain entry and exit systems designed to facilitate safe ingress and egress. These devices may include portable staircases, ladders, or personnel hoists that provide access to and from the confined space. Although these devices are essential for facilitating rapid evacuation or extraction of victims from confined spaces, unfortunately, not all confined spaces have them.

Basket stretchers or rescue slings are used to immobilize and safely extract victims from confined spaces. In these systems, rescuers secure a victim onto a stretcher or sling and then carefully maneuver them out of the confined space. Basket stretchers or rescue slings are suitable for victims who are injured, unconscious, or unable to move independently. They provide support and stability during extraction and minimize the risk of further injury.

Rope access and rope rescue techniques involve using ropes and specialized equipment to access and extract victims from confined spaces. Rescuers may employ techniques such as rope rigging, rappelling, or high-angle rescue to reach and retrieve the victim. Rope access and rescue techniques require specialized training, equipment, and expertise to ensure safe and effective execution. They are particularly useful in confined spaces with vertical or difficult-to-access entry points (**FIGURE 6-21**).

FIGURE 6-20 An SRL is designed for fall protection, automatically locking to stop a fall and providing safe working conditions at elevated heights.

Courtesy of Guardian Fall.

FIGURE 6-21 A grain elevator, used for storing and transferring large quantities of grain, is a confined space with vertical entry points that are difficult to access.

© Oliver Strewe/The Image Bank/Getty Images

TABLE 6-3 Methods of Rescue

Direct lifting	▪ **Description:** Manually lifting or pulling the victim out using physical strength ▪ **Application:** Suitable for conscious, cooperative, and not seriously injured victims ▪ **Limitations:** Requires sufficient physical strength and may be limited by the weight and condition of the victim ▪ **Equipment:** Harnesses, straps, or clothing to secure the victim
Mechanical advantage systems	▪ **Description:** Using pulley systems or winches to provide leverage and mechanical advantage for lifting or pulling the victim ▪ **Application:** Effective for lifting heavy loads or overcoming obstacles within the confined space ▪ **Requirements:** Proper setup, rigging, and operation by trained personnel ▪ **Equipment:** Pulleys, winches, ropes
Self-retracting lifelines (SRLs)	▪ **Description:** Fall arrest devices with built-in winches or retrieval systems that allow for self-rescue or assisted rescue ▪ **Application:** Particularly useful for vertical or elevated confined spaces ▪ **Benefits:** Provides controlled descent or ascent, minimizing risks associated with free fall or uncontrolled movement ▪ **Equipment:** SRLs with rescue capabilities
Confined-space entry and exit systems	▪ **Description:** Systems designed to facilitate safe ingress and egress ▪ **Application:** Essential for rapid evacuation or extraction of victims ▪ **Limitations:** Not all confined spaces have these systems ▪ **Equipment:** Portable staircases, ladders, personnel hoists
Basket stretchers or rescue slings	▪ **Description:** Used to immobilize and safely extract victims ▪ **Application:** Suitable for injured, unconscious, or victims unable to move independently ▪ **Benefits:** Provides support and stability during extraction ▪ **Equipment:** Basket stretchers, rescue slings, immobilization devices
Rope access and rope rescue techniques	▪ **Description:** Using ropes and specialized equipment to access and extract victims ▪ **Application:** Useful in confined spaces with vertical or difficult-to-access entry points ▪ **Requirements:** Specialized training, equipment, and expertise ▪ **Equipment:** Ropes, harnesses, pulleys, mechanical advantage systems

Rescuing victims from confined spaces involves using one of six primary methods. Each method has specific applications based on the condition of the victim and the characteristics of the confined space. **TABLE 6-3** breaks down these methods, their descriptions, their applications, their limitations, and the equipment used in the method.

Horizontal Retrieval Methods for Confined-Space Rescue

Horizontal retrieval methods are used to extract victims along a horizontal plane and are capable of safely and efficiently moving individuals through narrow or confined spaces to safety. Horizontal retrieval lines are problematic in that the line of sight is usually abandoned; people can be pulled painfully across corners; ropes can become damaged to the point of destruction; and on occasion, the lift is at the end of the horizontal pulling operation, with the newly damaged rope becoming the primary line suspending the survivor load.

Other devices used for extraction include the **LSP Half Back** (a vertical extrication device combining the functionality of a full-body harness with the ability to immobilize a patient in cases of suspected cervical or spinal trauma), the MAST Sling-Link, and the LSP Cinch Collar, as well as diaper sling harnesses.

The standard equipment for a horizontal retrieval method includes the following:

- **Rescue harness:** A full-body rescue harness equipped with attachment points for securing the victim and connecting to rescue systems.
- **Rope or webbing:** Strong and durable rope or webbing to create anchor points, haul systems, and rigging for the retrieval operation.
- **Pulleys:** High-strength pulleys for creating mechanical advantage systems and redirecting ropes or webbing as needed.
- **Carabiners and connectors:** Locking carabiners and connectors for attaching equipment, creating anchor points, and securing the victim.
- **Rope grabs or ascenders:** Mechanical devices used for climbing or descending ropes safely and efficiently.
- **Descent control devices:** Commonly used devices that are designed to function as pulley and safety/locking mechanisms.
- **Rescue stretcher or basket:** A stretcher or basket designed for horizontal retrieval that provides support and stability for the victim during transportation.
- **Helmet and PPE:** Protective helmet, gloves, and other PPE used to ensure the safety of rescuers and victims during the operation.

Before initiating a horizontal retrieval method, you must first assess the condition of the victim and provide necessary medical treatment or stabilization before initiating the retrieval process.

To start, establish secure anchor points both inside and outside the confined space. These anchor points must be able to support the anticipated loads. Next, proceed to rigging all hauling systems, pulleys, and mechanical advantage devices together to facilitate horizontal movement and provide lifting power.

Once the method has been fully rigged, securely attach the victim to a rescue stretcher or basket using appropriate straps, harnesses, or immobilization devices. Ensure that the victim is positioned comfortably and securely within the stretcher to minimize movement during transportation.

When the victim is in place, use the hauling system, mechanical advantage device, or manual pulling to move the victim along the horizontal plane toward the exit point. Carefully coordinate each movement to ensure smooth and controlled progress, avoiding sudden jerks or movements that could endanger the victim or rescuers. As the system encounters obstacles or tight spaces, safely maneuver the victim around these obstructions using pulleys, redirects, or changes in rigging. Coordinate with team members to adjust movements and rigging as needed to overcome any challenges you encounter.

Once the designated exit point or egress location has been reached safely, with the victim secure and stable throughout the process, use additional personnel or equipment as needed to assist with lifting, lowering, or maneuvering the victim to safety. Now that the victim is free from the space, a postextraction assessment of the victim's condition can prompt any necessary medical treatment or further assistance.

Following the extraction, a careful audit of the effectiveness of the retrieval operation will help the team identify lessons learned and provide insights for future rescue efforts.

Keeping the Space Clear

To make sure the area for a rescuer retrieval system is clear and unobstructed, careful preplanning, equipment preparation, establishment of a safe work zone, and monitoring must take place.

Conduct a thorough assessment of the confined space to identify potential hazards and obstacles and a rescue plan that includes the roles and responsibilities of the rescue team, the equipment needed, and the procedures to be followed. The assessment should reveal which retrieval system (e.g., tripods, winches, harnesses) is best suited for the rescue.

Once the plan is in place, remove any unnecessary objects or debris from around the entrance to the confined space. Note any sharp edges or abrasive surfaces that could harm ropes or air lines. Set up barriers and signage to prevent unauthorized entry and delineate the rescue area. Blowers or fans ventilate the confined space to remove hazardous gases and maintain a supply of fresh air, and they also keep the space clear for rescuers.

Each rescuer entering the confined space must be properly harnessed and connected to the retrieval system. The retrieval system must be positioned directly over the entry point to ensure a straight and unobstructed path for retrieval. Before loading, conduct a test lift to ensure the retrieval system is functioning correctly and can operate smoothly and without obstruction. If rescuers encounter an obstacle or if the atmosphere becomes hazardous, the retrieval system can be used to remove them from the confined space. To better understand how to perform an evacuation, follow the steps in **SKILL DRILL 6-2**.

SKILL DRILL 6-2

Evacuation Methods for Confined-Space Operations

Step 1: Identify Confined-Space Types

Review different types of confined spaces that may require evacuation, such as tanks, sewers, and silos (seen here), emphasizing the specific challenges each type presents in evacuation scenarios. Discuss common evacuation complications and the necessary precautions for each confined space.

Step 2: Implement Accountability Systems

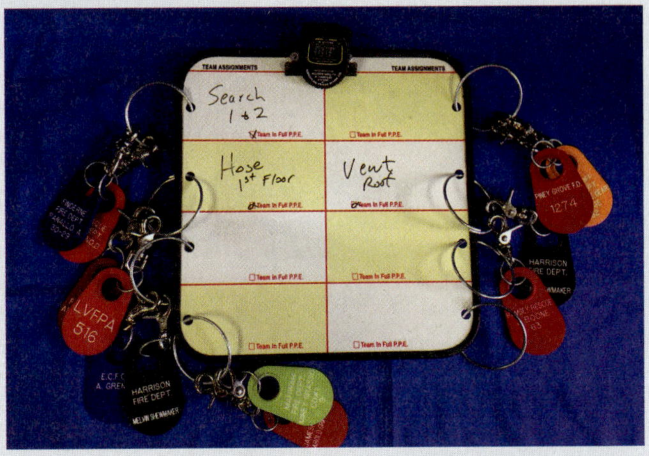

Set up an accountability board to track the personnel entering and exiting the confined space. Assign a safety officer responsible for maintaining accountability and ensuring everyone is accounted for during and after the evacuation.

Step 3: Provide Medical Assistance

Ensure rescue personnel are trained in basic first aid and emergency medical procedures, such as stabilizing injuries and providing immediate care. Have medical supplies readily available, and be prepared to treat common injuries related to confined-space evacuations, such as trauma or respiratory distress.

Step 4: Inspect Rescue Equipment

Conduct a thorough equipment check prior to confined-space entry, including self-contained breathing apparatuses (SCBAs), ropes, harnesses, and communication devices. Confirm that all equipment is in good working condition and that each team member knows how to use it properly.

Continues.

SKILL DRILL 6-2 CONTINUED

Evacuation Methods for Confined-Space Operations

Step 5: Signal Evacuation

Demonstrate the use of an air horn or other alarm system to signal an evacuation. Use the standard evacuation signal of three short blasts, ensuring all team members recognize and respond to the signal immediately.

Step 6: Conduct a Debrief

After the evacuation drill, gather all team members for a debrief session. Discuss what went well, areas for improvement, and any observations regarding the evacuation process. Incorporate feedback into future training and operations.

Immobilization Devices

Spinal immobilization is a critical aspect of confined-space rescue operations, aiming to prevent further injury to the spinal cord and surrounding structures during extrication. Immobilizing and securing a victim to a retrieval device involves carefully stabilizing the victim's body to prevent further injury and ensure their safe removal from a confined space. To better understand how to immobilize and secure a victim to a retrieval device, follow the steps in **SKILL DRILL 6-3**. Various methods and equipment can be employed to achieve spinal immobilization effectively.

Spinal immobilization devices are specifically designed to stabilize the spine and prevent movement during extrication. Common examples include cervical collars, head immobilization devices, KEDs, spine boards, LSP Half Backs, scoop stretchers, and vacuum mattresses.

Cervical collars (**FIGURE 6-22**) come in various sizes and are adjustable to fit different neck circumferences. A cervical collar should fit snugly, but it should not be restrictive.

Head immobilization devices, such as padded blocks or head blocks (**FIGURE 6-23**), are used to stabilize the victim's head and prevent lateral movement. They are often integrated into spinal immobilization devices or can be used separately.

A KED (**FIGURE 6-24**) consists of a vest, padded straps, and a head immobilization device, allowing rescuers to secure the victim's entire body while maintaining spinal alignment.

An LSP Half Back (**FIGURE 6-25**) is a vertical extrication device designed for rugged use, combining the functionality of a full-body harness with the ability to immobilize a patient in cases of suspected cervical or spinal trauma. It consists of two main components: a durable, padded full-body harness and a removable aluminum back stay with a head immobilizer. This design allows for quick and effective immobilization and vertical extrication of patients from confined or elevated spaces, such as high-rise buildings, storage tanks, ships, and automobiles.

SKILL DRILL 6-3

Immobilizing and Securing a Victim to a Retrieval Device Using Two Rescuers

Step 1: Ventilate the Confined Space

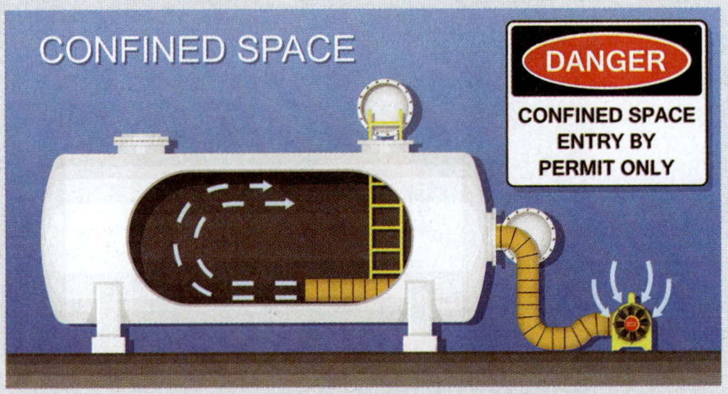

Before entering, ensure the confined space is properly ventilated to remove hazardous gases and provide a safe environment. Use appropriate ventilation techniques based on the type of confined space, such as mechanical or natural ventilation.

Step 2: Monitor Air Quality

Utilize monitoring equipment to check for harmful gases, oxygen levels, and other potential hazards before and during entry. Ensure the atmosphere remains safe throughout the rescue operation.

Step 3: Enter with Proper Line Management

Two rescuers should enter the confined space using horizontal entry techniques, ensuring proper line management to avoid entanglement or complications. One rescuer leads the line while the other follows, maintaining communication and coordination.

Step 4: Prepare the Victim for Extrication

Once the rescuers reach the victim, begin packaging them for extrication using either a long or short spine board, depending on the situation and space constraints. Ensure the victim is secured tightly to prevent movement during extraction.

Continues.

SKILL DRILL 6-3 CONTINUED

Immobilizing and Securing a Victim to a Retrieval Device Using Two Rescuers

Step 5: Immobilize the Spine

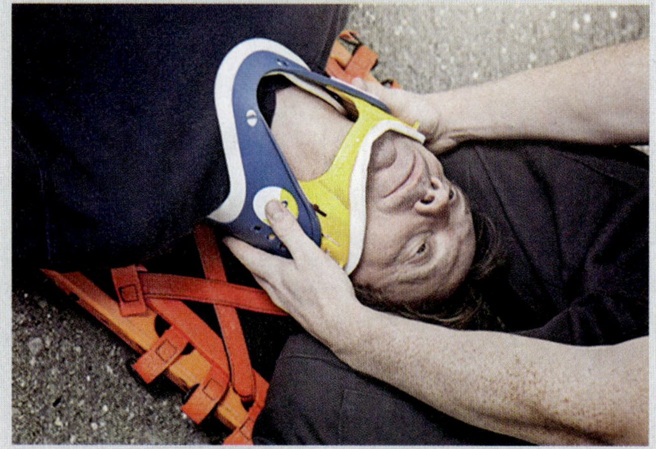

Apply spinal immobilization techniques, including positioning the head and neck in alignment and securing a cervical collar to prevent further injury. Ensure the victim is stable before attempting to move them.

Step 6: Transfer Victim to Retrieval Device

Once immobilized, carefully move the victim to the retrieval device, such as a Sked® or other stretcher designed for confined-space rescue. Ensure the victim remains secure and stable during the transfer.

Step 7: Coordinate with Haul Team for Extraction

Maintain communication between the rescuers and the haul team as they begin moving the victim out of the confined space. Rescuers should continue to guide and support the victim while the haul team operates the retrieval system, ensuring a smooth and coordinated extraction.

There are two types of spine boards: A long spine board (**FIGURE 6-26**) is used to immobilize the entire body, and a short spine board immobilizes only the upper half of the body (the head, neck, and spine). The short spine board's advantage is that it allows more versatility in movement. Straps should be padded to prevent pressure injuries, and harnesses should distribute weight evenly across the body.

Scoop stretchers (**FIGURE 6-27**) are medical devices specifically designed to safely lift and transport patients with minimal movement, making them ideal for victims with suspected spinal injuries. A scoop stretcher consists of two halves that can be separated and slid under a victim from either side, then reconnected to form a complete stretcher. This design allows victims to be moved with minimal disruption to their body position.

A vacuum mattress (**FIGURE 6-28**) is designed to provide customized support that conforms to the victim's body shape in addition to immobilization. A vacuum pump is required to extract the air from the vacuum mattress. Vacuum mattresses offer superior comfort and stability to other immobilization methods.

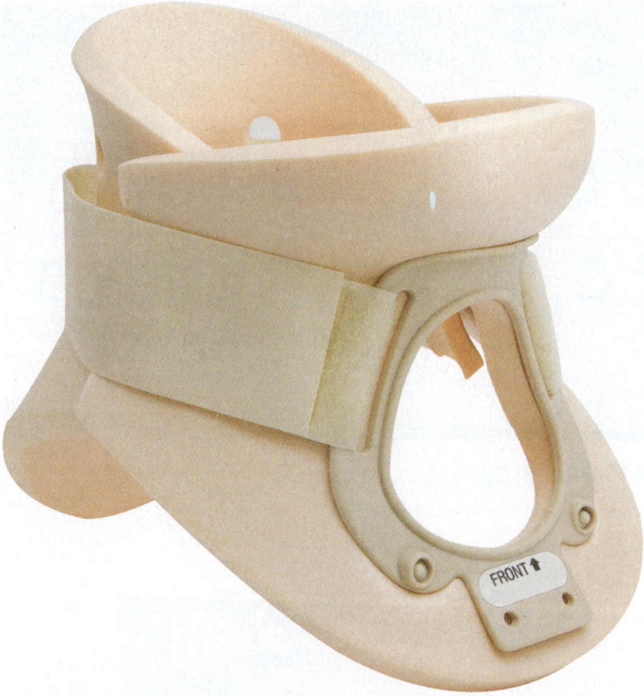

FIGURE 6-22 A cervical collar stabilizes a victim's neck during a confined-space rescue operation, ensuring safe extraction.

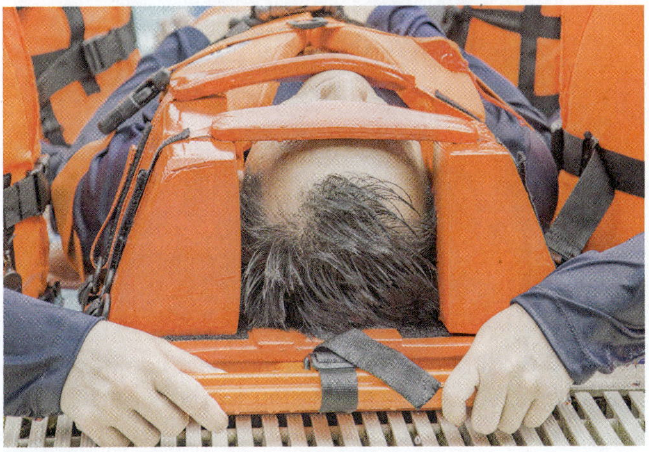

FIGURE 6-23 A head immobilization device.

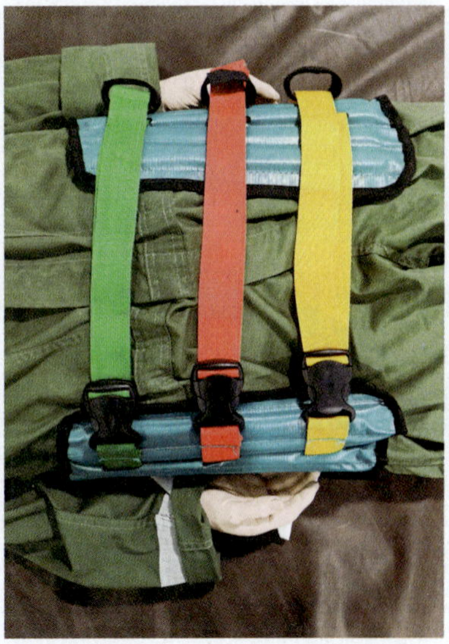

FIGURE 6-24 A KED is used to immobilize a victim's spine during a confined-space rescue for safe and secure extrication.

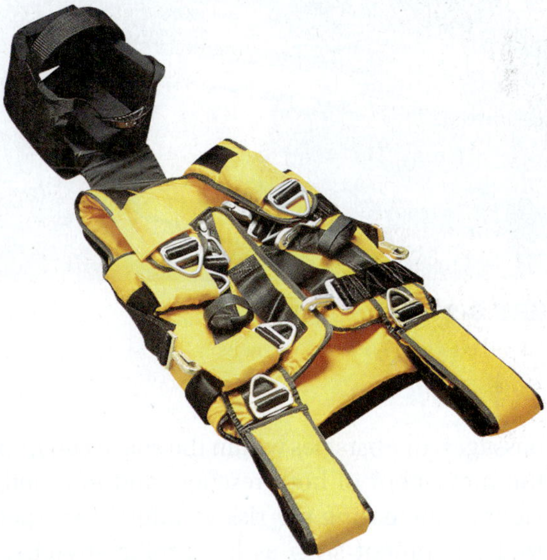

FIGURE 6-25 A LSP Half Back extrication device is used to safely secure and extract a victim from a confined-space scenario.

Hazards and Immobilization Devices

Removing victims on immobilization devices from a confined space presents several hazards that rescuers must address to ensure the safety of both the victim and the rescue team, such as the following:

- **Environmental hazards:** Confined spaces may contain hazardous atmospheres, such as toxic gases, low oxygen levels, or flammable substances. Rescuers must assess and mitigate these hazards before initiating the removal process to prevent exposure to harmful conditions. This can involve air monitoring, ventilation, and the use of appropriate PPE.
- **Physical constraints:** The confined-space environment may pose physical constraints that complicate the removal process. Limited space, narrow

FIGURE 6-26 A long spine board.

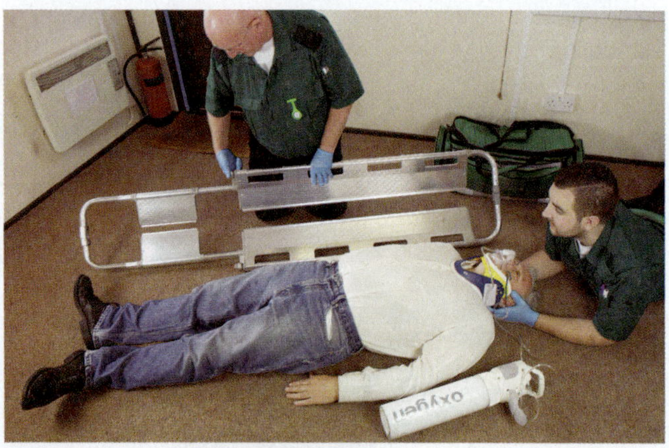

FIGURE 6-27 A scoop stretcher.

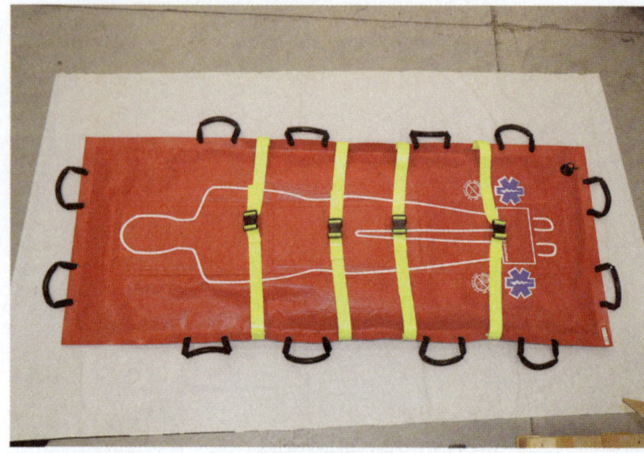

FIGURE 6-28 A vacuum mattress.

FIGURE 6-29 A rescuer using a tag line to communicate in a confined space. One pull = stop, two pulls = haul.

passages, or obstacles within the space can impede the movement of both rescuers and immobilized victims, increasing the risk of injury. Use specialized equipment such as low-profile stretchers or flexible immobilization devices to navigate tight spaces.

- **Spinal movement:** Improper handling or movement of the victim during removal can exacerbate spinal injuries, especially if the victim's spine is not adequately immobilized. Rescuers must ensure that the victim remains in a stable and neutral position throughout the extraction process to minimize the risk of further injury.
- **Equipment failure:** Malfunction or failure of immobilization devices or retrieval systems can compromise the safety of the removal operation. Rescuers should inspect all equipment thoroughly before use and have contingency plans in place to address any unexpected failures or malfunctions. Be sure to carry spare equipment and tools.
- **Communication challenges:** Effective communication between rescuers is essential during the removal process to coordinate movements, anticipate hazards, and respond to changing conditions. However, communication may be challenging in confined spaces because of limited visibility, noise, or communication barriers (**FIGURE 6-29**). Ensure all team members are briefed on the communication plan.

- **Psychological stress:** Confined-space rescue operations can be physically and mentally demanding for both rescuers and victims. Stress, anxiety, or panic reactions may affect decision-making and performance, increasing the risk of accidents or errors during the removal process. Provide psychological support, conduct regular training and drills to build confidence, and ensure rescuers are well rested and prepared for the operation.

To mitigate these hazards, rescuers must undergo thorough training in rescue procedures, hazard assessment, and victim-handling techniques. They must also use appropriate PPE, follow established protocols and guidelines, and maintain clear communication throughout the operation. By addressing these hazards proactively, rescuers can safely remove victims on immobilization devices from confined-space incidents while minimizing the risk of harm to all involved parties. To better understand how to negotiate hazards and remove all victims from a confined-space incident, follow the steps in **SKILL DRILL 6-4**.

TABLE 6-4 contains further details on the process of spinal immobilization.

SKILL DRILL 6-4

Removing All Victims from a Confined-Space Incident

Step 1: Don Operational PPE

Rescuers should properly don PPE, including helmet, gloves, and harness, to ensure safety during the confined-space operation. Verify that all equipment is secured and operational before proceeding.

Step 2: Inspect Rope Systems and Anchor Points

Inspect all rope systems, anchor points, and hardware for integrity and proper placement. Confirm that each anchor point is secure and capable of supporting the rescue loads.

Continues.

SKILL DRILL 6-4 CONTINUED

Removing All Victims from a Confined-Space Incident

Step 3: Conduct a Safety Briefing

The rescue unit should hold a safety briefing to review the incident plan, roles, hazards, and communication protocols. Ensure all team members are clear on their assignments and aware of the situational risks.

Step 4: Set Up the 4:1 System with Belay and Backup

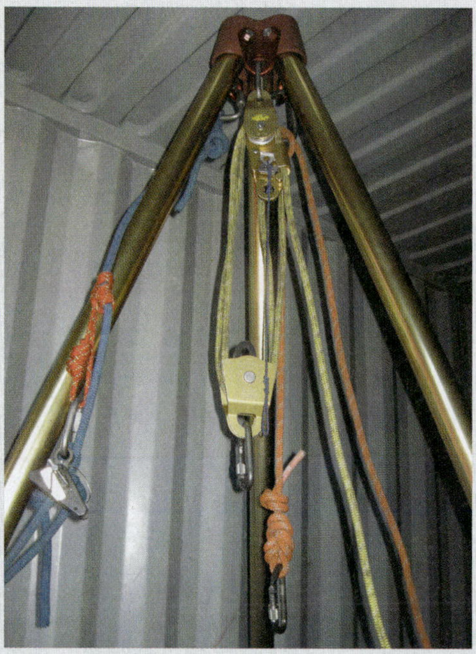

Rig a 4:1 mechanical advantage system with a belay line and backup safety measures in place to control descent and ascent during rescue operations. Ensure the system is properly secured and capable of functioning under load.

Step 5: Lower Rescuer into Confined Space

Use the 4:1 system to safely lower the rescuer into the confined space, ensuring constant communication between the haul team and the rescuer. Monitor progress closely to avoid any obstructions or hazards.

SKILL DRILL 6-4 CONTINUED

Removing All Victims from a Confined-Space Incident

Step 6: Convert from Lowering to Hauling System

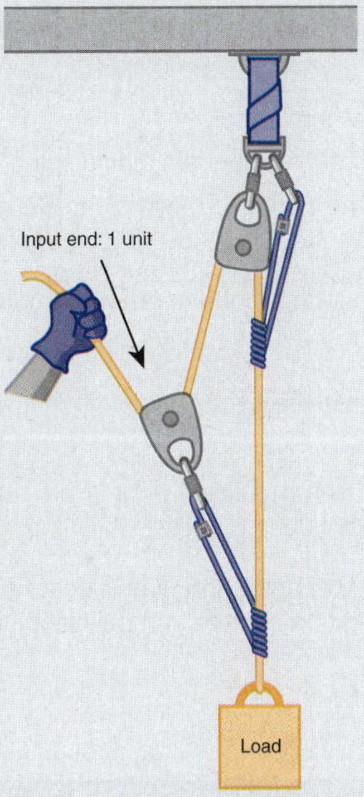

Convert the system from a lowering to a hauling configuration to facilitate victim retrieval. Build and implement a 3:1 or similar hauling system, if necessary, for greater mechanical advantage during the extraction process.

Step 7: Connect Victim to Retrieval System

Once the victim is packaged or if ambulatory, connect them to the retrieval system for safe extraction. Double-check connections and harnessing to ensure the victim is secure.

Step 8: Coordinate with Haul Team to Extricate Victim

The haul team should work in coordination with the rescuers inside the confined space to smoothly and safely extract the victim. Maintain control of the ropes and communicate throughout the operation to ensure efficiency and safety.

Steps 3, 4, and 5: © Courtesy of Illinois Fire Service Institute.

TABLE 6-4 The Process of Spinal Immobilization

Step	Description
Assess the victim's condition	▪ Evaluate the extent of the victim's injuries and potential spinal trauma to determine the appropriate immobilization technique.
Stabilize the head and neck	▪ Maintain neutral alignment to prevent movement that could exacerbate spinal damage; manual stabilization can be used initially.
Select and apply immobilization devices	▪ **Cervical collar:** Place the collar around the victim's neck. ▪ **Spine board or other device:** Transfer the victim while maintaining spinal alignment. ▪ **Padding:** Fill voids between the victim and the device to minimize movement and pressure points.
Secure the victim	▪ Use immobilization straps or other securement methods to firmly but gently secure the victim to the device.
Maintain communication and monitoring	▪ Explain each step to the victim, providing reassurance; continuously monitor their vital signs and neurological status.
Remove the victim	▪ Use a retrieval system to carefully remove the immobilized victim from the confined space, ensuring smooth and controlled movement.
Transfer to medical personnel	▪ Hand over the victim to medical staff for further assessment and treatment, providing all gathered information about the victim's condition and the immobilization measures taken.

CHAPTER FEATURE: AFTER-ACTION

IN SUMMARY

- The steps involved in establishing search parameters are as follows:
  1. Assess the confined space.
  2. Define the search areas.
  3. Mitigate risks and establish safety measures.
  4. Communicate the search parameters.
  5. Continuously assess and adjust the search parameters.
  6. Document and report.
- To properly assess a confined space for the purposes of establishing search parameters, you must evaluate the space's layout, identify the hazards present, and determine the considerations that must be taken for the weather and temperature conditions.
- Quickly identifying key witnesses and prioritizing interviews based on their potential knowledge of the incident is essential to get the most updated and critical information as soon as possible. During interviews, ask open-ended questions. Clearly identify yourself, your role, and the purpose of the interview. Be sensitive to cultural differences and communication styles when conducting your interviews, and be sure to maintain your professionalism and exhibit empathy throughout the process. Reassure witnesses that you will respect their confidentiality and privacy rights, both during and after their interviews. Build rapport by establishing a respectful and supportive atmosphere that encourages them to feel comfortable and share information.
- Types of technical rescues that may be applicable to confined-space incidents include nonentry rescue, entry rescue, complex rescue, and rescue versus recovery.
- Establishing the rescue area involves creating a designated zone where rescue operations will be conducted, controlling access to the incident scene, and implementing measures to protect personnel from hazards present in the confined-space environment. A thorough risk assessment of the area must take place first.

- Space-specific hazards may include atmospheric issues (oxygen enrichment or deficiency, toxic vapors, airborne contaminants); physical objects (uneven surfaces, sharp edges, falling objects); or chemical (corrosive substances, irritants, carcinogens), biological (mold, bacteria, viruses), or radiological (ionizing radiation) contaminants. These hazards can be identified through visual inspection, atmospheric monitoring, a review of SDSs, and consultation with subject-matter experts.
- Both immediate and ongoing monitoring needs consist of continuous atmospheric monitoring, repeated periodic visual inspections of the space, and regular victim assessments.
- Each component of a victim removal or retrieval system must have sufficient load capacity to safely accommodate all anticipated loads without compromising the system's integrity.
- Mechanical advantage systems enable rescuers to exert less force while achieving greater lifting or pulling power, making it easier to manage heavy loads effectively.
- Six primary methods are relied on to retrieve victims during confined-space emergencies: direct lifting, mechanical advantage systems, SRLs, confined-space entry and exit systems, basket stretchers or rescue slings, and rope access and rope rescue techniques.
- Spinal immobilization devices are specifically designed to stabilize the spine and prevent movement during extrication. Common examples include cervical collars, head immobilization devices, KEDs, spine boards (also known as *backboards*), LSP Half Backs, scoop stretchers, and vacuum mattresses.

KEY TERMS

1:1 mechanical advantage system Also known as a *direct pull system*, a system that provides a mechanical advantage of 1:1, meaning that the force applied to the system equals the force exerted on the load. This system involves a single rope or cable directly attached to the load, with no additional pulleys or mechanical devices to amplify force.

2:1 mechanical advantage system A system that provides a mechanical advantage of 2:1, meaning that for every unit of force applied to the system, two units of force are exerted on the load; this system typically involves a pulley arrangement where the load is attached to one end of the rope, and the rope runs through a pulley that redirects the force. The other end of the rope is then pulled, requiring half the effort needed to lift the load directly.

3:1 mechanical advantage system A system that provides a mechanical advantage of 3:1, meaning that for every unit of force applied to the system, three units of force are exerted on the load; this system typically involves multiple pulleys arranged in a block-and-tackle configuration, with the rope threaded through the pulleys to create a mechanical advantage.

4:1 mechanical advantage system A system that provides a mechanical advantage of 4:1, meaning that for every unit of force applied to the system, four units of force are exerted on the load; this system typically involves a combination of pulleys in a block-and-tackle arrangement.

5:1 mechanical advantage system A system that provides a mechanical advantage of 5:1, offering even greater amplification of force compared to a 3:1 system; this system utilizes additional pulleys and ropes arranged in a compound pulley system, with multiple stages of mechanical advantage to achieve a 5:1 ratio.

ascender A rope grab device used to ascend a fixed rope.

basket stretcher A type of stretcher designed for transporting patients, particularly in rugged or outdoor environments; its design features a sturdy, elongated, and open frame, typically made of metal or composite materials, that is covered with a durable, often waterproof, fabric mesh or canvas. Also known as a *Stokes basket* or *rescue stretcher*.

belay A rope fixed around an object intended to secure it.

belay device A braking mechanism through which a secondary line is rigged.

belay line A secondary (backup) line to protect against a fall or system failure. Also called a *safety line*.

carabiner A load-bearing metal connector with self-closing gates used to link the elements of a high-angle system.

cervical collar A rigid or semirigid device placed around the victim's neck to restrict movement and maintain alignment of the cervical spine.

KEY TERMS CONTINUED

descent control device (DCD) A rappel device that creates friction by means of a rope running through it.

haul line A type of tag line that a haul team uses for pulling on a load.

haul system An arrangement of ropes, pulleys, and connectors that is arranged in a manner to provide mechanical advantage to raise a load under controlled circumstances.

head immobilization device A device, such as padded blocks or head blocks, used to stabilize the victim's head and prevent lateral movement.

hitch A method of tying a rope in such a way that when the object or rope is removed, the knot falls apart.

Kendrick extrication device (KED) A multipart immobilization device that provides full-body support and stabilization. It consists of a vest, padded straps, and a head immobilization device, allowing rescuers to secure the victim's entire body while maintaining spinal alignment.

kernmantle rescue rope A rope design consisting of an interior core (the kern), which usually supports the major portion of the load on the rope, and an outer sheath (the mantle) that serves to protect the core but also may support a minor portion of the load.

lifting bridle A safety device used in rescue operations and lifting scenarios to securely attach a load or patient to a lifting mechanism, such as a helicopter, hoist, or crane; it is designed to evenly distribute the weight of the load, ensuring stability and safety during the lifting process.

litter basket A stretcher or basket designed for horizontal retrieval that provides support and stability for the victim during transportation.

LSP Half Back A vertical extrication device designed for rugged use, combining the functionality of a full-body harness with the ability to immobilize a patient in cases of suspected cervical or spinal trauma; it consists of two main components: a durable, padded full-body harness and a removable aluminum back stay with a head immobilizer.

progress capture device (PCD) A device that holds a rope in place to prevent a load from sliding back down when the pulling force is released.

Prusik hitch A type of friction hitch used in ascending and belaying.

Prusik loop A continuous loop of rope in which a Prusik hitch is tied.

pulley A device with free-turning, grooved metal wheels used to reduce rope friction.

Reeves™ Sleeve A medical device used in emergency situations, particularly in rescue and transport scenarios. It is a type of extrication device designed to immobilize and protect a patient's spine during transport.

safety/communication rope A specialized rope used in various rescue, climbing, and industrial applications to provide both a safety line and a means of communication between team members during operations; this type of rope is designed to ensure the safety of individuals working at heights or in hazardous environments while allowing them to communicate effectively.

scoop stretcher A medical device specifically designed to safely lift and transport patients with minimal movement, making them ideal for victims with suspected spinal injuries.

self-retracting lifeline (SRL) A fall arrest device that contains a drum-wound line that pays out from and automatically retracts back onto an enclosed drum during use and that automatically locks at the onset of a fall.

sheave A free-turning, grooved metal wheel used to reduce rope friction on a pulley.

spine board A rigid, flat surface made of plastic or wood to which a victim is secured using straps or a harness.

tag line A non-load-bearing line that connects to a litter to help it move from side to side in order to avoid obstacles, prevent swing-fall, or attain a more advantageous position.

vacuum mattress A mattress-like structure made of a durable, flexible material, typically reinforced with plastic or metal inserts to maintain rigidity, that conforms to the shape of the victim's body when air is extracted, providing customized support and immobilization.

Yates Spec Pak A victim immobilization and evacuation system that also functions as a lifting harness, eliminating the need for a separate victim harness; its self-adjusting (sliding) cervical restraint addresses the challenges associated with lifting patients vertically while they are secured to a rigid backboard.

REFERENCES

Ferno. n.d. "Vacuum Mattress Range." https://www.ferno.com/uk/ems/vacuum-mattresses?hl=en-us.

National Fire Protection Association. 2021. *NFPA 1006: Standard on Technical Rescue Personnel Professional Qualifications, 2021 Edition.* Quincy, MA: National Fire Protection Association.

"Permit-Required Confined Spaces." 1993. 29 C.F.R. 1910.146. https://www.ecfr.gov/current/title-29/part-1910/section-1910.146.

CHAPTER QUESTIONS

1. What is the first step in establishing the search parameters of a confined space?
   a. Define the search areas.
   b. Mitigate risks and establish safety measures.
   c. Assess the confined space.
   d. Monitor the confined space for atmospheric hazards.
2. What is one of the primary objectives of a confined-space rescue size-up?
   a. setting up ventilation equipment
   b. determining the number of victims and last known locations
   c. developing a sketch of the confined space prior to entry
   d. setting up retrieval and haul lines
3. What are the three types of technical rescue operations for confined spaces?
   a. nonentry rescues, entry rescues, and complex rescues
   b. horizontal rescues, vertical rescues, and self-rescues
   c. nonentry rescues, horizontal rescues, and vertical rescues
   d. nonentry rescues, entry rescues, and horizontal rescues
4. What is the first step before initiating a horizontal retrieval method?
   a. Remove the victim from the confined space.
   b. Immobilize the victim for transport and start necessary medical treatment.
   c. Put the victim in a rescue harness.
   d. Assess the victim's condition and start necessary medical treatment.

CHAPTER FEATURE: ON SCENE

A ROPE RESCUE FATALITY

In a rural town in the north-central United States, a volunteer fire department was conducting routine training exercises focused on rope rescue techniques. The training session aimed to enhance the skills of the volunteer firefighters in handling emergency situations involving vertical rescue scenarios.

One of the participants, a dedicated volunteer firefighter with several years of experience, was tasked with rappelling down a cliff to simulate a rescue operation. Equipped with safety gear and harness, the firefighter descended the rope smoothly under the supervision of the training officer. However, halfway through the descent, Firefighter A suddenly lost his grip and fell from the rope, landing on the rocky terrain below.

The Emergency Response, Rescue Operations, and Extrication and Treatment

Fellow firefighters immediately rushed to the victim's aid, assessing his condition and providing emergency medical treatment. Despite their efforts, it became evident that he had sustained critical injuries from the fall. Emergency medical services were summoned

A ROPE RESCUE FATALITY CONTINUED

to the scene, but tragically, the victim succumbed to his injuries before he could be transported to the hospital.

The Outcome

Following the incident, an investigation was launched to determine the cause of the accident. The investigation team comprised representatives from the fire department, occupational safety authorities, and relevant regulatory agencies. Their inquiry focused on examining various factors, including equipment failure, human error, training protocols, and adherence to safety procedures.

The investigation revealed several key findings:

- **Equipment inspection:** It was determined that the safety equipment, including the harness and ropes, had undergone regular inspections and maintenance. No defects or malfunctions were identified in the equipment used during the training exercise.
- **Human error:** Analysis of the incident pointed to a potential error in judgment or technique on the fallen firefighter's part, leading to the loss of control and subsequent fall. Despite his experience, the firefighter may have encountered difficulties in maintaining proper form or grip during the descent.
- **Training protocols:** The investigation scrutinized the training protocols and procedures followed during the rope rescue exercise. Although the training session adhered to established guidelines, there were discussions about the need for more rigorous training and evaluation of participants' proficiency in rope rescue techniques.
- **Safety culture:** The incident prompted a reassessment of the safety culture within the fire department, with a focus on reinforcing the importance of situational awareness, risk assessment, and continuous training to prevent similar accidents in the future.

The fire department committed to enhancing its training programs, conducting regular safety audits, and fostering a culture of accountability and continuous improvement. The memory of the fallen firefighter served as a poignant reminder of the sacrifices made by volunteer firefighters in service to their communities and reinforced the department's unwavering dedication to ensuring the safety and well-being of its members.

This was not a confined-space event, but it highlights an important point: Although training is critically important, training fatalities are unacceptable. Regardless of the nature of a technical rescue event, redundancy, including the use of backup lines, safety lines, and other safety protocols, must always be practiced. Even in training, lives are at stake.

1. What was the primary focus of the training exercise where the fatality occurred?
   - **a.** confined-space rescue for complex scenarios
   - **b.** water rescue techniques
   - **c.** rope rescue techniques for vertical scenarios
   - **d.** fire suppression
2. What happened to Firefighter A during the descent?
   - **a.** He lost his grip and fell from the rope.
   - **b.** His equipment failed.
   - **c.** His rope broke.
   - **d.** He sustained injuries from rappelling.
3. Which of the following factors was NOT examined by the investigation team after the incident?
   - **a.** human error
   - **b.** equipment failure
   - **c.** location hazards
   - **d.** training protocols
4. Which of the following was identified as a potential contributing factor to the accident?
   - **a.** lack of supervision by the training officer during descent
   - **b.** human error in judgment or technique during descent
   - **c.** inadequate equipment inspection
   - **d.** incorrect training protocols
5. What key lesson did the fire department learn from this incident?
   - **a.** Training fatalities are always inevitable in dangerous exercises.
   - **b.** There must always be redundancy in safety measures.
   - **c.** Training exercises should only be practiced by skilled firefighters.
   - **d.** Rescue equipment needs to be replaced more frequently.

CHAPTER

7

Operations

Ingress, Egress, and Victim Needs

KNOWLEDGE OBJECTIVES

After studying this chapter, you should be able to:

- Discuss the different methods used for entry into confined spaces. (**NFPA 1006 7.2.9** [horizontal spaces] and **7.2.13** [vertical spaces], pp. 180–182)
- Recognize ingress and egress routes and the process for measuring them. (**NFPA 1006 7.2.4**, pp. 182–183)
- Explain the process for controlling access and egress for confined spaces. (**NFPA 1006 7.2.10** [horizontal spaces] and **7.2.14** [vertical spaces], pp. 183–185)
- Recognize and identify all visible victims. (**NFPA 1006 7.2.1**, p. 185)
- Explain the process for developing a victim profile. (**NFPA 1006 7.2.1**, p. 185)
- Describe the processes involved when rendering patient care during a confined-space incident. (**NFPA 1006 7.2.10** [horizontal spaces] and **7.2.14** [vertical spaces], pp. 185–186)
- Describe the capabilities and limitations of self-contained breathing apparatus (SCBA) or supplied-air respirators (SARs) for victims. (**NFPA 1006 7.2.7**, pp. 190–194)
- Determine the expected air usage for victims. (**NFPA 1006 7.2.7**, pp. 192–194)
- Discuss the processes used to package victims. (**NFPA 1006 7.2.11** [horizontal spaces] and **7.2.15** [vertical spaces], pp. 186–188)
- Describe the differences and similarities between packaging for confined-space events versus others. (**NFPA 1006 7.2.11** [horizontal spaces] and **7.2.15** [vertical spaces], p. 188)
- Describe the processes used to maneuver low-profile packaging devices. (**NFPA 1006 7.2.11** [horizontal spaces] and **7.2.15** [vertical spaces], p. 188)
- Describe how to secure a breathing apparatus to a packaged victim attached to a retrieval system. (**NFPA 1006 7.2.7**, p. 190; p. 192)
- Explain the process for donning face pieces on victims with helmets or injuries. (**NFPA 1006 7.2.7**, p. 192)
- Discuss the steps needed to use a rapid victim harness technique. (**NFPA 1006 7.2.16**, pp. 188–189)

SKILL OBJECTIVES

- Ensure the point and route of entry are clear of hazards. (**NFPA 1006 7.2.9** [horizontal spaces] and **7.2.13** [vertical spaces], pp. 180–182)
- Discover entry and exit points. (**NFPA 1006 7.2.13** [vertical spaces], p. 183)
- Demonstrate emergency escape procedures while operating in a confined-space entry portal. (**NFPA 1006 7.2.1**, p. 184)
- Critique the scene and choose appropriate medical protocols to determine treatment priorities. (**NFPA 1006 7.2.10** [horizontal spaces] and **7.2.14** [vertical spaces], pp. 185–186)
- Establish an air supply on a victim. (**NFPA 1006 7.2.7**, p. 190; p. 192)

- Maintain a patent air supply on a victim throughout the incident. (**NFPA 1006 7.2.7**, p. 192)
- Demonstrate immobilization and patient treatment methods in confined spaces. (**NFPA 1006 7.2.11** [horizontal spaces] and **7.2.15** [vertical spaces], pp. 186–187)
- Demonstrate equipment decontamination procedures. (**NFPA 1006 7.2.17**, p. 191)

Introduction

Methods of entry to confined spaces differ depending on their type—horizontal or vertical—and rescuers must use different skills (e.g., crawling or climbing) to enter and exit the space.

Recognizing, measuring, and controlling ingress and egress routes is essential for effective confined-space rescue operations. By conducting thorough assessments, measuring dimensions, identifying hazards, and implementing control measures, responders can ensure the safety and accessibility of routes for both rescuers and victims in confined spaces.

Victims of confined-space emergencies require more than just rescue: Many need medical attention before, during, and after evacuation. A variety of techniques and equipment exist to effectively and safely package and extract victims from a confined space, but just as important as the extraction of victims is ensuring that the air those victims breathe during their removal from the space is safe and that none of their injuries is aggravated.

Methods of Entry

Not surprisingly, methods of entry to confined spaces vary depending on whether they are horizontal or vertical spaces. Horizontal entry methods involve entering confined spaces through side openings or passages. These methods are commonly used in scenarios such as tanks, pipelines, and underground vaults. Vertical entry methods involve entering confined spaces from above or below, such as manholes, shafts, and wells. These methods often require specialized equipment to ensure safe descent and ascent.

Horizontal Entry Methods

Direct entry into a horizontal confined space involves rescuers entering the space through a side opening (**FIGURE 7-1**). This method is suitable for spaces with large, easily accessible openings. In this method, rescuers secure the entry point, perform atmospheric monitoring and ventilation, and enter the space using appropriate personal protective equipment (PPE) while maintaining continuous communication with the safety team outside.

FIGURE 7-1 A horizontal confined space with a direct entry opening.

When a victim needs to be evacuated horizontally, stretchers or rescue litters are used to immobilize them and safely transport them from the space. The victim is strapped into a flexible litter, and rescuers carefully maneuver the device through the horizontal passage to safety.

In horizontal confined spaces like pipelines, small tunnels, and ducts, rescuers must crawl or crouch to navigate these narrow or low-height spaces. These entries require special protective gear, such as kneepads and helmets with headlamps. See Chapters 1 and 3 for details on appropriate PPE.

Vertical Entry Methods

Rescuers commonly use rope systems for controlled descent and ascent during rope rescues, including lower/haul systems (e.g., a pre-rigged haul safe system with progress capture) when entering vertical confined spaces that are deep or narrow. For this method of entry, rescuers set up an anchor system outside the confined space, secure their harnesses to it, and use descenders for controlled entry (ascenders are used to climb and exit the space).

Key Components of Rope Rescue Systems

- **Anchor system:** A secure anchor point is established outside the confined space, typically using strong and stable structures, to ensure the safety of the rescuers during their descent and ascent.

- **Harnesses:** Rescuers wear specialized harnesses that connect to the anchor system, providing safety and stability throughout the rescue operation.
- **Descenders:** These devices allow rescuers to control their descent into the confined space, providing a means to manage speed and stop as needed. They ensure a smooth and controlled entry while reducing the risk of free falls.
- **Ascenders:** Used during the ascent or climb out of the confined space, ascenders enable rescuers to move upward efficiently. They grip the rope securely, allowing the rescuer to pull themselves upward while maintaining safety.
- **Pre-rigged haul safe system:** This system includes progress capture mechanisms that allow for lifting or hauling operations. It ensures that if a load becomes unbalanced or slips, the system will automatically catch and hold the load, preventing accidents.

Safety Considerations

- **Training and certification:** Rescuers must be properly trained in rope rescue techniques, including the use of descenders and ascenders, to ensure their safety and the safety of those they are rescuing.
- **Equipment inspection:** Regular inspection of all rope rescue equipment is crucial to identify any wear or damage that could compromise safety during operations.
- **Communication:** Clear communication among team members is essential for coordinating movements, managing the rescue operation, and ensuring everyone is aware of their roles and responsibilities.

Tripod and Winch Systems and Ladder Systems

In a tripod and winch system, a tripod with a mechanical winch or mechanical advantage rope system is set up over the space's entry point to lift or lower victims or rescuers (**FIGURE 7-2**). These systems are commonly used for rescues in manholes, wells, and similar vertical spaces. In this type of entry, the tripod is positioned over the entry point, and the winch or rope system is used to lower rescuers or lift victims. The system allows for precise control and can quickly retrieve personnel in an emergency.

In shallow vertical spaces with stable entry points, ladder systems (**FIGURE 7-3**) can be used. In situations in which a fixed ladder is not present, portable ladders are set up for entry and exit. Rescuers descend and ascend using the ladder(s) while wearing harnesses connected to a fall protection system.

Hybrid Entry Methods

In some situations, a combination of horizontal and vertical entry methods may be required. These hybrid

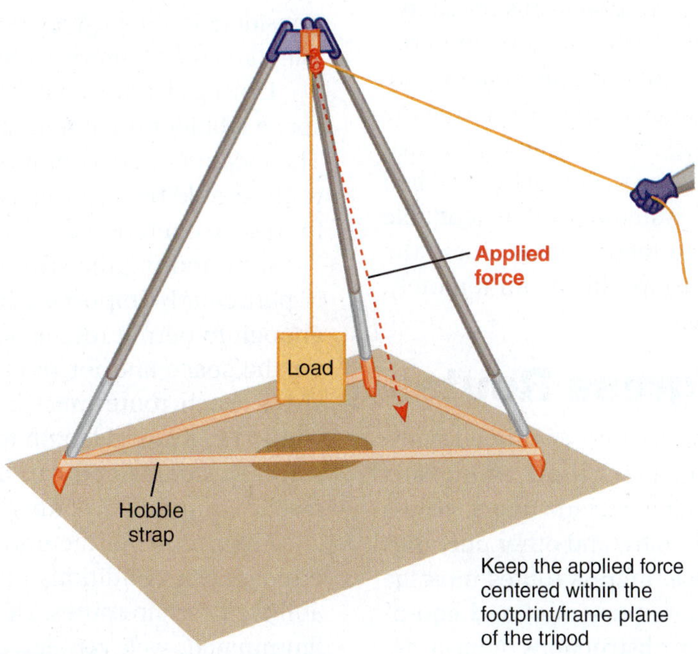

FIGURE 7-2 An Arizona vortex tripod in a directional frame configuration.

FIGURE 7-3 A shallow vertical confined space with ladder access.

FIGURE 7-4 A confined-space access hatch.

methods involve multiple techniques and equipment to navigate complex confined spaces.

In a multistage entry, rescuers use a combination of horizontal and vertical techniques to access and navigate the space. Multistage entries are generally used in complex spaces with multiple entry points or passages. A combination of rope systems, tripods, stretchers, and ladders may be necessary for rescuers to successfully navigate horizontal sections of the space and then transition to vertical techniques (or vice versa). Coordination and continuous monitoring are crucial to ensure safety.

In a remote access and retrieval entry method, remote-controlled devices and retrieval systems are used to assess and retrieve victims without entering the space. This method is appropriate for extremely hazardous or inaccessible confined spaces. Remote-controlled cameras, robotic retrieval systems, and extendable poles with retrieval hooks are deployed to assess the situation and, if possible, retrieve the victim remotely, minimizing the risk to rescuers.

Ingress and Egress Routes

The access to and means of entry into a confined space for rescuers and equipment are known as **ingress routes**. Common ingress routes include doors, access hatches (**FIGURE 7-4**), manholes, and other openings designed for entry to a space. Ingress routes must be wide enough to accommodate personnel and equipment and should be free from obstructions or hazards.

The Occupational Safety and Health Administration (OSHA) defines four different types of ingress into a confined space ("Permit-Required Confined Spaces" 1993):

- **Elevated:** A permit space where the entrance portal is above grade by four or more feet (1.2 or more meters).
- **Nonelevated:** A permit space where the entrance portal or opening is located fewer than four feet (1.2 m) above grade.
- **Restricted:** A permit space with a portal that is 24 or fewer inches (61 or fewer centimeters) wide in the least dimension.
- **Unrestricted:** A permit space with a portal that is greater than 24 inches (61 cm) wide.

Conversely, **egress routes** are the means of exit from a confined space for rescuers and victims. Egress routes should provide a clear and unobstructed pathway to safety, allowing for rapid evacuation in the event of an emergency. Considerations for egress routes include accessibility, visibility, and the presence of obstacles or hazards.

During size-up, a visual inspection of the confined space will identify potential ingress and egress routes. The location, size, condition, and accessibility of each route should be reviewed to determine their suitability for rescue operations.

Determining the size of ingress and egress routes is particularly important because they must be large enough to permit rescuers and their equipment to enter the space and for rescuers and victims to exit the space. Each route must be measured in terms of its width, height, and length to ensure it meets minimum safety standards and can accommodate all necessary rescue equipment, victims, and rescuers.

Environmental factors like lighting, ventilation, and atmospheric conditions may affect the safety and usability of certain routes. The best routes are adequately illuminated, well ventilated, and free from hazardous gases or contaminants.

The clearance around ingress and egress routes must also be checked carefully to ensure no obstructions or obstacles are present that could impede the movement of people or equipment into or out of the space. Potential hazards associated with ingress and egress routes, such as trip hazards, sharp edges, or unstable surfaces, must be identified and documented, and strategies and control measures must be developed to mitigate their risks and ensure the safety of rescuers and victims. Control measures to address these hazards may include the installation of handrails or guardrails, securement of loose debris, or the provision of additional lighting.

Discovering Entry and Exit Points

Determining the location of a confined space's entry and exit points requires a thorough external visual inspection for visible openings, manholes, hatches, or access panels. Following that, an examination of facility blueprints, schematics, or any available diagrams that detail the structure and layout of the confined space could reveal hidden or less obvious entry and exit points. Thermal imaging cameras can be used to detect hidden entry points or hazards. Remote-controlled cameras or drones can be used to explore and document the exterior and interior of the space to find entry or exit points—especially those that are difficult to access or hazardous. Extendable probes or poles should be used to check the stability and clearance of each identified entry and exit point before use.

Coordination with facility personnel before and after an event is crucial and helps to build understanding, trust, and knowledge about one another. Conversations with facility personnel or workers familiar with the confined space can yield valuable information about access points and potential hazards. Look for any labels, signs, or tags that indicate the presence of entry or exit points, such as emergency exit markers.

Clearing the Point and Route of Entry of Hazards

Ensuring the preferred route of entry to a confined space is clear of hazards requires a systematic assessment of all entry points to the space. Visual inspections, atmospheric monitoring devices, and gas detectors help you determine whether hazards are present and, if so, where they are located. Such factors as accessibility, visibility, and proximity to hazards must be considered when selecting the preferred route.

OSHA defines two confined-space access criteria ("Permit-Required Confined Spaces" 1993):

- **Open:** There are no obstacles, barriers, or obstructions within the space.
- **Obstructed:** The space contains some type of obstruction that a rescuer would need to maneuver around.

A rescue team should have procedures in place to remove physical obstructions, debris, and equipment from entry points to ensure unimpeded access. Ventilation can be used to clear any atmospheric hazards. Once the entry point is clear of all known hazards, barricades, warning signs, and control zones should be erected to prevent unauthorized access and protect rescuers from external hazards. Clear pathways to the entry point should be well marked and adequately illuminated, and if any slip hazards are present along the pathway or at the entrance, traction controls such as anti-slip mats or coatings can be implemented.

Controlling Access and Egress

Controlling access and egress is a critical aspect of confined-space operations to ensure the safety of both rescuers and victims. In most cases, a combination of personnel stationed at the confined space's access points and barrier tape is used to control access to the area. To better understand how to enact emergency escape procedures while operating in a confined-space entry port, follow the steps in **SKILL DRILL 7-1**.

An **entry permit system**—a formal system used to control and manage access to confined spaces, hazardous areas, or specific work sites—that outlines procedures for accessing and working in confined spaces is a good first step. A personnel accountability system, such as a tag-in/tag-out board, can help track who is inside the confined space at all times (see Chapter 5 for more on personnel accountability systems). The procedures such system covers include authorization, hazard identification, PPE requirements, and emergency procedures.

In horizontal confined spaces, all access points (e.g., manholes, hatches, doors, openings) should be secured to prevent unauthorized entry during rescue operations. Physical barriers (e.g., barricades) also help secure the space. Lockout/tag-out procedures can be used to de-energize and secure equipment or machinery associated with the confined space to prevent accidental activation.

Controlling access to and egress from vertical confined spaces can be considerably more involved. Appropriate entry systems for vertical confined spaces include ladders, stairways, personnel hoists, and mechanical lifts, each of which must be secured to prevent inadvertent access by unauthorized people. Fall protection systems, such as guardrails, personal fall arrest systems, or

SKILL DRILL 7-1

Emergency Escape Procedures

Step 1: Assess the Emergency Situation

Upon encountering an emergency in a confined space, immediately stop all work to assess the severity and nature of the situation. Remain calm and analyze potential hazards to avoid further escalation.

Step 2: Activate Emergency Communication

Quickly locate and activate the designated communication device (radio, alarm system, etc.) to signal an emergency. Ensure the signal is received by external teams to initiate rescue or evacuation support.

Step 3: Perform Self-Rescue Techniques

If conditions allow, use self-rescue techniques, such as moving carefully to the escape route while remaining tethered to safety lines. If a retrieval system (e.g., safety harness) is available, properly engage with the system to assist in extraction.

Step 4: Execute Safe Exit Using Escape Routes

Follow the established emergency escape route and procedures to exit the confined space as safely and quickly as possible. Ensure all team members are accounted for and that the exit process is controlled to prevent accidents during the evacuation.

Step 3: Courtesy of Illinois Fire Service Institute; **Step 4:** Courtesy of Ken Fowler.

positioning devices, are required for personnel working at heights or exposed to fall hazards during vertical entry and egress.

Identifying Victims

Recognizing and identifying victims in a confined-space emergency can be challenging because of factors such as poor visibility and inadequate lighting. A systematic search process starting from known entry points and proceeding methodically through the space is essential. Here are key steps to enhance this process:

- **Enhance visibility:** Use portable lighting, headlamps, and other equipment to improve visibility within the confined space.
- **Conduct a systematic search:** Conduct a thorough search, looking for signs of human presence or movement, such as limbs, clothing, or PPE. Also search for tools or items the individual was using in the work area.
- **Listen for sounds:** Pay attention to any calls for help, labored breathing, or other signs of life.
- **Establish communication:** If possible, establish verbal or visual communication (e.g., hand signals) with visible victims to obtain information about their condition and needs.
- **Provide immediate medical assistance:** If needed, provide immediate medical assistance to stabilize the victim's condition.
- **Document findings:** Record the location and condition of each visible victim, including any visible injuries or distress, and any relevant information provided by the victims.
- **Brief the incident commander (IC):** Inform the IC about the number and condition of visible victims to help coordinate resources and equipment for a rescue plan.

Victim Profiles

The development of victim profiles in confined-space rescue generally begins as ICs receive initial information about the emergency, including the following:

- The location and type of the confined space
- The nature of the emergency (e.g., entrapment, hazardous atmosphere)
- The number of potential victims involved
- A preliminary assessment of hazards and risks

After this information is collected, ICs interview witnesses, co-workers, and anyone else who may be familiar with the confined space or the victims to collect more details. These interviews can help ICs better understand the tasks that were being performed before the incident; provide information about victims' identities, medical histories, training, and experience; and identify specific hazards or challenges within the confined space. A full victim profile includes the following:

- The victim's name, age, and physical description
- The victim's medical conditions, allergies, or special considerations
- Details of the confined space, such as entry points, dimensions, and potential hazards
- A timeline of events leading up to the emergency

ICs analyze this information to develop a comprehensive understanding of the victim's situation; assess the severity of the confined-space emergency and the level of risk to victims; identify potential challenges or obstacles that may affect the rescue operation; and account for any environmental factors, such as temperature, air quality, and structural stability. This analysis will inform the IC's process of selecting the most appropriate rescue strategy.

Rendering Care to Victims

Once contact with victims is established, rescuers must conduct a rapid assessment known as **triage** to determine their condition and prioritize care. Triage principles guide rescuers in allocating resources and determining the order of treatment. The most critical areas a triage will cover are assessment of the victim's airway and need for breathing support; evaluation of pulse rate, blood pressure, and other elements of the victim's circulatory system, including checking for blood loss; and immobilization.

The first step is to establish whether the victim has airway patency. To establish patency, rescuers may need to perform airway maneuvers, such as head-tilt chin-lift or jaw thrust, to open the airway and facilitate breathing. Some victims experience respiratory distress as a result of atmospheric hazards or physical trauma; in these cases, rescuers must provide supplemental oxygen and assist as needed to maintain adequate oxygenation and ventilation.

Visible bleeding can be assessed and treated during triage, but close monitoring of patient vital signs (e.g., pulse rate and blood pressure) may indicate internal bleeding, other damage to the circulatory system, or the presence of physiologic **shock**, a life-threatening condition that occurs when the body and organs do not receive sufficient blood flow. Hemorrhage control and intravenous access may be necessary to stabilize a victim's circulatory system and prevent shock.

As discussed later in this chapter, victims with suspected spinal injuries require careful immobilization to prevent further injury during extrication. During triage, injuries may be detected that affect the victim's extremities (arms, hands, legs, and feet). If any injuries are found, a **splint**, which is a device used to shore up and protect injured limbs, should be used during the rescue.

Once triage is complete and victims have been stabilized, they can be safely extricated from the confined space. The extrication process must be smooth to minimize further injury to the victim. Throughout the retrieval, rescuers must continue to monitor vital signs and provide supportive care until victims are transferred to emergency services.

Packaging Victims for Retrieval

Packaging a confined-space victim for extrication ensures the victim's safety and minimizes the risk of exacerbating injuries during extraction. To better understand how to immobilize and treat a victim, follow the steps in **SKILL DRILL 7-2**. Before packaging begins, conduct a thorough assessment of the scene to identify hazards, assess injuries, and determine the best approach for extrication. Ensure that all necessary equipment, including medical supplies, immobilization devices, and communication tools, are readily available.

The head, neck, and spine must be immobilized before moving a victim to prevent further injury. Use manual stabilization techniques or cervical collars as needed. If victims are conscious and able to follow commands, instruct them to remain still and avoid unnecessary movement. That said, not every victim needing

SKILL DRILL 7-2

Treating an Immobilized Victim

Step 1: Conduct Patient Assessment

Begin by assessing the immobilized patient for any visible injuries, checking for consciousness, responsiveness, and breathing. Assess the head, neck, spine, and extremities to identify any potential trauma.

Step 2: Apply Spinal Immobilization

If spinal injury is suspected, apply a cervical collar (C-collar) to immobilize the neck and prevent further movement. Ensure that the C-collar is correctly sized and securely fastened.

SKILL DRILL 7-2 CONTINUED

Treating an Immobilized Victim

Step 3: Prepare for Vertical Lift with a Basket

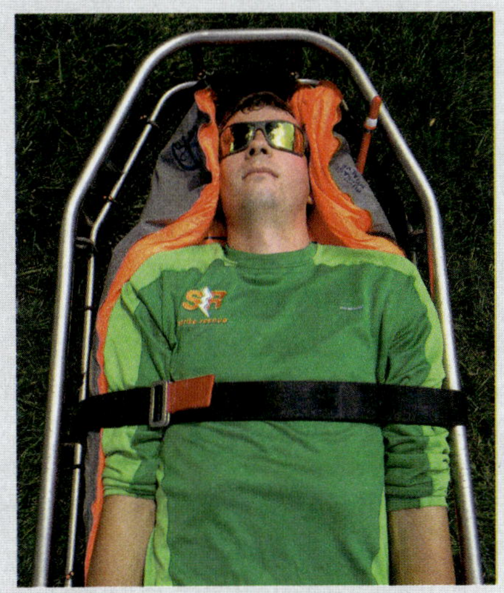

Once the patient is immobilized, carefully transfer the patient into a vertical lifting basket stretcher. Ensure the patient remains secure during the transition to prevent any shifting or additional injury.

Step 4: Attach the Bridle for Vertical Lifting

Attach the appropriate bridle system to the vertical lifting basket for hoisting the patient. Ensure that the bridle is securely fastened and correctly positioned to support a safe and balanced lift.

Step 1: Courtesy of Illinois Fire Service Institute; **Step 3:** Courtesy of Jason Williams.

removal from a confined space requires spine immobilization. The type of injuries sustained and determining what happened to the victim inside the space will dictate the level of care needed.

The most appropriate packaging method must be selected based on the victim's condition, the nature of the confined space, and the extrication plan. If using a vacuum mattress or spinal board, ensure it is properly positioned and padded to provide adequate support and immobilization. Check the equipment for any defects or damage that could compromise its effectiveness.

Carefully lift the victim and position them on the packaging device, taking care to maintain spinal alignment throughout the process. Secure the victim to the device using straps, harnesses, or immobilization devices, ensuring that they are snug but not too tight to restrict circulation or breathing. If necessary, apply additional immobilization devices, such as C-collars or headblocks, to further stabilize the victim's head and neck. Ensure that all immobilization devices are properly applied and adjusted to prevent movement during extrication.

As the packaging process is ongoing, maintain constant communication with the victim, reassuring them and providing instructions as needed. Monitor the victim's vital signs and level of consciousness, and be prepared to adjust the packaging or extrication plan accordingly.

Once the victim is securely packaged, ensure that the packaging device is properly secured to the extrication equipment, such as stretchers, litters, or rope systems. Double-check all connections and attachments to prevent the packaging device from becoming dislodged during extrication.

Once victims have been extricated from the confined space, reassess their condition and provide appropriate medical care or transport to definitive care facilities as necessary. Document all interventions, observations, and communications related to the packaging and extrication process for postincident analysis and reporting.

Unique Needs for Victim Packaging in Confined-Space Rescues

Patient packaging methods for confined-space events share some similarities with packaging methods for other types of emergencies, but there are also some distinct differences. For example, all methods involve immobilizing the patient to prevent further injury during transport. Prior to packaging a victim in any emergency situation, rescuers assess the victim's condition and injuries to determine which packaging method is most appropriate. The differences arise largely for five reasons:

- Space constraints
- Equipment accessibility
- Vertical versus horizontal movement requirements
- Environmental hazards
- Communication challenges

In most confined-space environments, there is little room to move around. A space's irregular shape or obstacles may require specialized packaging techniques for safe extraction. Accessing and maneuvering equipment inside a confined space can be particularly challenging, so compact and versatile packaging devices are specifically designed for these environments.

Patient packaging is the same for horizontal and vertical rescues, except that packaging for vertical rescues must be rated for vertical lifting. Some confined spaces contain hazardous atmospheres, limited oxygen, or other environmental dangers that require additional precautions and protective measures to be taken during patient packaging and rescue operations. A confined space's potential for limited visibility, noise, and other barriers can compromise communication between victims and rescuers during the packaging process.

Maneuvering Low-Profile Packaging Devices

Maneuvering low-profile packaging devices in confined-space rescue requires careful planning, coordination, and adherence to safety protocols outlined by NFPA and OSHA standards. Before a low-profile packaging device is deployed, a thorough assessment of the confined-space environment, including its dimensions, obstructions, and potential hazards, is necessary to determine the most suitable packaging device based on the space constraints, the victim's condition, and the extrication plan. It is worth noting that not all packaging devices are designed for vertical lifting.

Several types of low-profile packaging devices are specifically designed for use in confined spaces, including LSP Half Backs, Yates Spec Paks, **rescue sleds** (specialized equipment designed for transporting patients in emergency situations), and **flexible stretchers** (medical transport devices designed to safely and comfortably carry patients in various emergency situations, particularly when navigating through narrow or confined spaces). All low-profile packaging devices have benefits and liabilities, so no single option is right for all situations.

If using a vacuum mattress or similar device, ensure that it is properly inflated and positioned for optimal support and immobilization. Check all straps, harnesses, and attachments to ensure they are secure and in good condition. Carefully maneuver the packaging device into the confined space, accounting for any obstacles or tight passages that may impede movement. Position the device as close to the victim as possible while still allowing enough space for rescuers to work safely.

Once the packaging device is in position, carefully lift and transfer the victim onto the device, ensuring that they are properly aligned and supported. Secure the victim to the device using straps, harnesses, or immobilization devices, taking care to avoid any unnecessary movement or discomfort. Use coordinated teamwork to maneuver the packaging device through the confined space. Communicate effectively with team members to ensure smooth movement and avoid accidents or injuries.

Throughout the maneuvering process, monitor the victim's condition closely, paying attention to vital signs, comfort level, and any signs of distress. Be prepared to adjust the positioning of the packaging device or provide additional support as needed to ensure the victim's safety and well-being. Once the victim has been securely packaged and immobilized, coordinate with the rescue team to execute the extrication plan. Be sure to communicate any special considerations or hazards related to the packaged victim as you work together to safely maneuver the device and the victim out of the confined space.

Confined-Space Rescue Harnesses

A **rapid victim rescue harness** (**FIGURE 7-5**), or *webbing harness*, can be improvised using a single piece of webbing that is typically 20 to 25 feet long (6–7.5 m) and 1 inch (2.5 cm) wide. To create the harness, locate the middle of the webbing and form a large loop. Twist the

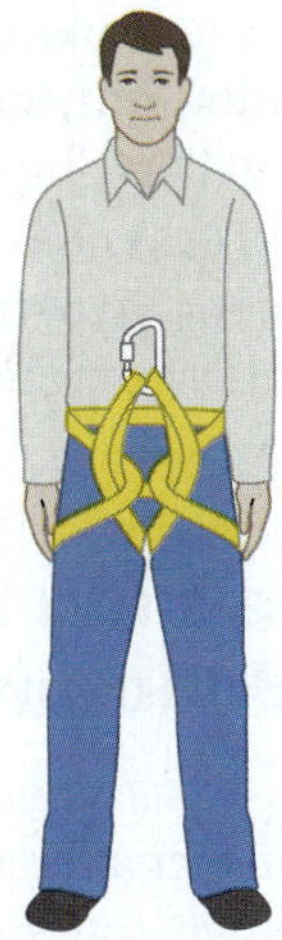

FIGURE 7-5 A rapid victim harness (or webbing harness).

loop once or twice to create two smaller loops, which will serve as footholds for the victim. Place these loops over the victim's feet and pull them up to their thighs, ensuring they fit snugly but comfortably. The two loose ends of the webbing should be wrapped around the victim's waist or chest, depending on their condition, and secured with a sturdy knot, such as a water knot. If vertical lifting is required, additional steps are necessary to ensure the victim is properly supported. After the webbing is wrapped around the victim's torso, the loose ends should be brought over the shoulders, crossed behind the victim's back, brought forward again, and securely tied. This harnessing method distributes the victim's weight more evenly and prevents the victim from tipping backward or slipping out of the harness.

A commercial confined-space rescue harness is the preferred type of harness; a rapid victim rescue harness should only be selected as a temporary solution in emergencies when proper equipment is unavailable. Extensive practice and training are required to create and safely transport a victim in an improvised harness because improper knotting, webbing placement, or weight distribution could lead to failure or injury.

A rapid victim rescue harness does not offer the safety features found in a commercial confined-space rescue harness, such as the **dorsal attachment point** located at the back of a commercial harness. This attachment point serves as the primary connection point for rigging systems (e.g., ropes, carabiners, and pulleys used to hoist or lower the victim). At the front of a commercial confined-space rescue harness, a **sternal attachment point** provides an additional anchoring option for secondary rigging connections or for securing the victim in a seated position. The sternal attachment point is primarily reserved for belay lines in ropes. Leg loops and a waist belt help to stabilize the victim and prevent slippage or movement during extraction. These are typically adjustable to ensure a snug and secure fit. Some commercial confined-space rescue harnesses feature a chest strap or **chest ascender attachment point** to facilitate vertical movement and positioning of the victim during rescue operations. Reflective materials or trim may be incorporated into the design of the harness to improve visibility in low-light conditions and enhance rescuer safety when visibility is poor.

Prior to using any commercial harness, it should be inspected for any signs of wear, damage, or degradation that could compromise its integrity or safety. Regular inspection and maintenance of the harnesses involve checking for signs of wear, fraying, or damage to the webbing, stitching, and hardware components. Harnesses should be cleaned according to the manufacturer's recommendations and stored in a clean, dry, and well-ventilated environment when not in use. Any worn or damaged components must be replaced immediately to maintain the integrity and safety of the harness.

The same holds true for any webbing that is used to create rapid victim harnesses. Webbing used in these environments is typically made of durable synthetic materials, such as nylon or polyester, that offer strength, flexibility, and resistance to abrasion and environmental factors. Webbing must be inspected regularly for signs of wear, such as fraying, cuts, burns, or chemical damage, because compromised webbing can fail under load. Cleaning should be performed using only mild soap and water because harsh chemicals can weaken the fibers. The webbing should be air-dried away from direct sunlight to prevent degradation from ultraviolet (UV) exposure.

Proper storage is essential to prolong the life span of webbing. It should be stored in a cool, dry place; kept free from dirt, oil, and other contaminants; and placed in a bag or container to protect it from physical damage and UV exposure. Maintaining a strict documentation process for inspection, cleaning, and usage helps ensure webbing remains reliable and safe for use in confined-space rescue operations.

Protecting Victims from Environmental Hazards

Using SCBA and SAR for Victims

SCBAs allow rescuers to enter environments where the air quality may be compromised as a result of factors such as toxic fumes, smoke, or oxygen deficiency; they can also be used to provide victims with access to safe, clean air, as described in Chapter 1. An SCBA typically consists of compressed air cylinders that provide an extended air supply for the wearer, allowing for prolonged operations in hazardous environments. The duration of the air supply depends on factors such as cylinder capacity, breathing rate, and workload.

SCBA units are designed to be portable and lightweight, allowing rescuers to maneuver freely within confined spaces and navigate obstacles while maintaining respiratory protection. This mobility is crucial for accessing victims in hard-to-reach areas and performing rescue operations efficiently.

SCBAs can be used in a variety of rescue scenarios, including confined-space rescues, hazardous material incidents, firefighting operations, and industrial emergencies. They are adaptable to different environments and hazards, making them a valuable asset for multipurpose rescue teams.

Some SCBA models feature integrated communication systems, allowing rescuers to maintain contact with team members and the IC during rescue operations. This facilitates coordination, enhances situational awareness, and ensures effective communication in challenging environments.

Despite providing an extended air supply, SCBA units have finite air capacities determined by cylinder size and pressure. Rescuers must closely monitor their air consumption and manage their workload to ensure they do not deplete their air supply before completing rescue operations. NFPA 1981 (NFPA 2019) outlines requirements for minimum air supply durations for SCBA units.

Although SCBA units provide respiratory protection, they can restrict mobility and agility because of their size and weight. Rescuers may encounter difficulties navigating tight spaces, climbing rope ladders, or performing intricate tasks while wearing an SCBA unit, and victims who are provided with SCBAs may find it more difficult to maneuver out of a space.

Wearing SCBAs can lead to heat stress and discomfort for rescuers and victims alike, particularly during prolonged operations or in high-temperature environments. Although some SCBAs include integrated communication systems, not all do. A victim wearing an SCBA mask may be difficult to communicate with, particularly in a noisy or chaotic confined-space environment. Rescuers must use alternative communication methods, such as hand signals or radio transmissions, to overcome communication barriers with victims wearing SCBAs.

SARs supply air to a face piece via a supply line that is in a stationary location outside the confined space. SARs are covered in detail in Chapter 1.

Environmental and other hazards can contaminate SCBA and SAR equipment. To better understand how to properly decontaminate equipment, follow the steps in **SKILL DRILL 7-3**.

Affixing a Breathing Apparatus to a Packaged Victim

Packaged victims who are connected to a retrieval system can be particularly challenging to fit with SCBA equipment. The "through-the-legs" method is an effective technique to ensure that the SCBA remains in place and provides a continuous air supply to the victim during extraction.

To perform the through-the-legs method, first ensure that the SCBA unit is fully functional and has been inspected according to the standard protocols. Ensure the air cylinder is adequately filled and the regulator is working. Clean and sanitize the face piece to prevent infection, especially if the victim has open wounds. Confirm that the retrieval system, including the harness and rope, is properly set up and ready for use. Prepare the SCBA, harness, retrieval system, and any additional securing straps or ropes.

Next, evaluate the condition of the victim, ensuring they are stable enough to be moved. Check for any visible injuries, particularly around the head and face. Assess the victim's consciousness and ability to cooperate. Position the victim in a manner that facilitates easy access to secure the SCBA and the retrieval system, and carefully place the SCBA mask on the victim's face, ensuring a secure fit and a proper seal. Once the mask is in place, verify that it is providing continuous airflow to the victim.

Now that the mask is on, secure the retrieval harness for the victim. The harness should be snug but not overly tight. Verify that the harness is properly adjusted and that all buckles and straps are fastened.

Place the SCBA unit on the victim's torso, ensuring that it is centered and aligned with the victim's body. A properly placed unit will appear as if the victim put the SCBA unit on backward, with the cylinder facing outward from the chest. Pass the SCBA's waist strap under one of the victim's legs. Raise it up and buckle it securely. Place the shoulder straps over the victim's shoulders and tighten them until the fit is snug. Depending on the type of packaging device that is used, the overall diameter of the packaged victim

SKILL DRILL 7-3

Proper Equipment Decontamination Procedures

Step 1: Inspect the SCBA Unit

Begin the decontamination process by conducting a visual inspection of the SCBA unit, focusing on any dirt, debris, or contaminants that need to be cleaned. Make sure the unit is turned off and properly secured before handling.

Step 2: Remove the Cylinder and Regulator

Detach the SCBA cylinder and regulator from the unit. This will allow for more thorough cleaning of individual components and ensure no water damage to sensitive parts.

Step 3: Prepare Cleaning Materials

Set up a cleaning station using a bucket of mild soap and water, a scrub brush, and a water hose. This is essential for gently scrubbing contaminants from the SCBA components without damaging the equipment.

Step 4: Clean the SCBA Cylinder

Use the scrub brush and soap solution to thoroughly clean the SCBA cylinder, removing any dirt, grime, or potential hazardous materials. Rinse the cylinder with clean water to ensure no soap residue remains.

Step 5: Dry the SCBA Unit

Once the SCBA components have been cleaned and rinsed, place the equipment in front of a fan or in a well-ventilated area to air-dry. Ensure that all parts are fully dry before reassembling the unit and storing it for future use.

could be too large for the opening. Be sure to consider this factor as the victim is packaged.

Connect a securing strap or rope to the retrieval system. This can be done using carabiners or other secure attachment methods. Double-check that all connections are secure and that the SCBA unit is tightly held in place against the victim's body. Ensure that the SCBA mask remains properly sealed on the victim's face and that the air supply is uninterrupted. Verify that the securing strap or rope does not restrict the victim's breathing or circulation. Confirm that the retrieval system is properly attached and that the victim is ready to be safely extracted.

Placing an SCBA Face Piece on a Victim with a Helmet or Facial Injuries

Helmets or injuries can complicate the process of using SCBAs with victims. If the victim is wearing a helmet that can be safely removed without exacerbating injuries, loosen the helmet strap and carefully remove it. If removal is not possible or advisable, proceed with the helmet on.

If the victim is wearing a helmet: Position the victim in a way that minimizes movement and discomfort. This could be lying down or sitting up, depending on the situation. Remove the helmet. Fully extend and loosen all the straps on the SCBA's face piece. Place the face piece on an angle and bring it up from below the victim's chin. Carefully position the face piece over the victim's chin and then move it upward until the nose and mouth are covered.

Gently pull the head harness and straps (there should be four to six straps, depending on the brand) over the victim's head. Once the face piece is in place, starting from the bottom and working upward, gently tighten the straps. Place the helmet back on the victim over the head harness and adjust the helmet's chin strap until it is secure.

If the victim has facial injuries: Fully extend and loosen all the face piece straps to minimize movement. Depending on the location and severity of the injury, either bring the face piece up from below the chin or gently from the side to avoid aggravating the injury. Carefully position the face piece over the chin and then move it upward until the nose and mouth are covered, taking care to avoid any injured areas. Starting from the bottom, gently tighten the straps, making sure not to apply excessive pressure to injured areas. Adjust the straps evenly to maintain a proper seal.

After the face piece is secure, check for any air leaks. If the victim is conscious, perform a negative-pressure test by covering the exhalation valve and asking the victim to inhale to see if the face piece pulls toward the face and does not leak. (If the victim is unconscious, perform the test with a fellow rescuer.) If leaks are detected, readjust the face piece and straps until a proper seal is achieved.

TABLE 7-1 shows the steps in maintaining a patent air supply on a victim, and **TABLE 7-2** shows air conservation techniques.

Determining Air Usage for Victims

A victim's level of physical exertion significantly affects their air consumption. Higher activity levels lead to increased breathing rates and, consequently, greater air usage. The larger and heavier victims are, the higher their metabolic rates will be, which means they will likely consume more air than others. Temperature,

TABLE 7-1 Maintaining a Patent Air Supply on a Victim

Establish Air Supply	Monitor Air Supply	Manage Air Supply	Respond to Air Supply Emergencies
Set up the air supply system.	Monitor the air pressure and the victim's consumption levels.	Implement air conservation techniques (see Table 7-2).	Be vigilant for any signs of malfunction of the self-contained breathing apparatus (SCBA).
Verify that the airflow is continuous.	Communicate the status of the victim's air supply to the rescue team.	Adjust the victim's breathing rates to extend the life of the air supply.	Implement emergency air supply measures if needed.
Ensure the air supply meets the durational needs of the operation.	Closely monitor the SCBA's gauges and indicators.	Manage the victim's stress and exertion levels to conserve air.	If air supply is compromised, execute an immediate evacuation plan.

TABLE 7-2 Air Conservation Techniques

Technique	Definition	Procedure
Controlled breathing	A technique that involves regulating the rate and depth of breaths to optimize oxygen intake while minimizing air consumption	▪ Inhale slowly and deeply through the nose, fully expanding the lungs. ▪ Exhale slowly and completely through the mouth, allowing the lungs to empty completely. ▪ Maintain a steady rhythm of breathing, focusing on smooth, controlled inhalation and exhalation. ▪ Avoid rapid or shallow breathing, which can increase air consumption and lead to hyperventilation.
Skip breathing	A technique that involves briefly holding the breath between inhalation and exhalation to reduce air consumption	▪ Inhale deeply through the nose, filling the lungs with air. ▪ Hold the breath for a brief pause (one to two seconds). ▪ Exhale slowly and completely through the mouth. ▪ Repeat the cycle, taking regular pauses between breaths to conserve air.
Box breathing	Also known as *square breathing*, a relaxation technique that involves equalizing the length of inhalation, breath hold, exhalation, and another breath hold	▪ Inhale deeply through the nose for a count of four seconds. ▪ Hold the breath for a count of four seconds. ▪ Exhale slowly and completely through the mouth for a count of four seconds. ▪ Hold the breath out for a count of four seconds. ▪ Repeat the cycle, maintaining a steady rhythm and focusing on relaxation.
Reilly Emergency Breathing Technique (R-EBT)	A method of conserving air during emergencies or when faced with limited air supplies	▪ Inhale deeply and hold the breath for as long as possible. ▪ Exhale slowly and completely through the mouth. ▪ Inhale again, but only partially, taking in a shallow breath. ▪ Hold the partial breath for a brief pause before exhaling. ▪ Repeat the cycle, alternating between full and partial breaths to conserve air.

humidity, and the presence of contaminants can also influence breathing rates and air consumption.

The length of time a victim will need to use an SCBA directly affects the victim's total air consumption. The performance and condition of the SCBA equipment, including its air cylinder capacity and regulator efficiency, are also critical factors in how much air will be available to the victim.

NFPA 1981: Standard on Open-Circuit SCBA for Emergency Services (NFPA 2019) specifies the minimum requirements for the design, performance, testing, and certification of SCBA units: SCBA units must provide a minimum rated service duration, which is typically 30, 45, or 60 minutes, depending on the cylinder size and air pressure. *NFPA 1404: Standard for Fire Service Respiratory Protection Training* (NFPA 2018) outlines the training requirements for SCBA users, emphasizing the importance of understanding air consumption rates and emergency procedures. It includes guidelines for calculating air consumption based on different activity levels and stress conditions.

To accurately determine the expected air usage for victims wearing SCBA during a confined-space rescue, you must do the following:

- Determine the potential hazards, temperature, humidity, and presence of contaminants in the space.
- Estimate the physical activity level of the victim (e.g., resting, moderate exertion, high exertion).

TABLE 7-3 Approximate Breathing Rates Based on Activity Level

Resting	Moderate Exertion	High Exertion
10 to 15 L/min	30 to 40 L/min	40 to 60 L/min

Calculating Air Consumption Rate for a Victim

Breathing rate (L/min) × Duration of rescue (min) = Total air consumption

Example:
A victim is operating at moderate exertion levels for a period of 30 minutes during a rescue in a confined space with a contaminated atmosphere.

35 L/m × 30 min = 1,050 liters of air

A 30-minute cylinder contains 45 ft^2 (1,274 L) of breathing air. A 45-minute cylinder contains 66 ft^2 (1,868 L) of breathing air. A 60-minute cylinder contains 88 ft^2 (2,491 L) of breathing air. Thus, a standard 45-minute SCBA cylinder would be sufficient for this scenario, but safety margins should be considered.

FIGURE 7-6 Calculating air consumption rate for a victim.

- Calculate the victim's breathing rate in liters of air per minute (L/min) based on their exertion level (**TABLE 7-3** and **FIGURE 7-6**).
- Estimate the length of time the SCBA must be worn by the victim.

Once the air consumption rate has been calculated, you will know whether the capacity of the SCBA cylinder you have access to will fall short of, meet, or exceed the victim's calculated air consumption.

CHAPTER FEATURE: AFTER-ACTION

IN SUMMARY

- Horizontal entry methods for entering confined spaces such as tanks, pipelines, and underground vaults include side openings or passages. Vertical entry methods involve entering from above or below a space, such as a manhole, shaft, or well. These methods often require specialized equipment to ensure safe descent and ascent. In some situations, a combination of horizontal and vertical entry methods may be required.
- Ingress routes are the access to and means of entry into a confined space for rescuers and equipment. Common ingress routes include doors, access hatches, manholes, or other openings designed for entry to a space. Egress routes are the means of exit from a confined space for rescuers and victims. Egress routes should permit rapid evacuation in the event of an emergency. Considerations for egress routes include accessibility, visibility, and the presence of obstacles or hazards. Ingress and egress routes must be wide enough to accommodate personnel and equipment and should be free from obstructions or hazards.
- Recognizing and identifying victims involves a systematic search process starting from known entry points and proceeding methodically through the space.
- ICs interview witnesses, co-workers, and anyone else who may be familiar with the confined space or the victims to collect more details. These interviews can help ICs better understand the tasks that were being performed before the incident; provide information about victims' identities, medical histories, training, and experience; and identify specific hazards or challenges.
- Once contact with victims is established, rescuers must conduct a rapid assessment known as *triage* to determine their condition and prioritize care.

- The most appropriate packaging method for a victim must be selected based on the victim's condition, the nature of the confined space, and the extrication plan.
- A rapid victim harness quickly and securely immobilizes and extracts victims from confined spaces. It provides a means of safely attaching the victim to a rigging system for hoisting or lowering operations.
- SCBAs and SARs allow a rescuer to enter environments where the air quality may be compromised because of factors such as toxic fumes, smoke, or oxygen deficiency; they can also be used to provide victims with access to safe, clean air.

KEY TERMS

chest ascender attachment point A specific location on a climbing or rescue harness designed for attaching a chest ascender, which is a device used to assist in ascending a rope; the chest ascender helps provide upward mobility while climbing—particularly in vertical environments—and plays a crucial role in ensuring safety and efficiency during ascents.

dorsal attachment point A primary connection point for a rigging system located at the back of a harness.

egress route The means of exit from a confined space for rescuers and victims.

entry permit system A formal procedure used to control and manage access to confined spaces, hazardous areas, or specific work sites; this system ensures that only authorized personnel are allowed entry and that all necessary safety measures are in place to protect workers from potential hazards.

flexible stretcher A type of medical transport device designed to safely and comfortably carry patients in various emergency situations, particularly when navigating through narrow or confined spaces; unlike traditional rigid stretchers, flexible stretchers are made from lightweight, flexible materials that allow them to conform to the contours of the patient's body and the surrounding environment.

ingress route The access to and means of entry into a confined space for rescuers and equipment.

rapid victim harness A piece of rescue equipment designed to quickly and securely immobilize and extract victims from confined spaces. Also known as a *webbing harness.*

rescue sled A specialized piece of equipment designed for transporting patients in emergency situations, particularly in environments where traditional stretchers may be impractical.

shock A life-threatening condition that occurs when the body and organs do not receive sufficient blood flow.

splint A device used to shore up and protect injured limbs during a rescue.

sternal attachment point A primary connection point for a rigging system located at the front of a harness.

triage A systematic process used to prioritize and categorize patients based on the severity of their condition and the urgency of their medical needs, especially in emergency situations or mass-casualty incidents; the goal of triage is to ensure that limited medical resources are allocated effectively, allowing for timely care to those who need it most.

REFERENCES

National Fire Protection Association. 2018. *NFPA 1404: Standard for Fire Service Respiratory Protection Training.* Quincy, MA: National Fire Protection Association.

National Fire Protection Association. 2019. *NFPA 1981: Standard on Open-Circuit SCBA for Emergency Services, 2019 Edition.* Quincy, MA: National Fire Protection Association.

"Permit-Required Confined Spaces." 1993. 29 C.F.R. 1910.146. https://www.ecfr.gov/current/title-29/part-1910/section-1910.146.

CHAPTER QUESTIONS

1. The first step to establish during triage is to ensure the victim has
   - a. no visible bleeding.
   - b. no compound fractures.
   - c. airway patency.
   - d. the ability to speak.
2. What is the primary connection point on a harness for rigging systems?
   - a. the side-shoulder attachment point
   - b. the dorsal attachment point
   - c. the lower-rear midsection attachment point
   - d. the sternal attachment point
3. Which of the following is an air conservation technique?
   - a. opening a mask bypass valve
   - b. only breathing through the mouth
   - c. only taking shallow breaths
   - d. practicing box breathing
4. *NFPA 1981: Standard on Open-Circuit SCBA for Emergency Services* (NFPA 2019) specifies the minimum requirements for which of the following?
   - a. design, performance, testing, and certification of SCBA units
   - b. design, color, testing, and certification of SCBA units
   - c. design, performance, testing, and cost of SCBA units
   - d. design, color, testing, and weight of SCBA units

CHAPTER FEATURE: ON SCENE

AN INCIDENT IN A RAILCAR

During routine loading operations at the chemical facility, a railcar containing hazardous chemicals experienced a mechanical failure, resulting in the release of toxic fumes. Several workers near the railcar were exposed to the fumes and required immediate rescue because of potential chemical exposure.

The Emergency Response

Upon discovering the incident, workers quickly activated the facility's emergency response protocol and notified the local fire department and hazardous materials (HazMat) team. HazMat personnel arrived at the scene and conducted an initial assessment of the railcar and surrounding area to identify the type and extent of the chemical hazard.

The Rescue Operations

Because the rescue was taking place in a unique type of confined space—a railcar—specialized entry procedures were implemented, including use of confined-space entry equipment. Highly trained technical rescue personnel equipped with chemical protective gear carefully entered the railcar to locate and assist affected workers. Continuous atmospheric monitoring was conducted to ensure safe conditions for rescue operations inside the confined space.

The Extrication and Treatment

The rescuers located and safely extracted the affected workers from the railcar and provided them with immediate medical attention. Following the rescue, a thorough decontamination process was carried out for both the rescued workers and the rescue team members to prevent secondary exposures.

The Outcome

The incident highlighted the importance of specialized training and equipment for confined-space and hazardous material rescue operations. Clear communication and coordination among all responding agencies and personnel played a critical role in the successful outcome of the rescue operation.

Risks Involving Railcars

Railcars pose specific dangers to workers. Workers may, for example, enter railcars to inspect or repair components such as valves, fittings, or internal linings. During loading or unloading operations, workers may need to access the interior of railcars to connect/disconnect hoses, check for leaks, or monitor product transfer. A tank car that is experiencing a problem may be loaded with toxic chemicals or contaminated with gases.

Under certain conditions, the interiors of railcars are considered to be confined spaces, so workers who enter them must be trained in confined-space entry procedures, which may include obtaining permits, conducting atmospheric testing, and implementing lockout/tagout procedures. They must also wear appropriate PPE, such as respiratory protection, chemical protective clothing, gloves, and eye protection, depending on the hazards present.

Various regulatory agencies, including OSHA and the Federal Railroad Administration (FRA), have specific regulations and standards governing confined-space entry in railcar operations. These regulations aim to protect workers from hazards associated with confined spaces and ensure compliance with safety protocols.

Communication protocols must be established between workers inside a railcar and those outside of it, and the car's atmospheric conditions must be constantly monitored. Rescue procedures and emergency response plans must be in place, including rapid access to rescue equipment and personnel trained in confined-space rescue.

This successful rescue exemplifies the challenges and complexities associated with confined-space rescue operations in industrial settings. Through effective emergency response, specialized training, and coordinated efforts, responders were able to safely rescue and treat the affected workers while mitigating further risks posed by the hazardous materials. This case underscores the importance of proactive safety measures and continuous improvement in emergency preparedness for similar incidents in the future.

1. What caused the need for rescue operations in the railcar incident?
   - **a.** A fire broke out in the railcar.
   - **b.** Workers were trapped after the railcar collapsed.
   - **c.** The railcar collided with another train.
   - **d.** A mechanical failure resulted in the release of toxic fumes.
2. Which specialized team was called to respond to the chemical release?
   - **a.** a search and rescue team
   - **b.** a hazardous materials team
   - **c.** the Environmental Protection Agency
   - **d.** the Federal Railroad Administration
3. What critical safety measure was implemented during the rescue operations?
   - **a.** continuous atmospheric monitoring inside the railcar
   - **b.** installation of fire-suppression systems
   - **c.** sealing off the entire rail yard
   - **d.** powering down all electrical systems inside the railcar
4. What was an essential component of the post-rescue process?
   - **a.** ventilation of the entire rail yard
   - **b.** a thorough debriefing process with workers and rescuers
   - **c.** a thorough decontamination process for workers and rescuers
   - **d.** evacuation of all nearby communities
5. What do regulatory agencies such as OSHA and the FRA require for confined-space entry in railcars?
   - **a.** Highly trained technical rescue personnel must be onsite.
   - **b.** Rescue procedures and emergency response plans must be in place.
   - **c.** All railcar components must be inspected.
   - **d.** A medical team must be onsite.

CHAPTER

8

Operations

Physical and Environmental Hazards

KNOWLEDGE OBJECTIVES

After studying this chapter, you should be able to:

- Describe how to identify space-specific hazards. (**NFPA 1006 7.2.4**, pp. 200–201)
- Identify and control energized and physical hazards. (**NFPA 1006 7.2.5**, pp. 201–204)
- Describe the hazardous atmospheres associated with confined-space incidents. (**NFPA 1006 7.2.9** [horizontal spaces] and **7.2.13** [vertical spaces], pp. 205–207)
- Discuss the process for monitoring the environment. (**NFPA 1006 7.2.3**, p. 205)
- Explain what documentation should be completed during air-monitoring operations. (**NFPA 1006 7.2.3**, pp. 214–216)
- Describe the types of operation of hazard-specific monitoring devices. (**NFPA 1006 7.2.9** [horizontal spaces] and **7.2.13** [vertical spaces], pp. 210–213)
- Discuss the importance of continued atmospheric monitoring. (**NFPA 1006 7.2.10** [horizontal spaces] and **7.2.14** [vertical spaces], pp. 207–210)
- Explain the process for managing exposures from hazardous materials. (**NFPA 1006: 7.2.5**, p. 218; pp. 220–221)
- Explain immediate and ongoing monitoring needs of the space, as well as victim conditions. (**NFPA 1006 7.2.4**, p. 218; pp. 220–221)
- Describe the capabilities and limitations of a self-contained breathing apparatus (SCBA). (**NFPA 1006 7.2.6**, pp. 220–221)
- Discuss the need for the rescue entrant to communicate their air levels to outside rescuers. (**NFPA 1006 7.2.6**, p. 221)
- Explain the procedures for operating ventilation equipment. (**NFPA 1006 7.2.5**, pp. 216–219)
- During a size-up, gather all pertinent information and relay it to the incident commander. (**NFPA 1006 7.2.3**, pp. 224–225)

SKILL OBJECTIVES

- Identify immediate hazards. (**NFPA 1006 7.2.4**, pp. 200–201)
- Identify and mitigate hazards. (**NFPA 1006 7.2.5**, pp. 200–204)
- Demonstrate the process of monitoring a confined space. (**NFPA 1006 7.2.13** [vertical spaces], pp. 205–210)
- Calibrate monitoring equipment. (**NFPA 1006 7.2.3**, pp. 213–214)
- Acquire and analyze air samples from a confined-space incident. (**NFPA 1006 7.2.3**, pp. 205–206 and **7.2.5**, pp. 205–206)
- While using SCBA, demonstrate conservative breathing techniques. (**NFPA 1006 7.2.6**, pp. 221–223)
- Employ ventilation for a confined-space incident. (**NFPA 1006 7.2.5**, pp. 216–219)

Introduction

Confined spaces, such as storage tanks, silos, or underground tunnels, pose significant dangers because of their limited size and restricted access. The inherent risks in these environments stem from both physical and environmental hazards that can severely affect health and safety. Physically, confined spaces often present challenges, such as the space's configuration, the limits to its access, and entrapment hazards. The restricted movement within these spaces also complicates emergency response efforts, making it difficult to quickly evacuate or assist individuals in distress.

Environmental hazards in confined spaces can exacerbate these dangers. For instance, the presence of flammable or explosive materials or gases in such areas can lead to fires or explosions if proper precautions are not observed. The combination of these physical and environmental factors creates a perilous situation where effective safety measures and proper training are essential to mitigate the risks and protect those who work in or enter these confined environments.

Monitoring and managing hazardous atmospheres in confined spaces is crucial to ensuring safety and preventing accidents. Regular and thorough atmospheric testing is essential to detect dangerous levels of toxic gases, oxygen deficiency, or flammable substances. Advanced detection equipment can provide real-time data on the presence of harmful gases, such as carbon monoxide, and measure oxygen levels to ensure they remain within safe limits. With continuous monitoring of these parameters, workers can be alerted to hazardous changes in the environment, enabling timely evacuation or corrective measures. Stringent protocols for pre-entry testing and continuous monitoring throughout the duration of work can significantly reduce the risk of exposure to hazardous atmospheres.

Effective management of hazardous atmospheres involves both proactive and reactive strategies. Proactively, safety measures such as ensuring proper ventilation and using personal protective equipment (PPE) can help control and mitigate atmospheric hazards. Reactive strategies include establishing emergency response plans and ensuring that workers are trained to recognize and respond to signs of atmospheric changes or contamination. Proper ventilation systems, such as mechanical fans or blowers, can be employed to dilute or remove harmful substances, and standby personnel should be prepared to assist in emergencies. By combining vigilant monitoring with robust management practices, the risks associated with hazardous atmospheres in confined spaces can be significantly minimized, protecting both workers and the integrity of the work environment.

Identifying and Controlling Hazards

Identifying Hazards

Examples of immediate hazards encountered in confined spaces include atmospheric hazards (e.g., toxic gases, oxygen deficiency, and combustible gases), physical hazards (e.g., the space's configuration, the limits to its access, and entrapment hazards), and biological hazards (e.g., mold, bacteria, and other pathogens). They can be identified using a variety of means, including visual cues (e.g., color changes or visible contaminants), sensory cues (e.g., unusual odors, labored breathing, dizziness, or nausea), and equipment readings (e.g., atmospheric monitor alarms, data from documents held by the facility [e.g., safety data sheets (SDSs) or previous entry permits], or other data generated by equipment). In addition to visual and sensory cues, advanced detection tools such as infrared cameras, photoionization detectors, and specific gas-detection tubes should be used to identify and quantify hazards more accurately. Prioritizing these hazards based on their immediacy and potential severity is crucial, with life-threatening conditions such as oxygen deficiency requiring immediate action.

A thorough assessment of the confined space is necessary to systematically identify and evaluate immediate hazards and their potential impact on rescuers and victims. Once they are identified, their mitigation must be prioritized based on their relative severity and urgency, taking into account the immediate safety needs of all personnel involved in the rescue operation.

Engulfment Hazards

Engulfment is defined as the surrounding and effective capture of a person by a liquid or finely divided (flowable) solid substance that can be aspirated and cause death by filling or plugging the respiratory system or that can exert enough force on the body to cause death by strangulation, constriction, or crushing, as described in Chapter 1. Engulfment hazards, such as soil, sand, grain, or flowing liquids, present unique challenges and require specialized training, equipment, and procedures.

A thorough pre-operation risk assessment of the confined space should yield information about engulfment hazards that may be present. Once they have been identified, consideration should be given to various factors about the hazards, such as the type of material that is involved, its depth and volume, the stability of the surrounding environment, and any other potential hazards that are present.

Engulfment hazard control measures include shoring, shielding, or trenching to prevent further collapse or movement of the material (**FIGURE 8-1**). Atmospheric monitors must also be deployed to monitor the space's atmospheric conditions during the efforts to mitigate the engulfment hazard.

Some rescues involving engulfment hazards require specialized rescue techniques, such as manual extraction or vacuum excavation equipment, to safely remove victims. It is not advised to attempt to remove a victim from an engulfing material using a mechanical advantage system because doing so can cause further injury or death, depending on the type of engulfing material.

FIGURE 8-1 A grain bin rescue kit being used to rescue an engulfed worker.

Controlling Energized and Physical Hazards

Energized hazards (also known as *electrical hazards*) are common in confined spaces, especially in industrial settings. They include exposed wires, energized equipment, and electrical machinery. Additional physical hazards include thermal hazards, such as steam-heating jackets; chemical hazards, such as the heat and off-gassing that can be associated with chemical reactions; and kinetic hazards, such as stored springs, pneumatic lids, or de-energized mixer blades.

NFPA 70E: Standard for Electrical Safety in the Workplace (National Fire Protection Association [NFPA] 2024a), which provides guidelines for identifying and mitigating electrical hazards in confined spaces, outlines safe work practices around energized hazards, including lockout/tagout procedures and PPE requirements, to protect rescuers against electrical shocks and **arc flash** incidents, which occur when light and heat are produced during an electrical explosion or discharge.

Lockout/tagout procedures for controlling energized hazards involve isolating energy sources, locking out equipment, and applying tags to prevent accidental start-up during rescue operations. "The Control of Hazardous Energy (Lockout/Tagout)" standard (1989) of the Occupational Safety and Health Administration (OSHA) in the *Code of Federal Regulations* (C.F.R.; 29 C.F.R. 1910.147) outlines requirements for implementing lockout/tagout procedures to control hazardous energy sources and protect workers from injury. More information on lockout/tagout can be found in Chapters 1 and 2.

Physical falling hazards, such as unguarded edges, open pits, and unstable surfaces, pose a significant risk in confined spaces. Rescuers must identify and mitigate these hazards to prevent falls and injuries. OSHA's "Duty to Have Fall Protection" standard (1979; 29 C.F.R. 1926.501) and *NFPA 2500: Standard for Operations and Training for Technical Search and Rescue Incidents and Life Safety Rope and Equipment for Emergency Services* (NFPA 2022c) provide guidance on fall protection measures, including the use of guardrails, safety nets, and personal fall arrest systems.

Guarding and barricading involve the use of barriers, cones, and signage to restrict access to hazardous areas and prevent unauthorized entry. OSHA's "Safety and Health Regulations for Construction" (1979; 29 C.F.R. 1926) and *NFPA 350: Guide for Safe Confined Space Entry and Work* (NFPA 2022a) provide recommendations for establishing barriers and signage to control access to confined spaces and protect workers from physical hazards. To better understand the process of identifying and mitigating hazards, follow the steps in **SKILL DRILL 8-1**.

SKILL DRILL 8-1

Identifying and Mitigating Hazards

Identify Hazards

Step 1: Cluttered Entrance

The entrance to this confined space is obstructed with equipment and debris that pose a trip hazard.

Step 2: Poor Ventilation

This confined space has no ventilation system. The absence of fans, ducts, and air circulation mechanisms leads to stagnant air and unsafe working conditions.

Step 3: Limited Visibility

Insufficient lighting makes it difficult to see all areas clearly. Shadows and dark corners increase the potential for accidents.

Step 4: Hazardous Materials

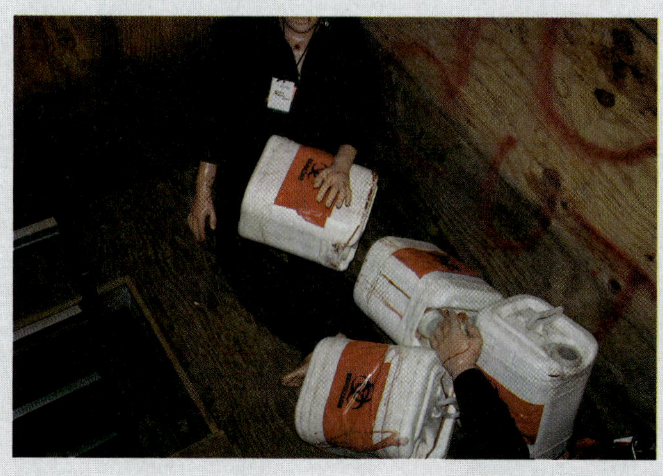

Chemical containers indicate the presence of hazardous materials. Proper handling protocols appear to be lacking.

SKILL DRILL 8-1 CONTINUED

Identifying and Mitigating Hazards

Step 5: Structural Hazards

Structural risks, such as unstable ladders, rusted beams, and exposed wiring, present a danger to personnel inside the space.

Steps 1, 2, 4, 5: Courtesy of Illinois Fire Service Institute.

Mitigate Hazards

Step 1: Clear Entrance

A confined-space entrance that is free of clutter and debris allows for easy and safe access. Warning signs are visible, and an emergency exit is unobstructed.

Step 2: Adequate Ventilation

A ventilation system with fans and ducts installed inside the space ensures constant airflow. An air quality monitor also improves worker safety.

Continues.

SKILL DRILL 8-1 CONTINUED

Identifying and Mitigating Hazards

Step 3: Enhanced Lighting

Bright lighting installed in the confined space eliminates dark corners and shadows, improving visibility. Portable lights are strategically placed for added safety.

Step 4: Safe Handling of Hazardous Materials

Hazardous materials are safely stored in marked containers, and any spillage has been cleaned up. Spill kits are easily accessible for quick response to future spills.

Step 5: Secured Structural Elements

Structural hazards have been addressed, with ladders properly secured, beams reinforced, and wiring covered with protective insulation. Safety checks are performed by trained personnel.

Steps 1 to 5: Courtesy of Illinois Fire Service Institute.

Monitoring and Managing Hazardous Atmospheres

Environmental Air Monitoring

Environmental air monitoring is the systematic process of detecting and measuring potential atmospheric hazards, such as toxic gases, flammable vapors, oxygen levels, and other contaminants. Rescuers create air-monitoring plans based on a space's identified risks and regulatory requirements.

The first step in an air-monitoring plan is to establish work zones—a "cold zone" for staging and command, a "warm zone" for staging of backup teams, and a "hot zone" for the entry point.

After this step, the air outside the confined space must be monitored to establish a baseline atmospheric condition. The air should be evaluated for its oxygen level, combustible gas concentrations, and toxic gas levels. Rescuers must assess what contaminants may be present and ensure gas monitors have the ability to detect those contaminants. Testing around the "hot zone" is especially important if gaining access to the space involves opening a hatch or manhole. These tests will reveal whether a flammable or explosive environment is present.

Next, air sampling with remote pick-up probes should be utilized to test varying points within the space. A full-volume atmospheric test must be completed; "Permit-Required Confined Spaces" (1993; OSHA 29 C.F.R. 1910.146) requires air monitoring to be completed in four-foot (1.2-m) increments. They should be positioned in areas where atmospheric hazards are most likely to accumulate, such as low spots, vents, or areas adjacent to potential emission sources. Each entry team should be provided with a monitor as part of the air-monitoring plan.

Any alarm activations or deviations from safe air quality thresholds must be acted upon immediately. However, alarms are not always the first warning of danger. For example, a 1 percent drop in oxygen levels indicates a 5 percent infiltration of another contaminant. Although such a shift in oxygen levels may not set off an alarm, it does indicate that a problem is present in the space.

If hazardous conditions are detected, personnel present in the space must be evacuated, and the space must be ventilated. If possible, the emission sources of any gases must be controlled. See **TABLE 8-1** for a description of various aspects of atmospheric contaminants and hazards found in confined spaces.

TABLE 8-1 Understanding Contaminants and Hazards

Aspect	Importance and Application
Density	Helps predict contaminant behavior within confined spaces. Heavier-than-air contaminants settle at lower levels, posing inhalation risks. Lighter-than-air contaminants accumulate in upper areas.
Toxicity	Critical for assessing health risks associated with exposure. Highly toxic substances can cause immediate health effects, requiring prompt protective measures.
Behavior under conditions	Contaminants exhibit varied behaviors based on temperature, pressure, and ventilation. Some chemicals become more volatile or reactive under specific conditions, increasing exposure risks.
Interpreting results	Understanding contaminant properties aids in accurate interpretation of monitoring data. Responders can assess the significance of readings and take appropriate control measures based on this knowledge.
Assessing risks	Combining contaminant properties with monitoring data allows for comprehensive risk assessment. This informs decision-making on entry procedures, protective gear, and emergency response strategies to safeguard personnel.

Gases Found Within Confined Spaces

Common gases can pose serious hazards in confined spaces. Atmospheric monitors are used to verify their presence; once they are found, strategies to remove or mitigate their risks must be implemented.

Perhaps the most common atmospheric hazards found in confined spaces are related to the presence of oxygen—whether the air is deficient in oxygen (perhaps because of displacement by other gases or chemical processes) or oversaturated with it (which increases the risk of burning if ignited). Explosion and fire risk can also be caused by the presence of flammable gases, such as methane, propane, hydrogen, and other hydrocarbons, which can ignite at specific concentrations. Toxic gases, such as hydrogen sulfide, carbon monoxide, ammonia, chlorine, and other chemicals present in industrial processes or decomposition, also pose atmospheric risks.

As mentioned previously, atmospheric monitors are used to detect the presence of flammable and toxic gases, and oxygen monitors are used to ensure the space's air maintains a safe oxygen level (typically 19.5 percent to 23.5 percent). If gases are detected or the oxygen level is determined to be unsafe, mechanical ventilation (e.g., blowers or fans) can be used to bring fresh air into the space and dilute the gases. Once a gas has been displaced or removed from the space, the source of its emission must be shut off or isolated to prevent its reappearance.

Another method, **purging**, involves using inert gases, such as nitrogen, to displace harmful gases from the space. However, just because any harmful gas has been displaced with nitrogen does not mean that the air is not safe. The oxygen has been displaced as well.

Workers and rescuers must have appropriate PPE for atmospheric hazards—specifically, respiratory protection—depending on the type and concentration of the gases present.

Isolating Confined Spaces for Gas Removal

The first step in isolating a confined space for gas removal is to identify all entry points leading into the confined space, including doors, hatches, or openings. All access points must be securely closed and sealed to prevent gases from entering. Physical barriers, such as barricades, locks, or covers, can seal off entry points and prevent unauthorized access. These barriers must be robust and capable of withstanding a variety of environmental conditions.

Disconnect or isolate external sources that could introduce gases into the confined space, such as pipelines, ventilation systems, or equipment that could be sources of hazardous vapors. Close and seal any vents, ducts, or openings connected to the confined space to prevent the circulation of air-carrying contaminants.

Warning signs or labels indicating the isolation status of the confined space must be posted to alert personnel and prevent unauthorized entry. Before allowing entry into the confined space, conduct thorough checks to verify that the isolation measures are effective and that the atmosphere inside the space is safe for human occupancy.

Throughout the isolation process, atmospheric monitoring equipment must be used to continuously monitor the air and ensure that hazardous gases are effectively isolated and safe conditions are maintained. If applicable, lockout/tagout procedures should be deployed to de-energize and secure equipment or machinery associated with the confined space to prevent accidental activation or release of hazardous substances.

Configuring a Confined Space for Optimal Environmental Monitoring

Configuring a confined space for optimal air monitoring involves consideration of various factors that influence the distribution of contaminants and atmospheric conditions within the space. These factors include the space's dimensions, entry points, and potential barriers that affect environmental-monitoring efforts.

Variances in the size, shape, and structural features of confined spaces directly affect the distribution and concentration of gases, vapors, and other contaminants present within them. The size and shape determine how gases disperse and accumulate. Larger spaces may have different airflow patterns than smaller ones, and ventilation fans can move different amounts of air per minute depending on their type. For these reasons, appropriate tools must be selected to create the correct levels of environment exchanges.

Entry points, including access hatches, doors, or openings, influence the flow of air and contaminants into and out of the confined space. Monitoring should prioritize areas near entry points because of the potential air movement. Structural barriers, such as equipment, machinery, or storage containers, can create stagnant zones where contaminants may accumulate.

Air-sampling locations should be strategically identified based on the space's dimensions and entry points to ensure that representative samples reflecting the actual conditions throughout the confined space are obtained. Monitoring efforts should be specifically focused on areas with potential barriers or stagnant zones to detect localized concentrations of contaminants. The airflow dynamics of the space's configuration, as well as the work and breathing zones of victims and rescuers, must be considered in order to monitor readings accurately and project where gases may be dispersed within the space.

According to OSHA standard 1910.146, "Permit-Required Confined Spaces" (1993), the atmosphere should be tested every four feet (1.2 m), both vertically and horizontally, in the direction of travel. Monitoring can be performed with a sampling hose every four feet (1.2 m) for vertical entries before the first rescuer enters (once inside, the rescuer will wear a monitor for ongoing monitoring purposes). All monitors are different; each has a set amount of time to pull a sample through the hose to get the measurement for each sample at these four-foot (1.2-m) intervals. Each monitor also has its own specifications for how much hose can be attached for sampling purposes. Rescuers must be familiar and proficient with the monitor they use.

Continuous Atmospheric Monitoring

In confined spaces, atmospheric conditions can fluctuate rapidly as a result of ventilation changes, chemical reactions, and the movement of personnel. Continuous monitoring enables rescuers to detect fluctuations in gas levels and take immediate action to address any deviations from safe levels. Real-time monitoring enables the team to assess the air quality and implement appropriate control measures.

Continuous monitoring provides early warning of hazardous conditions, allowing rescuers to evacuate the area if necessary to ensure their safety. Victims trapped in confined spaces are particularly vulnerable to atmospheric hazards. Ongoing atmospheric monitoring is essential to assess the air quality around the victim and determine the appropriate measures for their rescue and medical treatment.

Advances in technology have led to the development of reliable portable atmospheric monitoring devices. Rescuers should be trained in the use of these devices and understand their capabilities and limitations for effective monitoring. To better understand the process of monitoring a confined space, follow the steps in **SKILL DRILL 8-2**.

SKILL DRILL 8-2

Monitoring a Confined Space

Initial Setup and Equipment Check

Step 1: Prepare Equipment

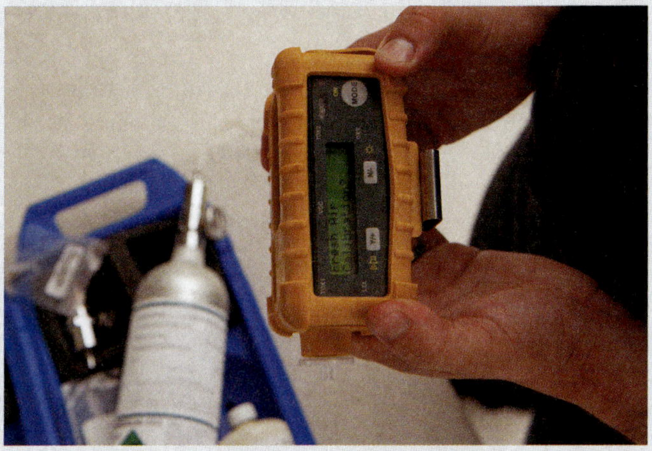

Prepare monitoring equipment, including single- and multi-gas monitors. Ensure that the batteries are charged and the sensors are properly calibrated.

Step 2: Calibrate Monitors

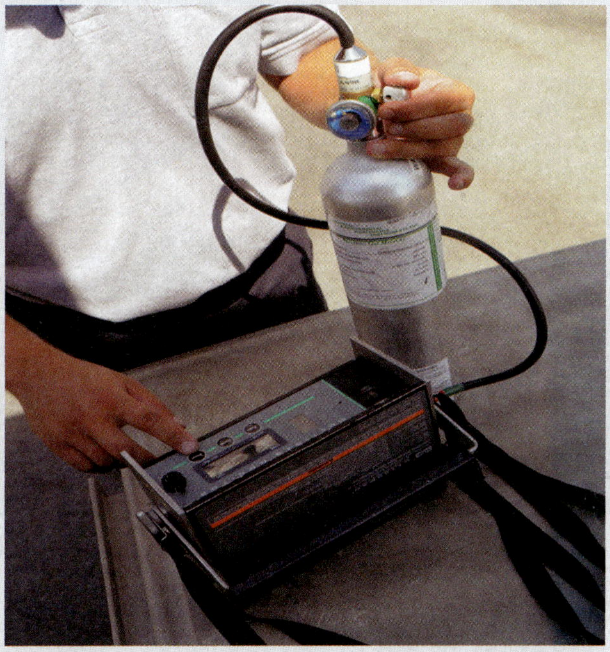

Calibrate your gas monitors using a calibration gas cylinder and a docking station. The monitor's display shows that the calibration process is in progress, ensuring the accuracy of readings.

Continues.

SKILL DRILL 8-2 CONTINUED

Monitoring a Confined Space

Single-Gas Monitor Usage

Step 1: Enter the Confined Space with Single-Gas Monitor

This monitor shows real-time concentration levels of a single gas as you proceed cautiously into the confined space.

Step 2: Check Oxygen Levels

The single-gas monitor display shows oxygen levels in the space. Compare the reading to acceptable safety standards to ensure that the oxygen levels are within safe limits for entry.

Step 3: Assess the Situation

If the single-gas monitor's display shows elevated levels of a toxic gas, pause to assess the situation and determine if it is safe to continue.

Step 1: © noomcpk/Shutterstock; **Step 2:** © Oil and Gas Photographer/Shutterstock; **Step 3:** © King Ropes Access/Shutterstock

SKILL DRILL 8-2 CONTINUED

Monitoring a Confined Space

Multi-Gas Monitor Usage

Step 1: Enter with a Multi-Gas Monitor

A multi-gas monitor shows real-time data for multiple gases detected simultaneously as you proceed cautiously into the confined space.

Step 2: Check the Monitor Display

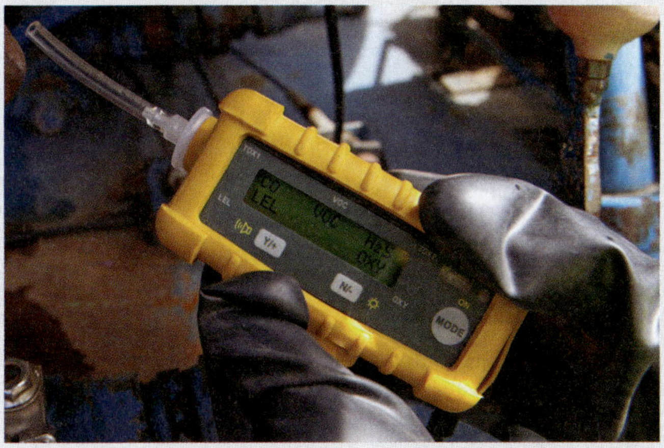

A multi-gas monitor display shows readings for oxygen, carbon monoxide, hydrogen sulfide, and combustible gases, ensuring awareness of all potential gas hazards in the space.

Step 3: Monitor Multiple Gases Simultaneously

Use the multi-gas monitor to check different areas of the confined space for potential hazards.

Continues.

SKILL DRILL 8-2 CONTINUED

Monitoring a Confined Space

Documentation and Safety Protocols

Step 1: Record Gas Monitor Readings

Record gas monitor readings in a log or an electronic device, documenting the data along with the time and specific location within the confined space for safety reference.

Step 2: Communicate Findings

Communicate gas-monitoring results to the safety team outside the confined space using a two-way radio. Based on the findings, safety protocols will be reviewed and updated as necessary.

Steps 1 and 2: © King Ropes Access/Shutterstock

Hazard-Specific Monitoring Devices

Atmospheric monitors are designed to detect and measure the concentration of gases in the atmosphere. There are several types, including the following:

- **Single-gas monitors:** Also known as *single-gas detectors*, these devices are designed to detect and measure the concentration of a single gas, such as oxygen, hydrogen sulfide, carbon monoxide, or combustible gases.
- **Multi-gas monitors:** Also known as *multi-gas detectors*, these devices can detect and measure the concentration of multiple gases simultaneously, providing comprehensive monitoring in confined spaces where multiple hazards may be present. They are the most widely used monitors because of their relative cost and their ability to monitor multiple gases simultaneously.

Oxygen monitors, which are also known as *oxygen-deficiency monitors* or *oxygen-deficiency indicators*, measure the concentration of oxygen in the atmosphere (**FIGURE 8-2**). These devices are used to identify oxygen-deficient (<19.5 percent) or oxygen-enriched (>23.5 percent) environments, which can pose significant risks to personnel.

Combustible gas detectors are specifically designed to detect the presence of flammable gases or vapors in the atmosphere (**FIGURE 8-3**). These devices utilize various detection technologies, such as catalytic bead sensors, infrared sensors, or semiconductor sensors, to detect combustible gases, such as methane, propane, and hydrogen, and **volatile organic compounds (VOCs)**, which are organic compounds that have a high vapor pressure and low water solubility at room temperature, which allows them to evaporate into air.

Atmospheric monitors (**FIGURE 8-4**) assess the overall air quality in confined spaces by measuring parameters such as **particulate matter (PM)** (a mixture of solid particles and liquid droplets found in air), VOCs, carbon dioxide, and other airborne contaminants. These

FIGURE 8-2 An oxygen monitor.

© Oil and Gas Photographer/Shutterstock

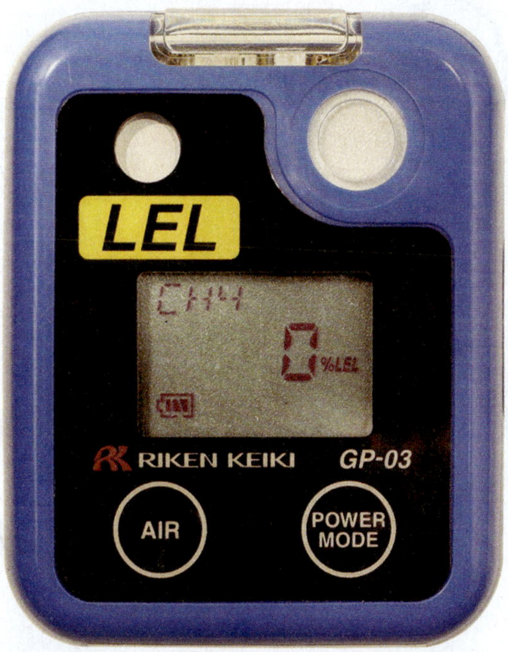

FIGURE 8-3 A combustible gas detector.

Courtesy of RKI Instruments, Inc.

devices provide valuable information about the presence of pollutants and potential respiratory hazards in the environment.

Single-gas monitors detect and measure the concentration of a single gas, such as oxygen, combustible gases (lower explosive limit [LEL]), hydrogen sulfide, or carbon monoxide. Multi-gas monitors are specialized monitoring systems designed for use during confined-space entry operations. These systems typically combine multiple sensors and monitoring capabilities, including gas detection, oxygen monitoring,

FIGURE 8-4 A four-gas atmospheric monitor.

© Jones & Bartlett Learning. Courtesy of MIEMSS.

and atmospheric monitoring, to provide comprehensive safety monitoring for confined-space work. Their compact and portable design makes them suitable for confined-space entry, and they feature a user-friendly interface, along with audible and visual alarms for quick notification of dangerous conditions. On the other hand, they are more costly than single-gas monitors, and they require regular maintenance and calibration to ensure the accuracy and reliability of their readings. They offer limited capabilities of detecting nonstandard gases, and their sensors may have specific environmental and operational limitations (e.g., temperature range, humidity). Never rely on an LEL sensor to determine the toxicity of a substance that is present.

Variances in readings for uncommon gases can be beneficial. Users should be aware of the baseline gas for the monitor they use. Pentane is a commonly used gas for calibration; a conversion chart is available for less common gases.

Personal gas detectors are wearable devices designed to provide individual personnel with real-time alerts and alarms in the event of hazardous atmospheric conditions (**FIGURE 8-5**). They are worn by personnel entering confined spaces and provide audible, visual, or vibrating alerts when hazardous conditions are detected. At least one monitor should be used with each entry team.

Photoionization detectors (PIDs) are used to detect VOCs and other hazardous substances based

FIGURE 8-5 A personal gas detector.

© Yevgeniysh/Shutterstock

FIGURE 8-6 A photoionization detector (PID).

© Red_Shadow/Shutterstock

on their ability to ionize when exposed to ultraviolet (UV) light (**FIGURE 8-6**). PIDs use UV light to ionize gas molecules, creating positively charged ions and free electrons. The resulting ionization current is proportional to the gas concentration. PIDs are sensitive to a wide range of VOCs and certain gases with ionization potentials below the energy of the UV lamp. They are commonly used for detecting and measuring VOCs in various industries, including environmental monitoring, industrial hygiene, HazMat response, confined-space rescue, and indoor air quality assessments. PIDs offer a rapid response time, real-time measurements, and sensitivity to a broad range of VOCs, but they have limited detection capability for non-VOC gases and are subject to potential interference from water vapor and other compounds.

Flame ionization detectors (FIDs) use a hydrogen flame to ionize carbon-containing compounds in a gas sample (**FIGURE 8-7**). The resulting ions produce a measurable electrical current. FIDs are highly sensitive to hydrocarbons and are often used for detecting and quantifying VOCs, including methane, ethane, benzene, and other hydrocarbons. They are commonly used in environmental monitoring, industrial emissions testing, and gas chromatography applications to analyze organic compounds. In specific circumstances, they can be used for confined-space rescue, especially if a known hydrocarbon is present. Their benefits include a high sensitivity to hydrocarbons, making them excellent for quantifying complex mixtures of VOCs. However, they require a hydrogen source for operation and are primarily intended to detect hydrocarbons.

pH paper, also known as *litmus paper* or *pH indicator paper,* is a simple and inexpensive tool used to measure the acidity or alkalinity (pH) of a solution (**FIGURE 8-8**). When pH paper is dipped into a solution, it changes color based on the pH of the solution, allowing users to quickly determine whether the solution is acidic, neutral, or alkaline. It can also detect the presence of corrosive gases when the paper is dampened with water.

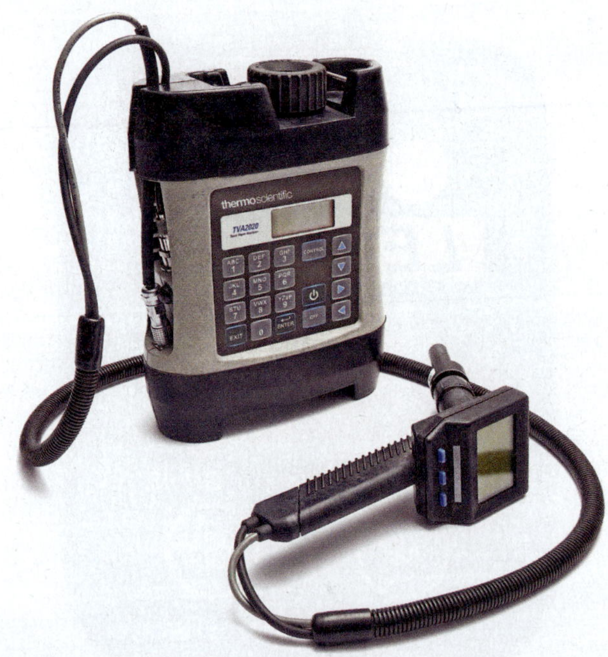

FIGURE 8-7 A flame ionization detector (FID).

Image courtesy of The Thermo Scientific™ TVA2020 Toxic Vapor analyzer.

Colorimetric tubes provide instantaneous measurement of specific gases by drawing air through a tube containing a chemical reagent that changes color in the presence of a specific gas or vapor (**FIGURE 8-9**). The length of the color stain correlates with the gas concentration. The tubes are available for a wide range

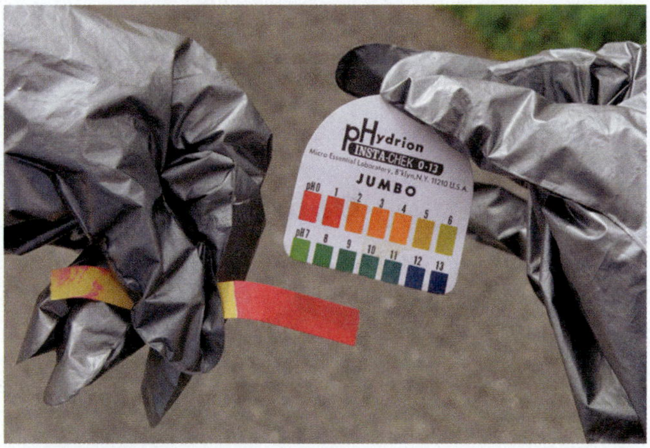

FIGURE 8-8 pH paper.

FIGURE 8-9 A colorimetric tube.

of gases, including toxic gases (e.g., hydrogen sulfide, chlorine), combustible gases (e.g., methane, propane), and specific chemicals (e.g., ammonia, formaldehyde). They are used for qualitative gas detection in field settings, emergency response, and industrial hygiene assessments where rapid, onsite measurements are required. Although they provide an immediate visual indication of gas presence and are simple to use, with no power source or calibration needed, they offer only semiquantitative measurements, and their use is limited to specific gases. Additionally, the tubes have a limited shelf life.

Colorimetric tubes and colorimetric chips are both used for detecting and measuring the presence of hazardous gases, but they differ in their design and application. Colorimetric tubes are glass tubes filled with a reagent that changes color when exposed to specific gases. They are typically used with a hand pump, drawing a known volume of air through the tube to measure gas concentrations based on the color change along a marked scale. Colorimetric chips, on the other hand, are compact digital sensor devices that use similar chemical reactions to detect gases but provide an electronic readout of the gas concentration. Unlike tubes, chips allow for continuous monitoring and are more user-friendly for ongoing assessments, whereas tubes are generally better suited for one-time spot checks.

Choosing the Right Air-Monitoring Equipment

PIDs are suitable for detecting a broader range of VOCs and have faster response times compared with FIDs. FIDs are more specific to hydrocarbons and are often used in environmental and industrial applications requiring precise measurement of organic compounds.

PIDs focus on VOC detection, whereas multi-gas monitors provide simultaneous monitoring of oxygen levels, combustible gases, carbon monoxide, and hydrogen sulfide, making them ideal for confined-space entry and industrial safety.

Colorimetric tubes offer immediate visual detection but are limited to specific gases and provide semiquantitative results. Electronic detectors such as PIDs and four-gas meters provide real-time measurements and continuous-monitoring capabilities.

The right choice of gas-detection equipment for a particular incident has many dependencies: the specific application, the targeted gases, and the desired level of measurement accuracy. PIDs and FIDs excel in VOC and hydrocarbon detection, multi-gas monitors provide comprehensive monitoring in confined spaces, and colorimetric tubes offer rapid qualitative screening for specific gases.

Calibrating Air-Monitoring Equipment

Calibration of air-monitoring equipment must be performed by trained personnel using certified calibration gases and following established procedures. This process is necessary to ensure the accuracy and reliability of measurements taken in confined-space rescue operations. Proper calibration involves adjusting the equipment to a known reference standard to verify its accuracy and performance. Regular maintenance and calibration schedules ensure that air-monitoring equipment remains accurate and reliable.

The accuracy of air-monitoring equipment can drift over time as a result of various factors, such as usage, environmental conditions, or exposure to contaminants. Regular calibration helps maintain the integrity of measurements, ensuring that the equipment provides reliable data for assessing atmospheric conditions.

The first step is to select a calibration gas that matches the target gas or gases to be measured by the monitoring equipment. Calibration gases are typically certified mixtures of specific concentrations of gases (e.g., oxygen, carbon monoxide) that are traceable to national standards.

Next, ensure the air-monitoring equipment is properly set up and connected to calibration gas cylinders according to the manufacturer's instructions. Confirm that all connections are secure and free from leaks.

The best practice is to follow the equipment manufacturer's instructions to initiate the calibration process. This usually involves activating the calibration mode or initiating a calibration sequence on the equipment. Generally, manufacturer recommendations also call for allowing the equipment to stabilize and warm up before it is put into use.

Once the equipment is warmed up and ready for calibration, introduce the calibration gas into the air-monitoring equipment through its calibration port or inlet. Adjust the flow rate of the calibration gas to ensure a steady and accurate supply to the equipment. The equipment's calibration controls are used to adjust settings based on the known concentration of the calibration gas. Some equipment may require adjustments to zero settings or span settings to match the calibration gas concentration.

As the calibration process proceeds, the equipment analyzes the calibration gas and adjusts its internal settings based on the detected concentration. Be sure to verify that the equipment displays the correct readings that correspond to the known concentration of the calibration gas.

Once the calibration process is complete, the results must be recorded. Proper documentation of the calibration process includes the calibration gas used, the concentration applied, and the equipment's response. The results should be recorded in a calibration log or maintenance record so that the equipment's performance can be tracked over time.

Typically, air-monitoring equipment used in confined-space rescue operations should be calibrated regularly, with more frequent instances performed for critical applications or in demanding environments. Best practices include performing bump testing at the beginning of each work day and before using monitors at a confined-space incident.

Air-Monitoring Documentation

Documentation during air-monitoring operations within a confined space is crucial for maintaining an accurate record of atmospheric conditions, monitoring results, corrective actions, and overall safety measures implemented during confined-space entry and work. It helps ensure compliance with safety regulations, facilitates incident analysis, and supports continuous improvement of confined-space safety practices. OSHA regulations require the documentation of the results of initial and periodic tests performed, accompanied by the names or initials of the testers and an indication of when the tests were performed. **FIGURE 8-10** shows a template for a confined-space air-monitoring report.

Standard pre-entry documentation includes a risk assessment report and a monitoring plan. The risk assessment report should include all results of the assessment, including identified hazards and the control measures that will be used to mitigate them. It should also outline what gases or contaminants are planned to be monitored based on the space's characteristics and potential risks. The monitoring plan should specify all monitoring locations, the parameters that will be measured (e.g., oxygen levels, LEL, specific toxic gases), the monitoring equipment that will be used, the contingency measures, and the response protocols.

The documentation must also include all data that will be collected, including baseline monitoring data taken from outside the confined space before entry, initial readings of oxygen levels, LEL, and any other relevant parameters. It must also capture all continuous-monitoring data captured during confined-space entry and work activities, time-stamped readings from atmospheric monitors placed inside the space, alarms, deviations from safe thresholds, and hazardous conditions observed during monitoring.

Incident reports, which describe any incidents or deviations from safe conditions identified during air-monitoring operations and the response actions taken, such as evacuation procedures, additional ventilation measures, or other corrective actions, are also key pieces of atmospheric monitoring data that must be collected.

Date of Monitoring:	[*Date completed*]
Location:	[*Specific location within the confined space*]
Monitoring Team:	[*Names and roles of individuals conducting the air monitoring*]
Atmospheric Conditions	
Oxygen Level:	Initial Reading: [*Initial oxygen concentration percentage*] Final Reading: [*Final oxygen concentration percentage*] Acceptable Range: [*Specify acceptable range for oxygen concentration*]
Toxic Gas Levels:	Contaminant 1: [*Specify the contaminant*] Initial Reading: [*Initial concentration*] Final Reading: [*Final concentration*] Acceptable Range: [*Specify acceptable range*]
	Contaminant 2: [*Specify the contaminant*] Initial Reading: [*Initial concentration*] Final Reading: [*Final concentration*] Acceptable Range: [*Specify acceptable range*] *(Add more contaminants as applicable)*
Combustible Gas Levels:	Gas 1: [*Specify the gas*] Initial Reading: [*Initial LEL percentage*] Final Reading: [*Final LEL percentage*] Acceptable Range: [*Specify acceptable range*]
	Gas 2: [*Specify the gas*] Initial Reading: [*Initial LEL percentage*] Final Reading: [*Final LEL percentage*] Acceptable Range: [*Specify acceptable range*] *(Add more combustible gases as applicable)*
Observations	
Ventilation Status:	Ventilation System: [*Specify and describe the type of system used*] System Efficacy: [*Specify the effectiveness of the ventilation system in terms of improving air quality*]
Contaminant Sources:	Sources: [*Specify all sources contributing to air contamination*] Potential Impact: [*Specify their potential impact on air quality and safety*]
Equipment and Conditions:	Equipment Used: [*Specify and describe the atmospheric monitoring equipment used*] Environmental Conditions: [*Specify the environmental conditions (e.g., temperature, humidity) that affect the monitoring*]
Actions Taken	
Ventilation Adjustments:	Adjustments Made: [*Detail any adjustments made to the system*] Impact on Air Quality: [*Describe any impact ventilation had on the air quality and atmospheric conditions*]
Personnel Safety Measures:	Measures Implemented: [*Describe the safety measures implemented based on the results of the atmospheric monitoring*] PPE Used: [*Specify the PPE and other precautions used by personnel*]
Recommendations	
Further Monitoring:	Recommendations: [*Recommendations for ongoing air monitoring based on initial findings*] Targets for Ongoing Assessment: [*Specific areas or contaminants requiring continued assessment*]
Risk Mitigation Strategies:	Suggestions: [*Suggestions for additional controls or interventions to minimize risks*] Training: [*Training or procedural enhancements to improve safety during confine-space operations*]
Additional Notes	
Incident Summary:	Overview: [*Overview of the confine-space rescue incident and its impact on air quality*]
	Lessons Learned: [*Lessons learned and insights gained from air-monitoring activities*]

FIGURE 8-10 Template for a confined-space air-monitoring report.

Equipment maintenance records, such as equipment calibration, maintenance checks, sensor calibrations for atmospheric monitoring devices, equipment inspections, repairs, and calibration dates, must be retained to ensure the accuracy and reliability of the monitoring data that are gathered.

Following entry into the confined space, the team must maintain records of all monitoring data collected throughout operations. This includes changes in atmospheric conditions, work processes, or environmental factors that may affect air quality. All monitoring data and incident reports must be completed and signed by responsible personnel, and the documentation must be accurate, legible, and stored securely for future reference and regulatory compliance.

Ventilating a Confined Space

Ventilation helps to maintain a safe atmosphere within a confined space. Proper ventilation helps remove hazardous gases, contaminants, and potentially harmful atmospheric conditions, reducing the risk of asphyxiation or exposure-related injuries. Ventilation is typically achieved using ventilation fans (battery powered, gas powered, and electric) and requires careful maintenance and equipment checks to ensure effectiveness and safety. **TABLE 8-2** shows varieties of ventilation fans and the pros and cons of each.

Positive-pressure ventilation (PPV) involves the use of a fan to create a higher pressure inside the confined space than outside of it, thus pushing contaminants out through designated exit points. PPV is preferred because the source can be checked with monitoring equipment.

Negative-pressure ventilation (NPV) is a ventilation method that creates a lower pressure inside the space, drawing fresh air in through designated entry points and exhausting contaminated air through a controlled outlet. Negative-pressure exhaust ventilation is useful in situations involving flammable or toxic atmospheres and contaminants that are heavier than air. It requires the use of explosion-proof equipment that eliminates sources of ignition.

Ventilation is primarily used in confined spaces to introduce fresh, clean air into the confined space, diluting and displacing potentially hazardous contaminants. By controlling the airflow, ventilation helps to create a safe and breathable environment for rescuers and victims. Effective ventilation helps to remove harmful gases, vapors, or airborne particles from the confined space, reducing the risk of exposure and ensuring safe working conditions. The air should be changed in the space six to eight times per hour during a confined-space operation (see **Appendix E** for details on air changing and calculating the turnover rate of ventilation). Proper air changes should be maintained even if rescuers are using supplied-air respirators (SARs) or self-contained breathing apparatuses (SCBAs) because carbon dioxide will be continuously exhausted into the space and displace fresh air.

Different types of ventilation systems may be used based on the specific requirements of the confined-space rescue. Some need only natural ventilation, which utilizes natural air currents or openings (e.g., doors, windows) to introduce fresh air and exhaust contaminated air from the space. Mechanical ventilation involves the use of mechanical equipment such as vent fans, blowers, or air movers to actively circulate air and remove contaminants. Forced-air ventilation combines natural and mechanical ventilation methods to achieve optimal airflow and air quality within the space.

Ventilation directly affects air-monitoring operations in the following ways:

- **Dilution of contaminants:** Proper ventilation dilutes the concentration of contaminants present in the confined space, making it easier to obtain accurate readings using air-monitoring equipment.
- **Stabilization of airflow:** Ventilation helps to stabilize the airflow within the confined space, ensuring a consistent and uniform distribution of air for monitoring purposes.
- **Enhanced monitoring accuracy:** With improved air quality resulting from ventilation, monitoring equipment can provide more reliable and accurate readings of atmospheric conditions, including oxygen levels, toxic gas concentrations, and combustible gas levels.
- **Facilitation of safe entry:** Ventilation creates a safer environment for rescuers to enter the confined space, enabling them to conduct air-monitoring activities more effectively and with reduced risk.

During an operation, it's essential to continuously monitor the effectiveness of the ventilation. Atmospheric monitoring equipment can assess air quality in areas affected by ventilation. Readings should not be taken directly in the initial path of the ventilation to see the effect of the ventilation within the space.

TABLE 8-2 Pros and Cons of Different Types of Ventilation Fans

	Battery Powered	Electrically Powered	Gas Powered
Pros			
Portability	Highly portable; easy transport to remote or inaccessible spaces	Limited mobility compared to battery- or gas-powered fans	Highly portable; easily transported to remote or inaccessible places without need for electrical outlets or cables
Versatility	Versatile; usable in various confined-space environments	Limited by availability of electrical outlets or generators	Versatile; suitable for large spaces or environments with high air resistance
Independence	Operate independently of external power sources	Continuous operation dependent on a reliable power source	Operate independently of external power sources
Cons			
Limits to run time	Limited run time based on battery capacity	N/A	Requires steady supply of fuel; refueling during operations may be necessary
Cost	Higher up-front costs due to batteries and chargers	Lower operating costs compared with battery-powered fans	Higher maintenance costs for engine tune-ups, oil changes, and fuel system inspections
Power output	Lower power output compared with electric or gas-powered fans	Higher power output suitable for larger spaces	Higher power output suitable for larger spaces or environments with high air resistance

Based on the results of the monitoring, adjustments to ventilation settings or equipment can be made to optimize air quality and ensure the safety of personnel inside the confined space. Ventilation strategies should be integrated into the overall air-monitoring plan for confined-space rescue operations.

Operating Ventilation Equipment

Before starting ventilation operations, assess the confined space to determine the type and extent of atmospheric hazards present, such as toxic gases, flammable vapors, or low oxygen levels. Select the appropriate ventilation equipment based on the size and configuration of the confined space, as well as the specific hazards identified during the assessment. The ventilation equipment must be in good working condition, properly calibrated, and free from defects or damage. Ambient noise factors must also be considered when selecting, using, and placing ventilation devices.

The ventilation equipment should be strategically placed outside the confined space to ensure adequate airflow and distribution, and the ventilation equipment must be placed in a clean and safe area to avoid the possibility of introducing outside hazards into the space. Ducting or hoses can be used to direct airflow into the space, taking care to avoid obstructions or obstacles that may impede ventilation. To prevent movement or dislodgement during operation, the equipment must be secured in place using straps, anchors, or other securing devices as necessary. Special ventilation equipment that is nonsparking or intrinsically safe may be required in spaces with high concentrations of flammable materials.

Once the equipment has been positioned and is ready for use, it should be powered up and checked to ensure that all components, including motors, fans, filters, and control panels, are functioning properly. Gradually increase the airflow rate to establish a positive pressure differential between the confined space and the surrounding environment, drawing clean air into the space and exhausting contaminated air out. The equipment's built-in gauges or other monitoring devices should be used to maintain a safe atmosphere within the confined space. Flagging tape may be used to indicate the presence of the airflow at exit and entry points.

As the operation continues, atmospheric conditions in the space must be continuously monitored to ensure that the ventilation remains sufficient. Atmospheric monitors can alert you to any changes in gas concentrations, oxygen levels, or other hazards. The ventilation equipment settings may need to be adjusted to maintain safe conditions, such as increasing airflow rates in response to elevated gas levels or decreasing airflow rates to conserve oxygen.

Once rescue operations are complete and all personnel have exited the confined space, gradually reduce airflow rates and shut down the equipment in a controlled manner. Conduct a final inspection of ventilation equipment to ensure that it is properly shut down and secured and that all components are in good working condition.

All ventilation operations, including start-up and shutdown procedures, airflow rates, and atmospheric monitoring results, must be documented for future reference and debriefing. This documentation will inform a postoperation evaluation to assess the effectiveness of ventilation in controlling atmospheric hazards and supporting rescue operations. To better understand the process of ventilating a confined-space incident, follow the steps in **SKILL DRILL 8-3**.

Managing Exposures to Hazardous Materials

Managing exposures to hazardous materials in a confined-space incident involves identifying, assessing, and controlling potential risks to responders and

SKILL DRILL 8-3

Ventilating a Confined-Space Incident

Step 1: Gather Ventilation Equipment

Prepare ventilation equipment for deployment.

Step 2: Position Equipment for Optimal Airflow

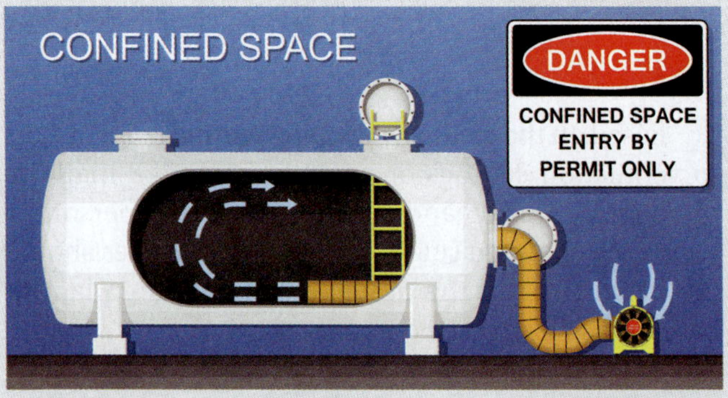

Strategically position equipment to maximize air circulation and distribution by drawing fresh air into the space.

SKILL DRILL 8-3 CONTINUED

Ventilating a Confined-Space Incident

Step 3: Prepare Battery-Powered Ventilation Equipment

Place battery-powered equipment (like this confined-space blower system) in areas where electrical connections are not available.

Step 4: Set Up Corded Vent Fans

Position corded electrical ventilation fans to maximize continuous air circulation within the space.

Step 5: Inspect Fan Components

Carefully inspect fan blades and housings for debris, dust, or obstructions that could affect airflow.

Step 6: Monitor Air Quality

Monitor the air quality inside the confined space using gas detectors or atmospheric monitoring equipment to verify that ventilation is successful.

Step 2: © Pepermpron/Shutterstock

victims. The types of substances present in a confined space may include gases, vapors, liquids, or solids that pose a risk to health and safety. *NFPA 470: Standard for Competence of Responders to Hazardous Materials/Weapons of Mass Destruction Incidents* (NFPA 2022b) provides guidance on identifying and classifying hazardous materials and understanding their properties and hazards.

Once hazardous materials are identified in a space, a risk assessment should be conducted to evaluate the potential health and safety risks to responders and victims. This involves assessing such factors as the toxicity, flammability, reactivity, and physical hazards associated with the materials. OSHA's "Hazard Communication" standard (1994; 29 C.F.R. 1910.1200) requires employers to assess the hazards of chemicals present in the workplace, communicate information about these hazards to employees, and implement appropriate control measures to protect workers. Control measures can be found within the *Emergency Response Guidebook* (ERG) or the SDS.

Based on the results of the risk assessment, exposure control measures should be implemented to minimize the risk of exposure to hazardous materials. These may include engineering controls (e.g., ventilation systems), administrative controls (e.g., work practices and procedures), and PPE. *NFPA 1500: Standard on Fire Department Occupational Safety, Health, and Wellness Program* (NFPA 2021b) outlines requirements for controlling exposures to hazardous materials, including the use of engineering controls, administrative controls, and PPE to mitigate risks to responders.

Proper PPE for protecting responders from exposure to hazardous materials may include respiratory protection, protective clothing, gloves, eye and face protection, and other specialized equipment. *NFPA 2500: Standard for Operations and Training for Technical Search and Rescue Incidents and Life Safety Rope and Equipment for Emergency Services* (NFPA 2022c) and *NFPA 1983: Standard on Life Safety Rope and Equipment for Emergency Services* (NFPA 2017) provide requirements for the selection, care, and use of PPE and equipment used in confined-space rescue operations, including protection against hazardous materials.

Continuous monitoring and detection of hazardous materials may involve using gas detectors, atmospheric monitors, radiation detectors, and other equipment that measures concentrations of hazardous substances. *NFPA 600: Standard on Facility Fire Brigades* (NFPA 2020) outlines requirements for the monitoring and detection of hazardous materials and atmospheres in industrial facilities, including confined spaces, to protect responders and facilitate effective emergency response.

A decontamination station and procedures must be set up before entry into the space occurs. In the event an exposure takes place, these decontamination procedures should be implemented to remove contaminants from personnel, equipment, and victims. This may involve decontamination showers, fans for dry decontamination, cleaning procedures, and disposal of contaminated PPE and materials. OSHA's "The Control of Hazardous Energy (Lockout/Tagout)" standard (1989; 29 C.F.R. 1910.147) requires employers to establish procedures for decontaminating equipment and work areas contaminated with hazardous materials, including confined spaces. More detail on decontamination procedures can be found in Chapter 5.

Always ensure that the rescue team coordinates extensively with HazMat specialists and other agencies during complex confined-space rescues that involve hazardous materials.

Using SCBAs to Manage Exposure to Hazardous Atmospheres

SCBAs provide a reliable source of breathable air that allows workers—and rescuers—to safely enter environments with potentially life-threatening contaminants, such as toxic gases, smoke, and low-oxygen conditions.

The capabilities of SCBAs are significant. SCBA units are equipped with air cylinders that supply breathable air, allowing the user to operate in environments where the air is contaminated or oxygen deficient. It provides a sealed system that prevents the inhalation of harmful substances like toxic gases, vapors, and particulate matter. This is essential in confined spaces where such hazards are present. An SCBA unit is designed to be portable, allowing rescuers to move freely within confined spaces, and the harness and face piece are ergonomically designed for comfort and ease of movement. Modern SCBA units include features such as low-air alarms, integrated communication systems, and heads-up displays that provide real-time information on air levels and other critical parameters.

SCBA units do, however, have their limitations. The duration of the air supply is finite and depends on the capacity of the air cylinder and the user's breathing rate. Typical SCBA units provide between 30 and 60 minutes of air, which means they require careful monitoring and planning. The units can be heavy and cumbersome, which may limit a wearer's agility and increase physical strain, particularly in tight or awkward spaces. Proper maintenance and regular training are crucial to ensure the effectiveness of SCBA units.

SAFETY TIP

Users must be familiar with donning, doffing, and operating the equipment, as well as conducting pre-use checks and maintenance, but best practices dictate that an SCBA should never be removed inside the space. The face piece and helmet can muffle speech, making communication difficult. Integrated communication systems can mitigate this factor, but they require additional training and may sometimes fail.

The Need for Rescue Entrants to Communicate Air Levels to Outside Rescuers

Rescue entrants must continuously monitor and communicate their air levels to outside rescuers. This allows for real-time assessment of the entrant's air supply, ensuring they have sufficient air to complete their tasks and exit the confined space safely. Communication of air levels is an important part of coordinating rescue operations. When outside rescuers have up-to-date information about the status of the entrant's air supply, they are in a much better position to make informed decisions about entry times, backup plans, and the deployment of additional resources.

By regularly updating outside rescuers on air levels, the risk of depletion of the air supply is minimized. If air levels approach critical thresholds, outside rescuers can initiate evacuation procedures promptly to prevent entrapment or exposure to hazardous atmospheres. Rescue entrants should exit the space before any low-air alarms activate. This will allow for any unexpected issues that may occur during the exit process and permits the entrant time to safely proceed through decontamination procedures, if necessary.

In the event of an emergency, immediate knowledge of the entrant's air levels allows outside rescuers to provide timely assistance. This could involve sending in additional SCBA units, deploying a secondary rescue team, or initiating a rapid extraction.

Many safety protocols and regulations mandate the monitoring and communication of air levels in confined spaces. Adhering to these requirements ensures compliance and enhances overall safety. To better understand how to use different breathing techniques for air conservation using an SCBA, follow the steps in **SKILL DRILL 8-4**.

SKILL DRILL 8-4

Breathing Techniques for Air Conservation Using an SCBA

Controlled Breathing

Step 1: Inhale Deeply Through the Nose

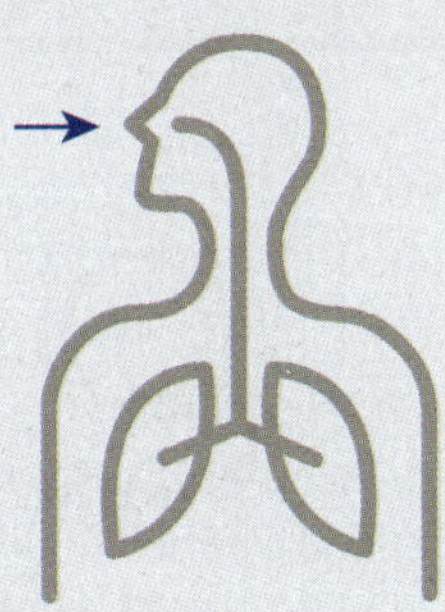

Inhale deeply through the nose, filling the lungs to maximize oxygen intake.

Step 2: Exhale Completely

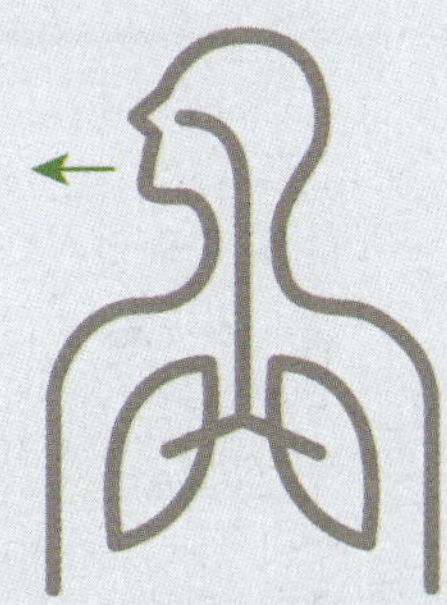

Exhale fully through the mouth to clear the lungs and reduce CO_2 buildup.

Continues.

SKILL DRILL 8-4 CONTINUED

Breathing Techniques for Air Conservation Using an SCBA

Step 3: Maintain Steady Rhythm

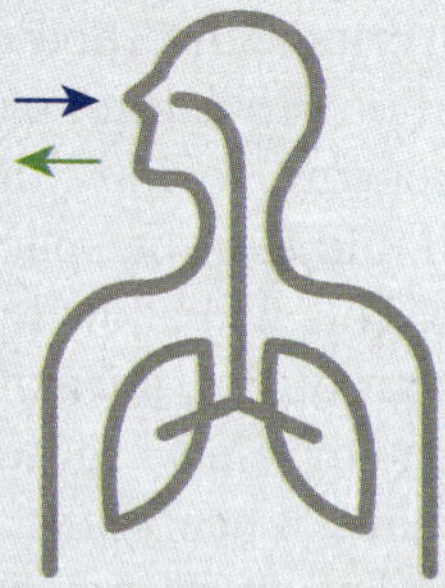

Maintain a steady breathing rhythm for optimal air conservation.

Step 4: Avoid Rapid Breathing

Avoid rapid, shallow breaths to prevent increased air usage.

Skip Breathing

Step 1: Inhale Deeply Through the Nose

Inhale deeply through the nose, taking in a full breath.

Step 2: Hold Breath for One to Two Seconds

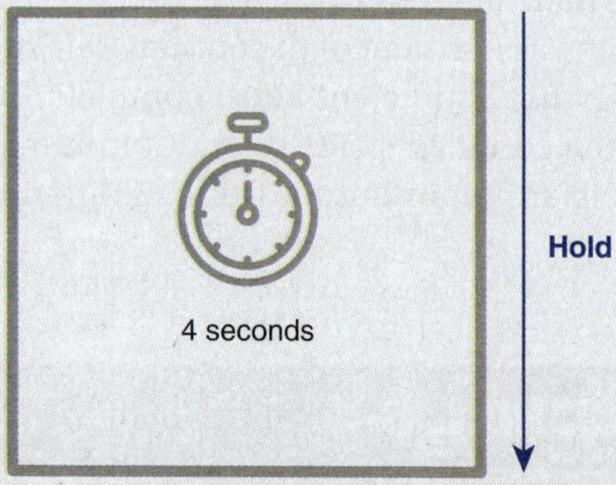

Hold breath for one to two seconds to conserve air.

Step 3: Exhale Slowly

Exhale slowly and fully through the mouth.

Step 4: Repeat the Cycle

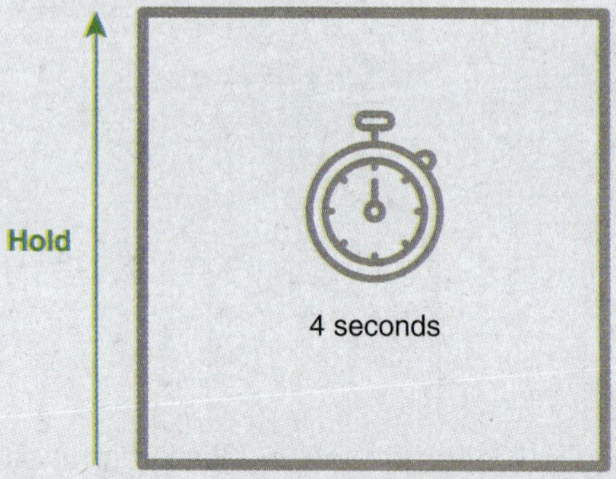

Repeat the skip breathing cycle to maximize air conservation.

SKILL DRILL 8-4 CONTINUED

Breathing Techniques for Air Conservation Using an SCBA

Box Breathing

Step 1: Inhale for Four Seconds

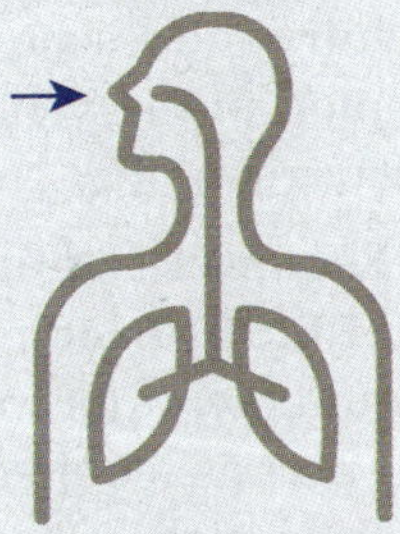

Inhale deeply over four seconds to control breathing rate.

Step 2: Hold Breath for Four Seconds

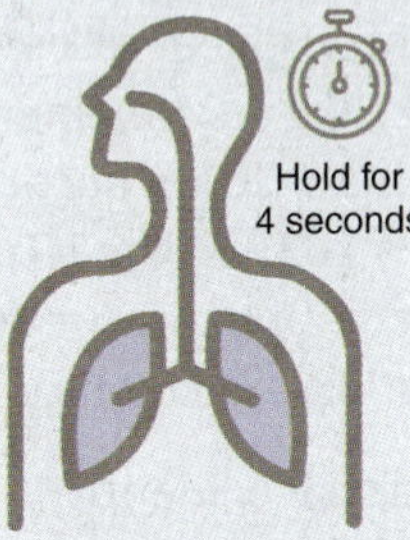

Hold breath for four seconds, maintaining calm focus.

Step 3: Exhale for Four Seconds

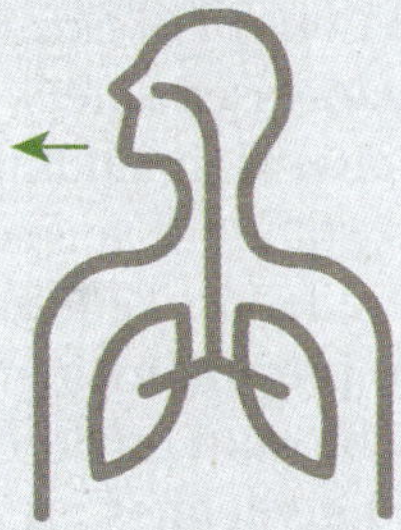

Exhale slowly over four seconds, conserving air with controlled release.

Step 4: Hold Breath for Four Seconds

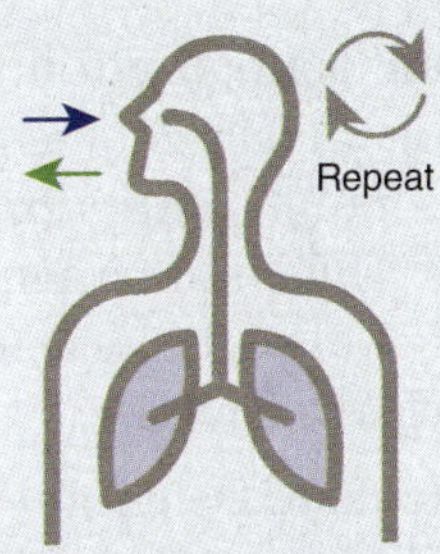

Hold breath for four seconds after exhaling. Repeat the four-step box-breathing cycle as necessary.

Reilly Emergency Breathing Technique (R-EBT)

Step 1: Inhale Deeply Through the Nose

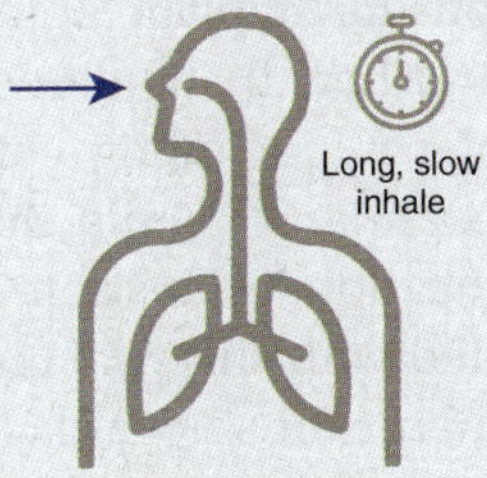

Inhale deeply through the nose, filling the lungs.

Step 2: Exhale While Humming

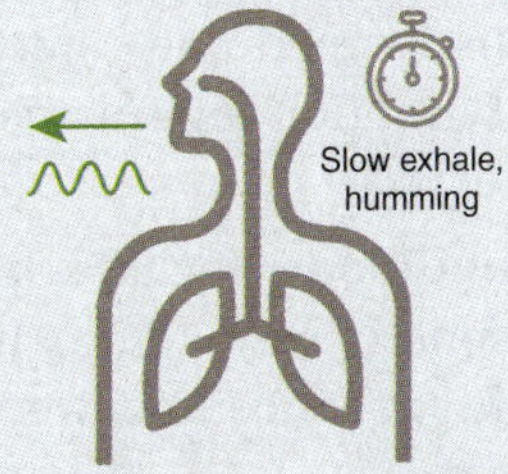

Exhale slowly while humming out the breath.

Step 3: Repeat the Cycle

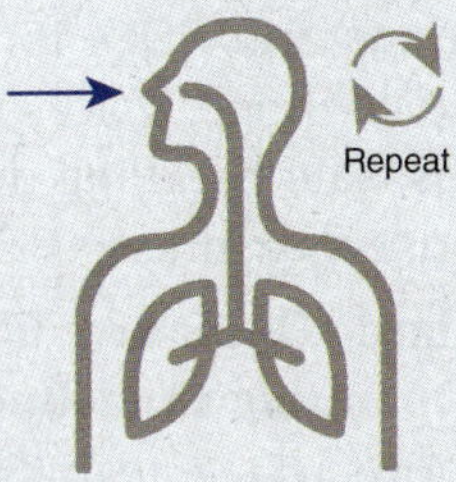

Repeat the R-EBT cycle to maximize air conservation.

Post–Size-Up Information Relay

In preparation for a size-up at a confined-space rescue emergency, responders gather critical background information from onsite personnel to help them assess the situation. This information can be an invaluable resource.

During a confined-space rescue size-up, teams gather, analyze, and relay all pertinent information to the incident commander (IC) in accordance with NFPA 1006 standards (NFPA 2021a). The IC uses this information to plan the rescue operation effectively. Example background information may include the following:

- **Type of confined space:** Identify the specific type of confined space (e.g., tank, silo, underground vault, vessel) where the incident occurred. Different types of confined spaces present varying hazards and challenges.
- **Dimensions and configuration:** Obtain dimensions and layout details of the confined space, including access points, entry/exit routes, and internal obstacles. This information helps in understanding the scope of the rescue operation and potential obstacles to navigate.
- **Nature of the incident:** Determine the nature of the incident that led to the rescue, such as a chemical spill, equipment malfunction, or structural collapse. Understanding the incident's cause provides insights into potential hazards and risks within the confined space.
- **Previous incident history:** Review any previous incidents or accidents related to the confined space, including lessons learned and safety recommendations. This information informs risk assessment and rescue strategies.
- **Atmospheric conditions:** Include specific details on initial atmospheric readings, including oxygen levels, flammable gas concentrations, and the presence of toxic gases.
- **Number and condition of victims:** Provide information on the number of known or suspected victims, their last known locations, and any information on their conditions or injuries.
- **Utilities and energy sources:** Identify any utilities or energy sources that may pose a hazard or need to be controlled during the rescue operation.
- **Weather conditions:** Include current and forecasted weather conditions that may affect the rescue operation, especially for outdoor confined spaces.

Reference materials play a crucial role in guiding responders during a confined-space rescue size-up. Examples of required reference materials may include the following:

- ***NFPA 350: Guide for Safe Confined Space Entry and Work* (NFPA 2022a):** This NFPA standard provides comprehensive guidelines for identifying, evaluating, and controlling hazards associated with confined-space entry and rescue operations. It covers atmospheric monitoring, ventilation, entry procedures, and rescue planning.
- **OSHA "Permit-Required Confined Space" Standard (1993; 29 C.F.R. 1910.146):** OSHA regulations outline safety requirements for working in confined spaces, including permit-required confined spaces. This includes information on entry permits, hazard assessment, training, and emergency response procedures.
- **NFPA Standard for Shipyard Employment (NFPA 306):** *NFPA 306: Control of Gas Hazards on Vessels* (NFPA 2024b) addresses safety measures related to shipyard operations, including confined spaces on vessels. This standard focuses on controlling gas hazards, ventilation requirements, and safe work practices to prevent incidents related to confined spaces in shipyard environments. NFPA 306 provides detailed guidelines on gas testing, atmosphere monitoring, ventilation systems, and emergency response procedures specific to shipyard confined spaces.
- **OSHA "Confined and Enclosed Spaces and Other Dangerous Atmospheres in Shipyard Employment" Standard (1994; 29 C.F.R. 1915, Subpart B):** OSHA's regulations for shipyard employment cover general safety and health standards applicable to shipyard workplaces, including confined spaces. Specific OSHA standards within this subpart address the identification, evaluation, and control of hazards associated with shipyard confined spaces. OSHA's requirements for shipyard confined spaces include entry procedures, permit systems, atmospheric testing, ventilation, PPE, rescue procedures, and employee training.
- **Manufacturer's Specifications and SDSs:** Obtain equipment specifications and SDSs for any substances present in the confined space. This information details hazards, safe handling procedures, and emergency response measures.
- **Rescue Team Standard Operating Procedures (SOPs) or Standard Operating Guidelines (SOGs):** Refer to the organization's rescue team SOPs/SOGs

for established protocols, roles, and responsibilities during confined-space rescue operations. SOPs/SOGs ensure consistent and coordinated response efforts.

- **Emergency Response Preplans or Plans:** Review facility-specific emergency response preplans or plans, including confined-space rescue protocols. Understand communication procedures, resource availability, and coordination with external agencies.
- **Confined-Space Entry Permit Checklist:** Review the confined-space entry permit checklist to ensure that all required data were collected and noted properly prior to the entry.

By leveraging background information and utilizing relevant reference materials, responders can conduct a comprehensive size-up to assess risks, identify hazards, and develop effective strategies for a confined-space rescue emergency.

CHAPTER FEATURE: AFTER-ACTION

IN SUMMARY

- Atmospheric, physical, and biological hazards can be identified using a variety of means, including visual cues, sensory cues, and equipment readings.
- Engulfment hazards, such as soil, sand, grain, or flowing liquids, present unique challenges and require specialized training, equipment, and procedures.
- Energized hazards include exposed wires, energized equipment, and electrical machinery. These also include pneumatic, hydraulic, and mechanical stored energy. Among the strategies used to mitigate energizing hazards in confined spaces are lockout/tagout procedures and PPE requirements.
- Environmental air monitoring is the systematic process of detecting and measuring potential atmospheric hazards, such as toxic gases, flammable vapors, oxygen levels, and other contaminants. These contaminants include hydrogen sulfide, carbon monoxide, ammonia, chlorine, fluorides, and other chemicals.
- Mechanical ventilation and purging, which involve using inert gases, such as nitrogen, to displace harmful gases from the space, are techniques used to remove toxins from the air of a confined space.
- Atmospheric monitors must be strategically placed inside a confined space to continuously monitor the air quality. They should be positioned in areas where atmospheric hazards are most likely to accumulate, such as low spots or adjacent to potential emission sources.
- Any alarm activations or deviations from safe air quality thresholds must be acted upon immediately. If toxic inhalation conditions are detected, personnel present in the space must be evacuated and ventilated, and if possible, the emission sources of any gases must be controlled.
- Mechanical ventilation of a confined space helps to maintain a safe atmosphere within it. Different types of ventilation fans are available, including battery-powered, gas-powered, pneumatic, and electrically-powered devices.
- Managing exposures from hazardous materials at a confined-space incident involves identifying, assessing, and controlling potential risks to responders, bystanders, and victims.
- SCBA provides a reliable source of breathable air that allows workers—and rescuers—to safely enter environments with potentially life-threatening contaminants, such as toxic gases, smoke, and low-oxygen conditions.
- In preparation for a size-up at a confined-space rescue emergency, responders gather critical background information to assess the situation. This information is then relayed to the IC, who uses it to plan the rescue operation effectively.

KEY TERMS

arc flash The light and heat produced during an arc fault, which is an electrical explosion or discharge.

colorimetric tubes Monitoring equipment that contains chemical reagents that undergo a color change when exposed to specific gases or vapors; the length of the color stain correlates with the gas concentration. Also known as *colorimetric chips.*

combustible gas detector A monitoring device specifically designed to detect the presence of flammable gases or vapors in the atmosphere.

KEY TERMS CONTINUED

energized hazard Exposed wires, energized equipment, and electrical machinery that may be present in a confined space.

environmental air monitoring The systematic assessment of pollutant levels in air by measuring the quantity and types of pollutants present.

flame ionization detector (FID) A monitoring device that uses a hydrogen flame to ionize carbon-containing compounds in a gas sample; the resulting ions produce a measurable electrical current.

multi-gas monitor A monitoring device that is capable of simultaneously detecting and measuring multiple gases (typically, oxygen, combustible gases, hydrogen sulfide, and carbon monoxide). Also known as a *multi-gas detector.*

negative-pressure ventilation (NPV) A method of ventilation that creates a lower pressure inside the space, drawing fresh air in through designated entry points and exhausting contaminated air through a controlled outlet.

oxygen monitor A monitoring device, also known as an *oxygen-deficiency monitor* or *oxygen-deficiency indicator,* that measures the concentration of oxygen in the atmosphere.

particulate matter (PM) A mixture of solid particles and liquid droplets found in air.

personal gas detector A wearable device designed to provide individual personnel with real-time alerts and alarms in the event of hazardous atmospheric conditions.

pH paper Also known as *litmus paper* or *pH indicator paper,* a simple and inexpensive tool used to measure the acidity or alkalinity (pH) of a solution.

photoionization detector (PID) A monitoring device that is used to detect volatile organic compounds (VOCs) and other hazardous substances based on their ability to ionize when exposed to ultraviolet (UV) light.

positive-pressure ventilation (PPV) A ventilation technique that uses a fan to create a higher pressure inside the confined space than outside, pushing contaminants out through designated exit points.

purging The removal of gases or particulate matter from air within an enclosed space by means of air displacement.

single-gas monitor A monitoring device that detects and measures the concentration of a single gas, such as oxygen, combustible gases (lower explosive limit [LEL]), hydrogen sulfide, or carbon monoxide. Also known as a *single-gas detector.*

volatile organic compounds (VOCs) Organic compounds that have a high vapor pressure and low water solubility at room temperature, which allows them to evaporate into air.

REFERENCES

"Confined and Enclosed Spaces and Other Dangerous Atmospheres in Shipyard Employment." 1994. 29 C.F.R. 1915, Subpart B. https://www.ecfr.gov/current/title-29/subtitle-B/chapter-XVII/part-1915/subpart-B.

"Duty to Have Fall Protection." 1979. 29 C.F.R. 1926.501. https://www.ecfr.gov/current/title-29/subtitle-B/chapter-XVII/part-1926/subpart-M/section-1926.501.

"Hazard Communication." 1994. 29 C.F.R. 1910.1200. https://www.ecfr.gov/current/title-29/subtitle-B/chapter-XVII/part-1910/subpart-Z/section-1910.1200.

National Fire Protection Association. 2017. *NFPA 1983: Standard on Life Safety Rope and Equipment for Emergency Services, 2017 Edition.* Quincy, MA: National Fire Protection Association.

National Fire Protection Association. 2020. *NFPA 600: Standard on Facility Fire Brigades, 2020 Edition.* Quincy, MA: National Fire Protection Association.

National Fire Protection Association. 2021a. *NFPA 1006: Standard on Technical Rescue Personnel Professional Qualifications, 2021 Edition.* Quincy, MA: National Fire Protection Association.

National Fire Protection Association. 2021b. *NFPA 1500: Standard on Fire Department Occupational Safety, Health, and Wellness Program, 2021 Edition.* Quincy, MA: National Fire Protection Association.

National Fire Protection Association. 2022a. *NFPA 350: Guide for Safe Confined Space Entry and Work, 2022 Edition.* Quincy, MA: National Fire Protection Association.

National Fire Protection Association. 2022b. *NFPA 470: Standard for Competence of Responders to Hazardous Materials/Weapons of Mass Destruction Incidents, 2022 Edition.* Quincy, MA: National Fire Protection Association.

National Fire Protection Association. 2022c. *NFPA 2500: Standard for Operations and Training for Technical Search and Rescue Incidents and Life Safety Rope and Equipment for Emergency Services, 2022 Edition.* Quincy, MA: National Fire Protection Association.

National Fire Protection Association. 2024a. *NFPA 70E: Standard for Electrical Safety in the Workplace, 2024 Edition.* Quincy, MA: National Fire Protection Association.

National Fire Protection Association. 2024b. *NFPA 306: Standard for the Control of Gas Hazards on Vessels, 2024 Edition.* Quincy, MA: National Fire Protection Association.

"Permit-Required Confined Spaces." 1993. 29 C.F.R. 1910.146. https://www.ecfr.gov/current/title-29/part-1910/section-1910.146.

"Safety and Health Regulations for Construction." 1979 29 C.F.R. 1926. https://www.ecfr.gov/current/title-29/subtitle-B/chapter-XVII/part-1926.

"The Control of Hazardous Energy (Lockout/Tagout)." 1989. 29 C.F.R. 1910.147. https://www.ecfr.gov/current/title-29/subtitle-B/chapter-XVII/part-1910/subpart-J/section-1910.147.

CHAPTER QUESTIONS

1. Which of the following helps predict contaminant behavior within confined spaces?
   - **a.** chemical density
   - **b.** assessment of risks
   - **c.** interpretation of results
   - **d.** vapor density
2. Air-sampling locations should be strategically identified based on which two characteristics of a confined space?
   - **a.** airflow dynamics and exit points
   - **b.** width and length
   - **c.** dimensions and configuration
   - **d.** depth and configuration
3. Within a confined space, ventilation helps to maintain a
   - **a.** safe atmosphere.
   - **b.** warm atmosphere.
   - **c.** cool atmosphere.
   - **d.** calming atmosphere.
4. Which of the following items contain chemical reagents that undergo a color change when exposed to specific gases or vapors?
   - **a.** Flame ionization detectors
   - **b.** Colorimetric tubes
   - **c.** Single-gas monitors
   - **d.** Oxygen monitors

CHAPTER FEATURE: ON SCENE

CARBON MONOXIDE EXPOSURE IN A CONFINED SPACE

A group of workers was tasked with performing maintenance inside a confined space at an industrial facility. The confined space, a storage tank, required routine inspections and repairs. To ensure a safe working environment, the workers deployed a ventilation fan at the entry point to provide fresh air circulation inside the tank, unaware that nearby equipment was emitting carbon monoxide as a result of combustion processes.

As the ventilation fan operated, it inadvertently drew surrounding air, including toxic carbon monoxide fumes, into the confined space. The workers, equipped with basic PPE and relying on the ventilation system to ensure a safe atmosphere in the space, entered and began their tasks.

Unfortunately, a short time after they entered the space, several workers began to experience symptoms of carbon monoxide poisoning, including dizziness, headache, and nausea. As carbon monoxide levels continued to rise within the space, some workers lost consciousness as a result of oxygen deprivation caused by carbon monoxide binding to hemoglobin in their blood.

The Emergency Response, Rescue Operations, Extrication, and Treatment

Upon discovering the emergency, the onsite safety team immediately initiated rescue operations. The safety team evacuated the affected workers from the confined space and administered first aid while calling for emergency medical services. The injured workers were swiftly transported to a nearby medical facility for further treatment and evaluation.

The Outcome

An investigation after the incident revealed the source of the carbon monoxide emissions and the failure of the ventilation strategy. The proximity of the ventilation fan to the emitting equipment and inadequate air

CARBON MONOXIDE EXPOSURE IN A CONFINED SPACE CONTINUED

monitoring contributed to the hazardous conditions within the confined space.

Lessons learned from this incident highlighted the importance of comprehensive risk assessment, including identifying potential sources of contaminants and implementing effective ventilation strategies. It underscored the critical need for continuous air monitoring and periodic reassessment of environmental conditions during confined-space operations.

After this incident, the company implemented stricter protocols for confined-space entry, including enhanced air-monitoring practices; increased awareness of carbon monoxide risks; and revised ventilation procedures to prevent similar incidents in the future. Training programs were also intensified to educate workers about the dangers of carbon monoxide exposure and the importance of vigilant safety measures in confined spaces.

This case study serves as a reminder of the serious risks associated with confined-space work and the critical importance of implementing robust safety protocols, including proper ventilation practices and comprehensive hazard identification, to protect the health and well-being of workers in confined spaces.

1. What was the primary source of carbon monoxide in the confined-space incident described?
   a. a poor ventilation system
   b. nearby equipment emitting carbon monoxide
   c. a leaking gas pipeline
   d. improper use of protective equipment
2. Which of the following symptoms did the workers experience as a result of carbon monoxide poisoning?
   a. skin rash, fever, and vomiting
   b. dizziness, headache, and nausea
   c. coughing, wheezing, and blurred vision
   d. muscle cramps, cold sweats, and confusion
3. Why did the ventilation strategy fail to protect the workers inside the confined space?
   a. The ventilation fan was turned off before the workers entered.
   b. The ventilation fan was too small for the confined space.
   c. The ventilation fan drew in surrounding air, including carbon monoxide.
   d. The ventilation fan malfunctioned and stopped working during the operation.
4. What safety measure was NOT in place and further contributed to the hazardous conditions?
   a. adequate air monitoring
   b. use of basic PPE
   c. multiple ventilation fans
   d. an onsite safety team

CHAPTER

9

Operations

Portable Anchor Systems, Lighting, and PPE

KNOWLEDGE OBJECTIVES

After studying this chapter, you should be able to:

- Discuss the proficiencies of different types of portable anchor systems. (**NFPA 1006 7.2.12**, pp. 230–236)
- Discuss the identified constraints of portable anchor systems. (**NFPA 1006 7.2.12**, pp. 230–236)
- Describe the process of assembling portable anchor systems. (**NFPA 1006 7.2.12**, p. 236)
- Explain the importance of stabilizing a portable anchor system to limit movement. (**NFPA 1006 7.2.12**, pp. 236–237)
- Explain the direction of force on a portable anchor. (**NFPA 1006 7.2.12**, p. 237)
- List the steps necessary to calculate force application on a portable anchor. (**NFPA 1006 7.2.12**, p. 237; pp. 240–241)
- Discuss the importance of portable lighting systems and the different methods used in setting them up. (**NFPA 1006 7.2.10** [horizontal spaces] and **7.2.14** [vertical spaces], pp. 245–246)
- Describe the different types of confined-space personal protective equipment (PPE). (**NFPA 1006 7.2.17**, p. 246; pp. 248–249)
- List the steps for limiting damage to confined-space equipment. (**NFPA 1006 7.2.11** [horizontal spaces] and **7.2.15** [vertical spaces], p. 249)

SKILL OBJECTIVES

- Using a portable anchor device, demonstrate how to remove a vertically oriented litter. (**NFPA 1006 7.2.12**, pp. 238–240)
- Deploy scene lighting. (**NFPA 1006 7.2.10** [horizontal spaces] and **7.2.14** [vertical spaces], pp. 247–248)
- Assess the scene for PPE and equipment requirements. (**NFPA 1006 7.2.10** [horizontal spaces] and **7.2.14** [vertical spaces], p. 246; pp. 248–249)
- Choose appropriate PPE and needed equipment. (**NFPA 1006 7.2.4**, p. 246; pp. 248–249)
- Demonstrate proper PPE selection. (**NFPA 1006 7.2.5**, p. 246; pp. 248–249)

Introduction

In confined-space rescue operations, assembling portable anchor systems is a critical task that ensures the safety of both rescuers and victims. From tripod-based systems to beam clamps and monopods, each type presents unique advantages and limitations. Understanding these nuances is essential for selecting the most suitable system for a given rescue scenario. Alongside their benefits, portable anchor systems also come with constraints that must be carefully considered. Factors such as weight capacity, stability on various surfaces, and compatibility with rescue equipment need to be thoroughly evaluated to ensure effective operation.

Stabilizing portable anchor systems to limit movement during rescue operations not only enhances the safety of the rescue team but also prevents potential injuries to the victim and minimizes the risk of system failure. Understanding the forces at play is crucial for ensuring the system can withstand the load imposed during rescue operations and maintain structural integrity. By comprehending how forces act on the system from various angles and directions, rescuers can make informed decisions regarding anchor placement and reinforcement.

Portable lighting is another essential aspect addressed in this chapter, with a focus on its significance in providing illumination in confined spaces. Different methods for setting up portable lighting are discussed, considering the specific requirements and constraints of both horizontal and vertical spaces.

Personal protective equipment (PPE) plays a critical role in safeguarding rescuers from the hazards encountered in confined spaces. The various types of PPE commonly used in confined-space rescue include respiratory protection, protective clothing, helmets, gloves, and footwear. Choosing PPE that is suitable for the specific hazards present in a confined-space environment and ensuring proper fit and functionality are critical factors in ensuring the safety of rescuers.

Portable Anchor Systems

Portable anchor systems (also called *high-directional systems*) are crucial in confined-space rescue operations because of their adaptability and reliability. These systems offer the flexibility to create secure anchor points in environments where permanent structures may be insufficient or unavailable. Their portability ensures that rescue teams can quickly set up anchors in various confined spaces, whether they are vertical shafts, tunnels, or tanks. This adaptability is essential for maintaining safety and efficiency in dynamic and challenging rescue scenarios. The most common portable anchor systems are monopods, tripods, davit arms, beam clamps, structural members to attach rigging points or preinstalled attachment points (e.g., rated eye bolts), and portable winch systems.

Many of these systems are designed to distribute loads evenly and can be adjusted to accommodate different types of confined spaces. This ensures that rescuers can work safely and effectively, minimizing the risk of falls or accidents. Additionally, portable anchors provide the strength required to support both the rescuer and the victim, ensuring a secure and reliable operation.

These systems are designed with ease of use and rapid deployment in mind. Time is critical, and the ability to quickly set up and dismantle anchor points can make a significant difference. Many portable anchor systems come with user-friendly features such as modular components and intuitive assembly instructions, which enable rescue teams to act swiftly without compromising on safety. This efficiency not only improves the effectiveness of the rescue operation but also helps in minimizing the exposure of both rescuers and victims to potential hazards within the confined space.

However, portable anchor systems also have limitations. They have specific load capacities that dictate their safe working limits. Exceeding these capacities can compromise system integrity and may pose significant risks to rescuers and victims. Rescuers must carefully assess the anticipated loads during confined-space rescue operations and select portable anchor systems with appropriate load ratings to ensure safety and prevent equipment failure.

The stability and integrity of anchor points are critical considerations when using portable anchor systems in confined-space rescue. Anchor points must be capable of supporting the anticipated loads and providing a secure attachment for rigging components. However, not all environments offer suitable anchor points, and rescuers may encounter challenges in finding stable and reliable attachment points within confined spaces. Rescuers must assess anchor point stability thoroughly and implement additional measures, such as anchor-stabilization techniques or alternative anchoring methods, to mitigate risks associated with unstable anchor points.

Confined-space rescue operations often occur in challenging and unpredictable environments, including hazardous atmospheres, extreme temperatures, and adverse weather conditions. These environmental factors can affect the performance and safety of portable anchor systems, potentially compromising rescuer and victim safety. Rescuers must account for environmental constraints when selecting and deploying portable anchor systems, ensuring compatibility with prevailing conditions and implementing appropriate precautions to minimize environmental hazards.

Confined spaces pose inherent challenges related to limited space and restricted access, which can affect the deployment and effectiveness of portable anchor systems. Rescuers may encounter difficulty positioning anchor points or rigging components within confined spaces, particularly in narrow or cluttered environments. Furthermore, space limitations may restrict the use of certain types of portable anchor systems, requiring rescuers to adapt their approach and utilize alternative anchoring methods or rigging configurations to accommodate spatial constraints.

Compatibility between portable anchor systems and other rescue equipment is essential for seamless integration and effective rescue operations. However, not all equipment may be compatible or interchangeable, leading to constraints in equipment selection and utilization. Rescuers must ensure compatibility between portable anchor systems, rigging hardware, harnesses, and other rescue gear to facilitate smooth operation and minimize risks associated with equipment mismatch or incompatibility.

Portable anchor systems used for confined-space rescue must comply with relevant standards and regulations to ensure safety and effectiveness. The Occupational Safety and Health Administration (OSHA) regulation "Permit-Required Confined Spaces" (1993) in the *Code of Federal Regulations* (C.F.R.; 29 C.F.R. 1910.146), National Fire Protection Association (NFPA) 1006 (NFPA 2021), and NFPA 2500 (NFPA 2022) provide guidelines for confined-space rescue operations, including portable anchor system equipment selection and usage. Regular inspections and maintenance of portable anchor systems are recommended and detailed in NFPA 1006, Chapter 7.1.2 (NFPA 2021).

Types of Portable Anchor Systems

Monopods

Monopods are single-leg support structures designed to provide vertical stability and load-bearing capacity (**FIGURE 9-1**). They typically consist of a telescoping mast or pole made of durable materials such as aluminum or steel, with adjustable height settings to accommodate different confined-space configurations. Monopods are portable and can be easily

FIGURE 9-1 A monopod portable anchor system.

transported to rescue sites, making them versatile tools for confined-space operations.

Monopods are rated for specific load capacities, which should be clearly marked by the manufacturer, and they require at least three tie-backs (four is the most stable). These ratings consider factors such as the weight of the victim, rescuers, and any equipment being used during the rescue operation. Rescuers must ensure that the selected monopod is capable of supporting the anticipated loads, and they must adhere to the manufacturer's guidelines for safe usage.

Monopods are typically deployed by extending the telescoping mast to the desired height and securing it in position using locking mechanisms or pins. Some models may feature adjustable legs or stabilizing feet to enhance stability on uneven or slippery surfaces. Proper installation is crucial to ensure the integrity and safety of the monopod during rescue operations.

Monopods may include built-in safety features to enhance stability and prevent accidents. These features may include nonslip foot pads, locking mechanisms to secure the mast in position, and secondary attachment points for backup systems such as safety lines or secondary anchors. Rescuers should inspect monopods for signs of wear or damage before each use and follow manufacturer recommendations for maintenance and inspection.

Monopod systems may have limited load-bearing capacity compared with other systems, restricting their use in heavy-duty rescue operations or with multiple rescuers. The stability and anchorage of monopod systems may be compromised in soft or unstable ground conditions, requiring careful assessment and reinforcement.

Tripods

Tripods, shown in **FIGURE 9-2**, are closely associated with confined spaces. Tripods are devices typically meant to support people as they are raised or lowered into and out of a space. Because the tripod will support people, it should not be used to support any other loads.

When you support a person on the tripod, its load capacity must not have been compromised by the potential overloading of equipment or prior stresses that could have caused unseen damage and may lead to unexpected failure. The footprint is the area between the legs of the tripod, where weight to the system is distributed. The resultant is the vector of where the force of the system is acting.

When choosing a tripod, be aware that there are differences based on the height of the unit, footprint, and lifting capacity. How you plan on using the tripod will affect the number of anchor points on the head of the tripod, attachment points for lifting devices, the adjustability of the height, and loads you can support or lift using the tripod.

Properly loading the tripod is just as important as not overloading it. The tripod has an axis that runs through its center to initially pick up the load, and it then transmits the load to the ground through the axis of each leg. To evenly load each leg, you must ensure that the load is applied along the center of all those axes. If you load the tripod in any other direction, the load will want to go in that direction. If a load is at an angle to the axis or perpendicular to the axis, the tripod can be easily pulled over. The greater the angle to the axis, the more unstable the tripod will become.

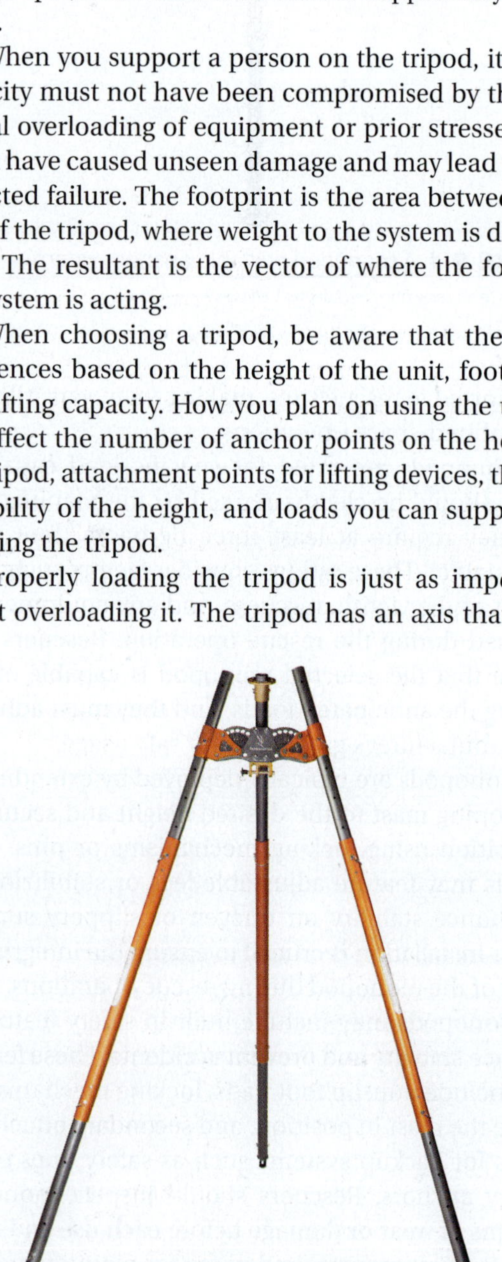

FIGURE 9-2 A tripod-based portable anchor system.

Used with permission from TerrAdaptor™ and Harken®, Inc.

> **TIP**
>
> Tripods are meant to lift and lower people. Lowering or raising equipment or other loads into and out of a confined space with a tripod should be forbidden.

To avoid tipping the tripod, you must use either equipment specifically designed for use with that tripod or an anchor point and a change-of-direction pulley to keep the direction of the load aligned with the tripod and then redirect the load using the change-of-direction pulley. This may sound like a simple process, but to achieve your desired outcomes, you must know the capacity and limitations of your equipment and how to properly use it. The equipment manufacturer is a good source of this information.

Even in situations where a ladder can be used to enter a confined space, a tripod is still valuable. A tripod allows you to connect the rescue entrants to a retrieval device and create an entry team rescue device that can be used from outside the space. If you need to rescue the rescuers from the space, you don't have to make entry or attempt to lift the injured rescuer out of the space—you can let the equipment do the work for you. Another advantage to using a tripod and ladder at the same time is that you can use the retrieval line for fall protection or as a fall arrester and limit the free-fall potential of a person climbing up or down the access ladder. When using a tripod as your sole means of entry and exit, each person entering the space should wear a retrieval line and a safety line while they are being raised or lowered. With two rescuers in the confined space, there should be two retrieval lines in use and a spare third line.

This requirement for the three lines (one retrieval line for each rescuer after they are in the space and a safety line for use during lifting) also requires three attachment points at the head of the tripod. Each line should be as independent of the others as possible.

When it is time to remove the victim from the space, one rescuer will have to exit the space using one retrieval line and the safety line. Both lines are then sent back into the space to be used to lift the victim. Per "Permit-Required Confined Spaces" (1993; 29 C.F.R. 1910.146), each entrant is required to wear a full-body harness and have a tag line with one end attached to the dorsal connection and the other end attached to an outside anchor. Once the victim has been removed, the safety line is lowered back in for use by the second rescuer as they exit the space. Although this may seem redundant, the use of a safety line ensures that if anything happens to either the retrieval line or the hoisting equipment, you have a backup means to safely recover your personnel and the victim. This process is illustrated in **FIGURE 9-3**.

There are three types of tripods: portable, heavy-duty, and confined-space entry tripods. Portable tripods are lightweight and easily transportable, making them ideal for mobile rescue operations. They typically feature adjustable legs, allowing for a stable setup on uneven or sloped surfaces. Portable tripods are available in various materials, such as aluminum or steel, with different load capacities and height adjustments to suit different rescue scenarios.

Heavy-duty tripods are designed for more demanding rescue applications, such as industrial or technical rescues involving heavier loads or extreme conditions (**FIGURE 9-4**). They are constructed from robust materials, such as steel or reinforced aluminum, providing enhanced stability and durability. Heavy-duty tripods often feature higher load capacities; greater height adjustments; and additional safety features, such as locking pins or anti-slip feet.

Confined-space entry tripods are specifically designed for use in confined spaces with limited access or restricted entry points. They are typically compact and lightweight, with adjustable legs and compact designs to facilitate setup in tight spaces. Confined-space entry tripods may feature specialized attachment points or accessories for connecting ropes, pulleys, or other rescue equipment. Modern tripods can also serve as high-point changes of direction, as long as they are rated for this capacity.

Only a rescue-type tripod should be used for any rescue that requires rope techniques. Rescue tripods can be tied off at the top to offset working outside of the center axis and/or the use of directional pulleys. Industrial tripods and davit arms are designed to be used with crank-type winches attached directly to the legs or davit arm.

Any type of overhead anchor system should be load tested by pulling down on the rope in the direction the load will be traveling. This will reveal any unwanted directional forces on the system prior to putting a life load on the system. If any unwanted directional forces are found, they must be corrected and then tested again.

See **Appendix F** for details on understanding and calculating mechanical advantage systems when using tripods.

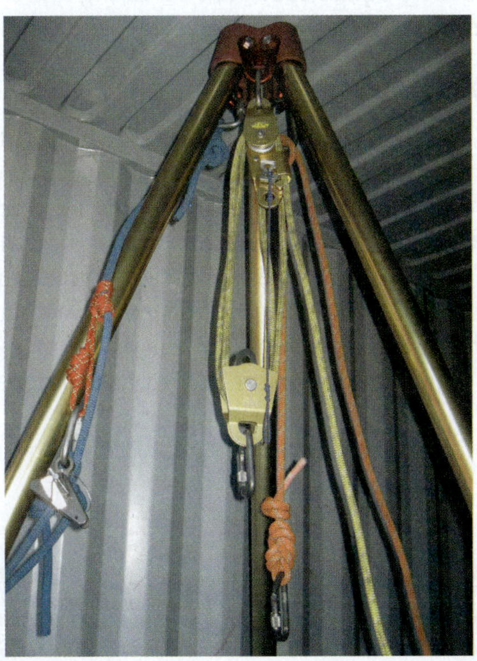

FIGURE 9-3 A tripod with a safety/retrieval system attached.

Courtesy of Illinois Fire Service Institute.

FIGURE 9-4 A heavy-duty tripod.

> **TIP**
>
> A tripod-based system consists of three legs connected at the top with a central point, resembling the shape of a traditional camera tripod. The legs are typically adjustable to accommodate different heights and terrain.
>
> The top of a tripod features a pulley system or attachment point where rescue ropes or other equipment can be secured. This allows rescuers to lower or raise victims safely within a confined space. Tripods are often equipped with safety features such as locking pins and stability braces to prevent accidental collapse during use.

Davit Arms

Davit arms provide a versatile and stable platform for accessing and retrieving victims from various types of confined spaces (**FIGURE 9-5**). They consist of a vertical post or mast with an extendable boom arm that can rotate horizontally. Davit arms are typically made of durable materials, such as steel or aluminum, that can withstand the loads encountered during rescue operations. Davit arms may be fixed or portable, with some models featuring adjustable height and reach to accommodate different confined-space configurations.

These systems can be installed either permanently or temporarily, depending on the specific application. Permanent installations involve securely mounting the davit arm to a stable structure, such as a concrete pad or building rooftop. Temporary setups may utilize bases or anchoring systems that can be quickly assembled and disassembled as needed. Proper installation is critical to ensure the stability and safety of the davit arm during rescue operations.

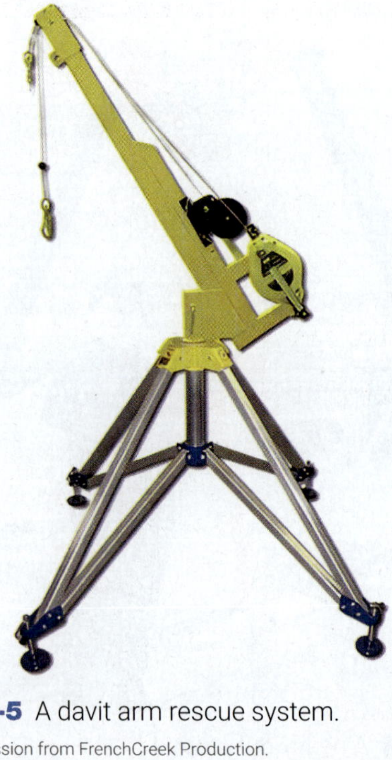

FIGURE 9-5 A davit arm rescue system.

Used with permission from FrenchCreek Production.

Davit arms are rated for specific load capacities, which should be clearly marked by the manufacturer. These ratings take into account factors such as the weight of the victim, rescuers, and any equipment being used. Rescuers must ensure that the selected davit arm is capable of supporting the anticipated loads during rescue operations and adhere to the manufacturer's guidelines for safe usage. Built-in safety features to enhance stability and prevent accidents may include locking mechanisms to secure the boom arm in position, secondary attachment points for backup systems, and warning labels or indicators to alert users of potential hazards.

Davit arm systems offer flexibility and adaptability for confined-space rescue operations, especially in areas with limited space or overhead clearance. They can be easily maneuvered and positioned to provide secure anchor points for rescue rigging. Davit arms can be equipped with accessories, such as winches and pulleys, to facilitate victim retrieval and hoisting.

These systems may require additional stabilization and support to prevent tipping or instability, especially when used in confined spaces with uneven terrain or high winds. The setup and deployment of davit arm systems may require specialized training and expertise to ensure proper operation and safety compliance.

Beam Clamps

Beam clamps are designed to securely attach to horizontal beams or structural members, providing a stable anchor point for various rescue operations (**FIGURE 9-6**). They are typically made of durable materials, such as steel, and are designed to clamp onto horizontal beams or structural members with a secure grip. They often feature adjustable jaws or bolts that can be tightened to ensure a firm attachment. Some

FIGURE 9-6 A beam clamp portable anchor system.

Courtesy of Pigeon Mountain Industries Inc.

beam clamps may also include swivel eye bolts or attachment points for connecting rescue ropes, pulleys, or other equipment.

Beam clamps are relatively easy to install and require minimal tools; they can be directly attached to overhead beams or structures using bolts or locking mechanisms. However, they must be properly aligned and tightened to prevent slippage during rescue operations.

Beam clamps are rated for specific load capacities, which should be clearly marked by the manufacturer. Rescuers must ensure that a beam clamp can support anticipated loads during rescue operations, including the weight of victims, rescuers, and any equipment being used. Safety features to enhance stability and prevent accidental disengagement may include locking mechanisms, safety pins, or secondary attachment points for backup systems.

Beam clamps are an effective way to secure lifelines and rigging systems to structural beams or overhead supports in confined spaces. They offer quick and easy attachment to beams without the need for drilling or permanent installation, minimizing setup time and labor. Although they are adjustable and adaptable to various beam sizes and configurations, beam clamps must be properly sized and rated to match the load requirements of the confined-space rescue operation.

Portable Winch Systems

Portable winch systems come in various configurations, offering different capacities and capabilities to suit different rescue scenarios. These systems typically consist of a winch unit, rigging attachments, and anchor points (**FIGURE 9-7**). Their capacities can vary widely depending on the specific model and manufacturer—anywhere from a few hundred to several thousand pounds of pulling force. The winch's capacity should be matched to the anticipated loads encountered during confined-space rescue operations to ensure safe and effective performance.

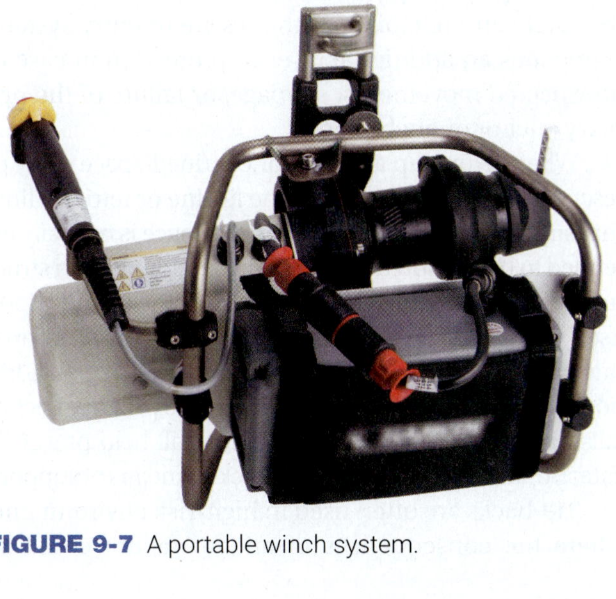

FIGURE 9-7 A portable winch system.

Portable winch systems provide mechanical advantage and pulling power for confined-space rescue operations, facilitating victim retrieval and hoisting. These systems provide mechanical advantage, allowing rescuers to multiply their pulling force and overcome resistance when retrieving victims. They offer versatility and adaptability for various rescue scenarios, including vertical and horizontal entries, as well as high-angle or low-angle rescues. Portable winch systems can be configured with different attachments and accessories to accommodate different rescue techniques and requirements. Many portable winch systems are compact and lightweight, making them easy to transport and maneuver in confined spaces.

Proper training and expertise necessary for the safe and effective use of portable winch systems include rigging techniques, load calculations, and equipment operation. Regular inspection and maintenance of winch components are essential to ensure reliable performance and compliance with safety standards and regulations. Some portable winch systems can be expensive to purchase and maintain, requiring budgetary considerations for procurement and ongoing operational costs.

Before deploying a portable winch system, you must identify potential hazards, determine load requirements, select appropriate rigging configurations, and identify suitable anchor points capable of supporting the anticipated loads and providing a stable attachment for the winch system. Anchor points should be secured to structural elements or rated anchor points certified for rescue operations.

To complete the installation, rig the portable winch system according to the manufacturer's guidelines and industry best practices, ensuring proper alignment, tensioning, and attachment of rigging components. Use appropriate rigging attachments, such as slings, carabiners, and pulleys, to optimize mechanical advantage and load distribution. When using winches or mechanical advantage systems, extra consideration should be given to the path of travel to avoid obstructions that could cause entanglements or pinch points. A winch will not give if something becomes snagged.

Winch systems must be operated in accordance with manufacturer instructions, including maintaining proper tension, monitoring load limits, and communicating effectively with the rescue team. Be sure to use appropriate PPE, such as gloves and eye protection, when handling winch components and rigging hardware.

Assembling Portable Anchor Systems

Assembling portable anchor systems for confined-space rescue requires following manufacturers' guidelines, as well as careful planning, adherence to applicable codes and standards, and attention to detail. Personnel who are responsible for assembling and using portable anchor systems must receive adequate training and demonstrate competency in their operation. These hands-on training sessions should cover assembly procedures, safety precautions, and emergency protocols to prepare rescuers for confined-space rescue operations. The team doesn't need to know how to assemble every portable anchor system on the market—just the equipment in your inventory.

Before assembly, thoroughly inspect all components of the portable anchor system for signs of wear, damage, or defects. Check for proper functioning of locking mechanisms, adjustable features, and structural integrity. Replace any damaged or worn components to ensure the safety and effectiveness of the anchor system.

Anchor points must be structurally sound and capable of supporting the anticipated loads during rescue operations. Avoid anchoring to fragile or unstable surfaces that may compromise the integrity of the system. Position the anchor points in strategic locations to provide optimal stability and support for the portable anchor system. Use appropriate anchoring methods, such as straps, clamps, or bolts, to secure the anchors to sturdy structural elements within the space. Ensure that anchors are firmly attached and properly tensioned to prevent movement or dislodgement during rescue operations.

If the portable anchor system includes a mast or support structure, assemble it according to the manufacturer's instructions. Extend the telescoping sections to the desired height and secure them in position using locking mechanisms or pins. Ensure that the mast is stable and vertical to provide a secure anchor point for rescue operations.

Once the anchor points and support structure are in place, attach lifelines, rigging, and other components necessary for rescue operations. Use appropriate connectors, carabiners, and knots to secure lifelines to anchor points and ensure proper alignment with the confined-space entry point.

Before using the system, perform function checks on the assembled system to verify its stability, load-bearing capacity, and overall integrity. Test each component and connection point to ensure that it functions as intended and can withstand the anticipated loads during rescue operations.

Keep detailed records of the assembly process, including the equipment used, anchor point locations, and any modifications made to the portable anchor system. Documenting the assembly process helps ensure accountability, traceability, and compliance with applicable codes and standards.

Stabilizing Portable Anchor Systems to Limit Movement

Unstable anchor systems can compromise rescuer and victim safety, leading to equipment failure, falls, and other serious incidents. Environmental conditions, such as wind, uneven terrain, or vibrations, can affect an anchor system's stability. Anchor systems subjected to excessive movement or instability are more prone to failure, jeopardizing the safety of rescuers and victims. Stabilization of the system helps prevent anchor failure by maintaining the integrity and stability of anchor points. Proper stabilization techniques, such as anchoring to stable structures or using anchor-stabilization devices, minimize anchor movement and enhance system reliability, reducing the risk of anchor failure during rescue operations.

In the context of confined-space operations requiring the setup of a tripod, the term **tie-back** refers to an additional safety measure where the anchor point or anchoring system used in the rescue setup is secured with an additional line or anchor. This is done to ensure redundancies and increase the overall safety of the system. The primary purpose of a tie-back is to enhance the safety and reliability of the rescue or entry system. It provides an additional layer of protection in case of unexpected movements, slippage, or failure of the primary anchor or anchor point.

When setting up a tripod for confined-space entry or rescue, the main line (such as the lifeline or retrieval line) that supports the person entering the space is typically intended to be anchored to the tripod or another solid structure. A tie-back secures this anchor point or the tripod itself to another anchor (e.g., a second tripod leg, a separate anchor point, or a structural element) using an additional line or rope. This ensures that if the primary anchor fails or shifts, the secondary tie-back will help prevent a catastrophic failure, providing a backup means of support.

Tie-backs are often used in high-risk environments where the consequences of anchor failure could be severe, such as in confined-space rescues where rescuers and entrants rely heavily on the integrity of the tripod and anchoring systems.

Unstable anchor systems can increase rescuer exposure to hazards within confined spaces, placing them at greater risk of injury or entrapment. Movement of anchor points or rigging components can disrupt rescue operations, hinder victim extrication efforts, and create unsafe working conditions for rescuers. Stabilization helps minimize rescuer exposure to these hazards by maintaining secure anchor points and preventing unexpected movements that could endanger rescuer safety.

Stabilization also ensures the effectiveness of rigging configurations during rescue operations. Rigging components attached to unstable anchor points may shift or dislodge during hoisting, lowering, or maneuvering, compromising the stability of the rescue system and posing risks to rescuers and victims. Stabilization mitigates these risks by providing a secure foundation for rigging hardware, enhancing rigging stability, and facilitating smooth and controlled rescue operations.

A stabilized system enables rescuers to maintain control over movement, preventing sudden shifts or unexpected motions that could endanger rescuer and victim safety. Controlled movement is essential for safely extricating victims from confined spaces, maneuvering them through tight passages, and positioning them for extraction.

An aerial apparatus can serve as an effective **high-point anchor** system in certain rescue or confined-space operations. A high-point anchor is a secure anchoring system positioned above the working area or entry point, providing a high point of attachment for ropes and allowing for efficient and controlled vertical movement of personnel and equipment. When it can be positioned close enough to the incident site, an aerial ladder or platform can provide a stable and elevated anchor point for rope systems. This setup is particularly useful when natural or structural high points are unavailable or inadequate.

However, it is crucial to remember the load capacity and operational limits of the aerial apparatus. The weight of the ropes, hardware, and any potential load must be within the apparatus's rated capacity to avoid overloading, which could compromise safety. Additionally, operators must ensure that the apparatus is properly stabilized, with outriggers fully deployed, to maintain balance and prevent tipping during the operation. Proper planning and adherence to the manufacturer's guidelines are essential when using aerial apparatus in this capacity.

Direction of Force on Portable Anchor Systems

The force exerted on a portable anchor can vary depending on such factors as the angle of pull, the weight of the load, and any dynamic forces involved in the operation. When a load is suspended directly below the anchor point, the force exerted on the anchor is primarily **vertical force**. Vertical force, which is typically the easiest to manage, is the most common scenario in confined-space rescue operations where victims or equipment need to be lifted or lowered vertically.

Horizontal force occurs when the load is pulled or pushed horizontally away from the anchor point. It can occur during operations where the direction of movement is not directly vertical, such as pulling a victim horizontally through an opening. Anchors must be capable of withstanding horizontal forces without tipping over or shifting to ensure the safety of rescuers and victims.

Angular force occurs when the load is pulled at an angle relative to the anchor point. Angular force can exert significant lateral pressure on the anchor and may cause it to shift or fail if not properly secured. Rescuers must consider the angle of pull and use appropriate rigging techniques to minimize angular force and ensure the stability of the anchor.

Dynamic force refers to the additional forces exerted on the anchor as a result of changes in velocity or direction of motion. Dynamic forces can occur during lifting, lowering, or movement of loads and can significantly increase the stress on the anchor. Rescuers must account for dynamic forces and select anchors capable of withstanding these additional loads to prevent accidents and injuries.

In many confined-space rescue scenarios, multiple forces may act simultaneously on a portable anchor. Rescuers must analyze the direction and magnitude of each force and ensure that the anchor is able to withstand combined loads without compromising safety. To better understand how to remove a vertically oriented litter using a portable anchor device, follow the steps in **SKILL DRILL 9-1**.

Calculating Force Application on a Portable Anchor

Force application refers to the distribution of forces exerted on a portable anchor system during rescue operations. These forces include the weight of the victim, tension from rigging components, dynamic forces from

SKILL DRILL 9-1

Removing a Vertically Oriented Litter

Step 1: Assess the Situation

Assess the entrance dimensions and litter positioning. Verify the litter's orientation inside the confined space.

Step 2: Prepare the Equipment

Set up the tripod for the lifting system at the entrance.

Prepare the pulley system with ropes for lifting.

SKILL DRILL 9-1 CONTINUED

Removing a Vertically Oriented Litter

Step 3: Attach the Litter to the Pulley System

Attach the lifting rope to the litter's top for balance and support.

Step 4: Start Lifting the Litter

Begin lifting the litter, maintaining a steady pace.

Step 5: Guide the Litter Out of the Confined Space

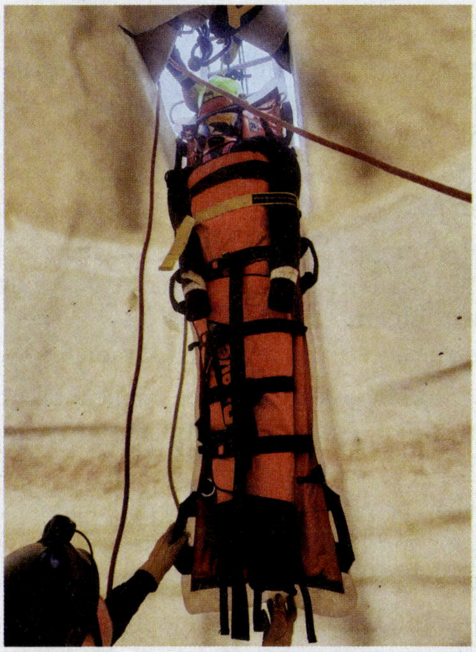

Position the litter carefully at the entrance for extraction.

Carefully guide the litter through the narrow opening.

Continues.

SKILL DRILL 9-1 CONTINUED

Removing a Vertically Oriented Litter

Step 6: Lower the Litter

Begin lowering the litter to ground level safely.

Step 7: Post-Operation Review

Conduct a thorough equipment inspection for any damage.

movement, and environmental factors such as wind or vibration. Properly assessing force application is essential for selecting suitable anchor points, determining equipment requirements, and ensuring the stability and safety of the entire rigging system.

Force application on portable anchors directly affects the safety and stability of rigging systems used for rescue and retrieval. Properly calculating force application ensures that portable anchors can withstand the forces exerted during rescue operations, minimizing the risk of anchor failure and ensuring the safety of rescuers and victims.

Several methods and factors can be used to calculate force application on portable anchors in confined-space operations, including the following:

- **Static analysis:** This method involves calculating the static weight of the victim and any additional loads, such as equipment or rescue personnel, to determine the total force applied to the anchor system.
- **Dynamic analysis:** Dynamic forces, such as those generated by movement or sudden impacts, must also be considered. These forces may exceed the static weight of the victim and require additional calculations to assess their impact on anchor stability.
- **Environmental factors:** External factors such as wind, vibration, or changes in load distribution can affect force application on portable anchors and must be accounted for in calculations.

To accurately calculate force application on portable anchors in confined-space operations, follow these steps:

1. Determine the total weight of the victim, including any equipment or additional loads.
2. Assess the distribution of weight and forces within the confined space, considering factors such as uneven terrain, movement, or changes in load distribution.

3. Calculate the static weight of the victim and any additional loads using appropriate formulas or weight measurements.
4. Consider dynamic forces that may be generated during rescue operations, such as sudden movements or impacts, and assess their potential impact on anchor stability.
5. Evaluate environmental factors, including wind, vibration, or changes in load distribution, and adjust calculations accordingly.
6. Select suitable anchor points capable of withstanding the calculated forces, and ensure that rigging components are properly configured and secured.
7. Regularly monitor anchor stability and reassess force application as conditions change during rescue operations.

Accurately calculating force application on portable anchors is essential for ensuring the safety and stability of rigging systems in confined-space operations. When calculating the load from an anchor that is at the point of hauling the total load, you must include the weight of the load and the amount of force that will be introduced (hauled) to move that load—for example, a 500-pound (226.7-kg) load requires more than 500 pounds (226.7 kg) of force to move it, so the anchor will experience a force of more than 1000 (453.6 kg) pounds. By understanding the principles of force application, utilizing appropriate calculation methods, and following the steps outlined in this chapter, rescuers can effectively assess anchor stability, select suitable anchor points, and mitigate the risk of anchor failure during rescue and retrieval operations in confined spaces.

Static and Dynamic Analysis for Force Application

Static analysis for force application involves calculating the total weight or force exerted on a structure or anchor point when all forces are in equilibrium, meaning there is no movement or acceleration. This method is commonly used in engineering and rigging to assess the stability and safety of anchor points and support structures.

The mathematical formula for static analysis is as follows:

$$F = m \times g$$

F is the force (in Newtons or pounds), m is the mass (in kilograms or pounds), and g is the acceleration due to gravity (32.2 feet per second squared [ft/s^2] or approximately 9.81 meters per second squared [m/s^2]). In this formula, $m \times g$ represents the weight of the object or load. When performing static analysis for force application, the total weight of the load, including any equipment, tools, or personnel, is calculated using this formula. This weight is then used to assess the load-bearing capacity of anchor points and support structures, ensuring they can safely withstand the applied forces without failure or deformation.

Static analysis is particularly useful in confined-space rescue operations for determining the maximum load capacity of anchor points, selecting appropriate rigging equipment, and ensuring the stability and safety of the rescue system. By accurately calculating force application through static analysis, rescuers can mitigate the risk of anchor failure and ensure the success of rescue and retrieval operations.

Dynamic analysis for force application involves calculating the forces exerted on a structure or anchor point when there is movement or acceleration involved. This method is essential in scenarios where loads are subjected to changes in velocity, such as during lifting or lowering operations in confined-space rescue.

The mathematical formula for dynamic analysis involves considering additional factors, such as acceleration, deceleration, and inertia. The formula to calculate force in dynamic situations is as follows:

$$F = m \times a$$

F is the force (in Newtons or pounds), m is the mass (in kilograms or pounds), and a is the acceleration in feet per second squared (ft/s^2) or meters per second squared (m/s^2). In dynamic analysis, the force (F) is the product of mass (m) and acceleration (a). This formula accounts for the additional forces exerted on a structure or anchor point resulting from changes in velocity or direction of motion.

Using Change-of-Direction Pulleys

Change-of-direction pulleys, also known as *redirection pulleys*, are used to redirect the direction of force in a rope system (**FIGURE 9-8**). By changing the angle of pull, change-of-direction pulleys enable rescuers to navigate around obstacles, traverse horizontal or vertical passages, and achieve optimal rigging configurations. These pulleys are commonly used in confined spaces to create rigging systems for hauling, lowering, and lifting operations. They allow rescuers to navigate complex environments, such as vertical shafts, horizontal passages, or confined chambers, by redirecting the rope around corners or obstacles.

If a change-of-direction pulley is needed based on the layout of the confined space and the desired rigging

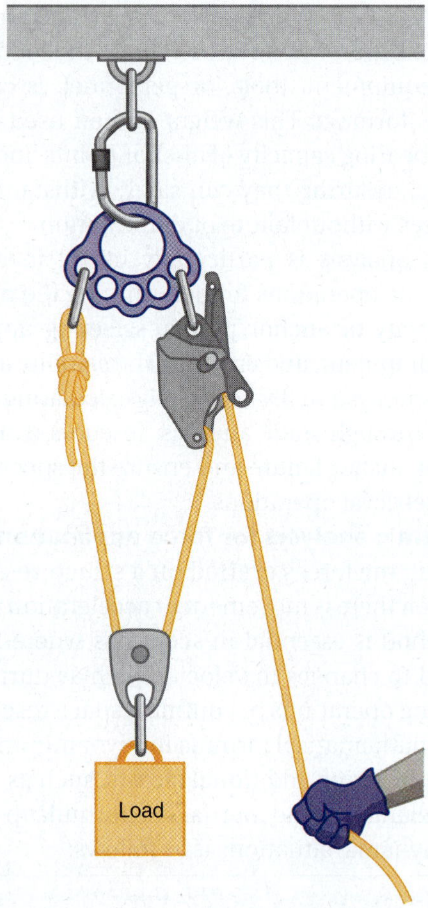

FIGURE 9-8 A change-of-direction pulley in action, redirecting the rope smoothly to minimize friction and aid in controlled lifting during the extraction process.

configuration, first select a suitable anchor point for the pulley and ensure that it is stable, secure, and capable of supporting the anticipated loads. Next, rig the pulley by attaching it to the anchor point using carabiners or connectors, and then thread the rope through the pulley to redirect the direction of force.

You can adjust the position and orientation of the change-of-direction pulley as needed to achieve the desired rigging configuration and ensure smooth operation and minimal friction on the rope.

Using Ascending and Descending Techniques

In confined-space operations, raising and lowering are the preferred and most often used methods for rescue. However, ascending and descending techniques can be useful in certain situations.

Ascending techniques are used in confined-space operations to access elevated or vertical areas within the space, such as vertical shafts, pits, or risers. They allow rescuers to reach victims, conduct assessments, or

FIGURE 9-9 A mechanical Prusik.

perform rescue operations in areas that are inaccessible by other means. These techniques involve using mechanical devices or climbing techniques to ascend a rope or vertical surface. Ascending devices, such as ascenders (see Figure 6-14 in Chapter 6) or **mechanical Prusiks** (a mechanical friction hitch used to attach a loop of cord around a rope; **FIGURE 9-9**), provide a grip on the rope and allow rescuers to ascend safely and efficiently.

Vertical rope work, whether it involves ascending or descending a rope, requires two rope lines, with one devoted to a safety function. To deploy an ascending technique, first select an appropriate ascending device based on the type of rope, environmental conditions, and personal preferences. Attach the ascending device to the rope and secure it to a harness or climbing system. Use proper climbing techniques, such as hand-over-hand or foot-locking methods, to ascend the rope safely and efficiently. Always maintain three points of contact with the rope, and periodically check the integrity of the ascending device and attachment points.

Descending techniques (often called *lowering techniques*) are used in confined-space operations to exit or enter elevated or vertical areas within the space or to lower victims to safety. Static-line descending is generally rare in confined-space rescue, but it does occur. The focus instead should be on lowering, especially when using retrieval systems. These techniques allow rescuers to retreat from elevated positions or evacuate victims from hazardous environments with controlled descent. They involve safely descending a rope or vertical surface using mechanical devices or controlled-descent methods. Descent control devices (see Figure 6-15 in Chapter 6) provide controlled friction on the rope and allow rescuers to descend safely and smoothly.

FIGURE 9-10 A self-retracting lanyard.

To deploy a descending technique, first select an appropriate descending device based on the type of rope, environmental conditions, and personal preferences. Attach the descending device to the rope and secure it to a harness or descent system. Use proper descent techniques, such as braking, feathering, or lowering methods, to control the speed and smoothness of descent. Always maintain control over the descent process and be prepared to adjust braking or positioning as needed to ensure a safe and controlled descent.

A **self-retracting lanyard** with rescue capabilities is used in confined-space operations to provide fall protection for workers and enable rapid rescue in the event of an emergency (**FIGURE 9-10**). Self-retracting lanyards are equipped with a retractable lifeline that automatically adjusts in length as the user moves, providing freedom of movement while maintaining fall protection. The self-retracting mechanism ensures that the lifeline remains taut and minimizes slack to reduce the risk of tripping or entanglement. Taking up slack also minimizes the effect of dynamic force by reducing the size of a potential fall.

TIP

The terms *self-retracting lifeline* (or SRL; see Chapter 6) and *self-retracting lanyard* are often used interchangeably, but there are slight distinctions, especially in specific industries or applications.

In the event of a fall, a self-retracting lanyard's internal braking mechanism activates, arresting the fall and absorbing energy to reduce the impact on the user. This fall arrest feature provides vital protection against falls from a height, mitigating the risk of serious injury or fatality in confined-space environments.

Some models are equipped with built-in rescue capabilities for self-rescue or assisted rescue of a fallen worker. Rescue features may include an integral winch or retrieval mechanism that enables the user or rescuer to raise or lower the fallen worker to safety.

Self-retracting lanyards are housed in robust and durable casings constructed from materials such as aluminum or thermoplastic, providing protection against impact, abrasion, and environmental hazards. The housing also contains internal components such as the lifeline, braking mechanism, and retrieval system, ensuring reliable performance in demanding confined-space environments.

Despite their rugged construction, these devices are designed to be lightweight and portable, allowing for easy transport and deployment in confined spaces. Their compact and ergonomic designs ensure that they do not impede the user's movement or add unnecessary bulk to their PPE.

A thorough inspection of a self-retracting lanyard to ensure it is in good working condition includes checking for signs of damage, wear, or corrosion on the lifeline, housing, and mechanical components, as well as verifying that the braking mechanism, retrieval system, and rescue features are functional and properly maintained.

To deploy a self-retracting lanyard, first attach it to a secure anchor point using a compatible anchorage connector, such as a D-ring or carabiner. Ensure that the anchor point can support the anticipated loads and is positioned to minimize swing hazards and fall distances.

Wear the self-retracting lanyard in accordance with manufacturer instructions and industry best practices, ensuring that the lifeline is properly routed and free from twists or tangles. Adjust the length of the lifeline as needed to maintain proper tension and minimize slack while allowing for unrestricted movement.

In the event of a fall, remain calm and activate the device's self-rescue capabilities, if they are available. Follow the manufacturer's instructions for engaging the retrieval mechanism or winch to raise or lower yourself to safety.

Exercise caution and control during self-rescue operations, maintaining awareness of potential hazards and maintaining communication with rescuers if possible. If self-rescue is impossible, communicate your situation to nearby rescuers and await assistance. Rescuers can use the self-retracting lanyard's rescue capabilities to perform a controlled descent or ascent to reach the fallen worker and provide assistance. However, these devices come with several limitations that must be considered to ensure safe and effective use. See **TABLE 9-1** for information on the limitations of self-retracting lanyards.

TABLE 9-1 Limitations of Self-Retracting Lanyards

Issue	Limitations
Limited mobility and range	▪ **Confined-space constraints:** Self-retracting lanyards are typically designed for vertical or near-vertical applications in which the worker has limited lateral movement. In confined spaces with complex layouts, the device may restrict the worker's movement, limiting access to certain areas. ▪ **Length of lifeline:** The length of the lifeline is fixed, which can be limiting in larger or more complex confined spaces where greater range or horizontal movement is required.
Anchoring challenges	▪ **Anchor point requirements:** Self-retracting lanyards require a solid and appropriate anchor point directly above the worker. In confined spaces, finding an adequate anchor point that is both strong and properly positioned can be challenging, especially in spaces with low ceilings or complex geometries. ▪ **Compatibility with tripods and davits:** Although self-retracting lanyards can be attached to tripods or davits, the setup needs to be carefully assessed to ensure the device functions properly and does not impede the rescue operation.
Retrieval mechanism limitations	▪ **Manual retrieval effort:** Although many self-retracting lanyards have a built-in retrieval winch, it often requires manual effort to crank the system, which can be time-consuming and physically demanding during a rescue. ▪ **Speed of retrieval:** The retrieval speed of self-retracting lanyards may not be fast enough in some emergency situations, especially when quick extraction is critical.
Potential for swing falls	▪ **Swing fall hazard:** If the anchor point is not directly above the worker, a swing fall can occur if the worker falls while moving laterally. This can result in the worker hitting the sides of the confined space, causing injury.
Environmental factors	▪ **Corrosive or harsh environments:** Confined spaces that involve exposure to corrosive substances, extreme temperatures, or other harsh environmental factors may degrade the self-retracting lanyard's components, affecting their performance and reliability. ▪ **Entrapment risk:** In confined spaces with narrow passages or obstacles, the line could become entangled, potentially hindering the worker's movement or the effectiveness of the rescue operation.
Weight limitations	▪ **Capacity limits:** Self-retracting lanyards have a maximum weight capacity. If the combined weight of the worker, equipment, and any other load exceeds this limit, the device may not function properly, increasing the risk of injury.
Training and misuse	▪ **User training:** Effective use of self-retracting lanyards in confined-space rescue requires specialized training. Improper use or lack of understanding of the equipment's limitations can lead to accidents. ▪ **Misapplication:** Using a self-retracting lanyard in situations for which it was not designed (e.g., horizontal applications without proper fall protection measures) can compromise its effectiveness.
Impact on rescue operations	▪ **Delayed response:** In an emergency, reliance on self-retracting lanyards alone may not provide the fastest means of rescue. Additional rescue techniques and equipment might be necessary to ensure a prompt and safe extraction. ▪ **Complexity of setup:** The need for correct setup and configuration of self-retracting lanyards can add complexity to the rescue operation, potentially leading to delays if not done properly.

Lighting the Scene

Effective illumination is essential for safe and efficient operations in confined spaces, especially during emergencies. Confined spaces often have limited natural light, making it challenging for rescuers to assess hazards, perform patient care, and execute rescue operations. Chemical light sticks are sometimes used to illuminate a space prior to entry. Portable lighting sources enhance visibility, allowing rescuers to navigate confined spaces safely, locate victims, and conduct critical tasks with precision. Types of portable lighting include the following:

- **Handheld flashlights:** Compact and versatile, handheld flashlights provide focused illumination for close-range tasks and inspections (**FIGURE 9-11A**).
- **Headlamps:** Head-mounted lights offer hands-free operation, allowing rescuers to maintain dexterity while illuminating their work area (**FIGURE 9-11B**).
- **Area lights:** Portable area lights provide broad, diffused illumination, ideal for illuminating larger confined spaces and work zones (**FIGURE 9-11C**).
- **Portable light towers:** Light towers equipped with high-intensity lamps offer elevated illumination for large-scale rescue operations and extended-duration tasks (**FIGURE 9-11D**).

FIGURE 9-11 Types of portable lighting include **(A)** a handheld flashlight, **(B)** a headlamp, **(C)** an area light, and **(D)** a portable light tower.

A: © Garna Zarina/Shutterstock; **B:** © Taras Vykhopen/Shutterstock; **C:** Courtesy of Nightstick c/o Bayco Products, Inc.; **D:** © Andyqwe/iStock/Getty Images Plus/Getty Images

Intrinsically safe lighting refers to lighting equipment that is designed to be safe for use in hazardous environments, such as confined spaces with a potential presence of flammable gases, vapors, or dust. This type of lighting is engineered to prevent the risk of ignition by ensuring that any electrical sparks or heat generated by the light are contained and do not ignite the surrounding atmosphere. The design typically includes low-energy circuits, encapsulated or sealed components, and materials that prevent sparks. Using intrinsically safe lighting in confined spaces is critical to protecting workers and maintaining safety in environments where explosive conditions may exist.

Several methods exist for setting up portable lighting. Adjustable tripods or stands can be used to position portable lights at optimal heights for maximum coverage and visibility. Mounting brackets, clamps, or magnetic mounts allow you to secure lights to walls, ceilings, or structural elements. Rigging systems, such as pulleys and ropes, can suspend lights from overhead structures or anchor points, enabling strategic positioning in confined spaces. Battery-powered lighting solutions offer flexibility and mobility, eliminating the need for external power sources and cords in remote or inaccessible locations.

Only lighting equipment designed for rugged environments that is resistant to impact, moisture, and hazardous conditions should be used in confined spaces. Lights with adjustable brightness levels and color temperatures are best because they can adapt to varying lighting conditions and improve visibility. Battery-powered units must have adequate battery life for a prolonged operation; rechargeable options can maintain continuous illumination throughout rescue operations. Redundant lighting systems mitigate the risk of equipment failure or power outages, ensuring uninterrupted visibility during critical tasks.

To better understand how to deploy scene lighting, follow the steps in **SKILL DRILL 9-2**.

PPE for Confined-Space Work

Confined spaces present unique hazards that require specialized PPE to ensure the safety of workers and rescuers. Specifics on personal-issue PPE and team-issue PPE, as well as details on different types of respiratory, clothing, eye and face, and hearing protection, can be found in Chapter 3 in the section "PPE's Role in the Development of a Safety Culture."

- Respiratory protection is crucial in confined spaces where air quality may be compromised because of the presence of hazardous gases or vapors or a lack of oxygen. Types of respiratory protection devices include a self-contained breathing apparatus (SCBA), which provides a portable air supply for the wearer, allowing them to breathe in hazardous atmospheres; a supplied-air respirator (SAR), which delivers breathable air from an external source to the wearer via a hose or tube connected to a clean air source; and air-purifying respirators (APRs), which filter contaminants from the air before the wearer breathes it in. Different cartridges or filters are available for specific contaminants.
- Protective clothing shields workers from physical and chemical hazards present in confined spaces, including cuts, abrasions, punctures, and exposure to harmful substances. Types of protective clothing include coveralls, chemical-resistant suits, gloves, boots, and hard hats.
- SCBA face pieces and other eye and face PPE safeguard against potential hazards like flying debris, splashing liquids, and chemical exposure. Types of eye and face protection include safety goggles, face shields, and chemical splash goggles.
- Hearing protection is essential in environments with high noise levels to prevent hearing loss or damage. Types of hearing protection include earplugs and earmuffs.

Guidance on PPE use for confined spaces can be found in the OSHA regulation "Respiratory Protection" (1998; 29 C.F.R. 1910.134), *NFPA 1981: Standard on Open-Circuit SCBA for Emergency Services* (NFPA 2019a), *NFPA 1989: Standard on Breathing Air Quality for Emergency Services Respiratory Protection* (NFPA 2019b), the OSHA regulation "Personal Protective Equipment General Requirements" (1974; 29 C.F.R. 1910.132), *NFPA 1992: Standard on Liquid Splash-Protective Ensembles and Clothing for Hazardous Materials Emergencies* (NFPA 2018), the OSHA regulation "Eye and Face Protection" (1994; 29 C.F.R. 1910.133), ANSI/ISEA Z87.1 (American Society of Safety Engineers 2020), and the OSHA regulation "Occupational Noise Exposure" (1974; 29 C.F.R. 1910.95).

Assessing PPE and Equipment Requirements in Confined Spaces

The first step in determining the need for PPE and equipment required for a confined-space rescue is to perform a thorough assessment of the space. This

SKILL DRILL 9-2

Deploying Scene Lighting

Step 1: Assess the Incident Scene

Assess the scene to determine optimal lighting positions, ensuring clear visibility for safety and operation.

Step 2: Prepare Lighting Equipment

Unpack and prepare portable lighting equipment for quick deployment to illuminate the scene effectively.

Assemble tripod-mounted lights for height and stability, allowing adjustable lighting angles for versatile coverage.

Step 3: Position Lighting for Optimal Coverage

Strategically place lighting at the confined space entrance to enhance visibility for rescuers. Adjust the lighting angle to reduce shadows and ensure the entrance area is fully illuminated, establishing a well-lit entry point.

Step 4: Deploy Additional Lighting Inside the Confined Space

Position additional portable light or lights inside the confined space to fully illuminate the immediate work zone, providing clear visibility for operations.

Continues.

SKILL DRILL 9-2 CONTINUED

Deploying Scene Lighting

Step 5: Conduct Power Management and Safety Checks

Use a ground fault circuit interrupter (GFCI) to prevent electrical hazards, ensuring that all lighting equipment operates safely throughout the operation.

Step 6: Monitor and Adjust Lighting as Needed

Monitor and adjust lighting to accommodate changing conditions, maintaining optimal visibility. As needed, deploy additional lights to provide sustained illumination for prolonged incident operations.

Step 7: Postincident Breakdown and Review

After the operation, turn off lighting equipment, disassemble it, and store it to ensure readiness for future incidents.

includes the identification of potential atmospheric, biological, chemical, or physical hazards that may be present, such as low oxygen levels, toxic gases, flammable atmospheres, physical obstructions, or engulfment hazards. You must also consider the specific requirements of the rescue scenario, such as vertical or horizontal access or the space's dimensions.

Once that has been completed, you are in a position to evaluate the PPE requirements based on the identified hazards and the nature of the space. The PPE that may be required includes respiratory protection, protective clothing, head protection, eye and face protection, hand protection, and foot protection, but you must consider how compatible the PPE may be with the tasks that must be performed and the other equipment that will be involved in the rescue.

This equipment may include rigging gear, communication devices, lighting, retrieval systems, tripods, winches, lighting equipment, and atmospheric monitoring instruments. The space may have specific challenges that will impair your ability to use this equipment, such as limited access or restricted mobility. Specialized tools or equipment may be necessary, depending on the conditions of the space.

Limiting Damage to Confined-Space Rescue Equipment

Limiting damage to confined-space rescue equipment helps ensure its effectiveness and longevity. It requires proper maintenance and regular inspections of all rescue equipment for signs of wear, damage, or malfunction. For each piece of equipment, the manufacturer's guidelines for maintenance and servicing must be followed to keep it in optimal condition. The procedures taken to limit damage to equipment must be performed in accordance with NFPA 1006 (NFPA 2021).

Rescue equipment must always be stored in a clean, dry, and well-ventilated area, away from direct sunlight and moisture. Use designated storage racks, shelves, or containers to prevent damage from stacking or crushing.

Always handle rescue equipment with care, avoiding rough handling, dragging, or dropping. Appropriate lifting and carrying techniques should be used to prevent strain or damage to the equipment and its accessories. During transport, equipment should be carefully stored to prevent shifting or damage en route to the incident scene.

Rescue equipment should only be used for its intended purpose and within its specified limits. Avoid overloading or exceeding weight capacities, load ratings, or operational limits. Use proper rigging and setup techniques to rig rescue equipment, and do so in accordance with manufacturer instructions and industry best practices. Use appropriate anchoring points, rigging configurations, and hardware to ensure stability and reliability during use.

To ensure the safe operation of all equipment, train personnel regularly in its use, including operating procedures, techniques, and safety precautions. During training or operations, monitor the equipment closely to detect any signs of stress, strain, or malfunction, and take corrective action as needed. While the equipment is in use, visual inspections can help identify any signs of wear, damage, or deterioration. Pre- and post-use inspection checklists can be a useful tool to catalog issues and the measures taken to abate them. Promptly address any issues or concerns to prevent further damage or compromise to the equipment's integrity.

Rescue equipment must be cleaned and decontaminated after each use to remove dirt, debris, or contaminants that could cause damage or corrosion (decontamination is covered in detail in Chapter 5 in the section "Decontamination Protocols"). After decontamination, ensure the equipment is completely dry before storage to prevent mold, mildew, or corrosion. Maintain detailed records of equipment inspections, maintenance, repairs, and replacements. Document any incidents of damage to equipment; when they occur, investigate their root causes to prevent recurrence.

CHAPTER FEATURE: AFTER-ACTION

IN SUMMARY

- Portable anchor systems are crucial in confined-space rescue operations because of their adaptability and reliability. These systems offer the flexibility to create secure anchor points in environments where permanent structures may be insufficient or unavailable.
- The most common portable anchor systems are monopods, tripods, davit arms, beam clamps, and portable winch systems.
- Stabilizing portable anchor systems to limit movement during rescue operations not only enhances the safety of the rescue team but also prevents potential injuries to the victim and minimizes the risk of system failure.
- Understanding the forces at play is crucial for ensuring that a portable anchor system can withstand the load imposed during rescue operations and maintain structural integrity. The forces are vertical force, horizontal force, angular force, and dynamic force.

IN SUMMARY CONTINUED

- *Force application* refers to the distribution of forces exerted on a portable anchor system during rescue operations. These forces include the weight of the victim, tension from rigging components, dynamic forces from movement, and environmental factors such as wind or vibration. Several methods can be used to calculate force application on portable anchors in confined-space operations, including static analysis, dynamic analysis, and environmental factors.
- Methods of lighting a confined-space rescue scene include handheld flashlights, headlamps, portable area lights, and portable light towers. Lighting can be secured using adjustable tripods or stands; mounting brackets, clamps, or magnetic mounts; or rigging systems. Battery-powered lighting solutions offer flexibility and mobility, eliminating the need for external power sources and cords in remote or inaccessible locations.
- Confined spaces present unique hazards that require specialized PPE to ensure the safety of workers and rescuers. The most essential forms of PPE for confined-space rescue are respiratory protection, protective clothing, eye and face protection, and hearing protection.
- Limiting damage to confined-space rescue equipment requires proper maintenance and regular inspections of all rescue equipment for signs of wear, damage, or malfunction. Rescue equipment must always be stored in a clean, dry, and well-ventilated area, away from direct sunlight and moisture. It must be handled with care, avoiding rough handling, dragging, or dropping. The equipment should only be used for its intended purpose and within its specified limits.

KEY TERMS

angular force The force exerted on an anchor point when the load is pulled at an angle relative to it.

ascending techniques Techniques used in confined-space operations to access elevated or vertical areas within the confined space, such as vertical shafts, pits, or risers. They allow rescuers to reach victims, conduct assessments, or perform rescue operations in areas that are inaccessible by other means.

beam clamp A portable anchor system device that is designed to securely attach to horizontal beams or structural members, providing a stable anchor point for various rescue operations.

change-of-direction pulley Also known as a redirection pulley, a system used to redirect the direction of force in a rope system.

davit arm A type of anchor system device that consists of a vertical post or mast with an extendable boom arm that can rotate horizontally and is typically made of durable materials, such as steel or aluminum, that can withstand the loads encountered during rescue operations. They may be fixed or portable.

descending techniques Techniques used in confined-space operations to exit elevated or vertical areas within the confined space or to lower victims to safety. They allow rescuers to retreat from elevated positions or evacuate victims from hazardous environments with controlled descent.

dynamic analysis for force application Calculation of the forces exerted on a structure or anchor point when there is movement or acceleration involved.

dynamic force The additional forces exerted on an anchor point resulting from changes in velocity or direction of motion.

force application The distribution of forces exerted on a portable anchor system during rescue operations.

high-point anchor A secure anchoring system used in rope rescue and confined-space operations, positioned above the working area or entry point. This anchor provides a high point of attachment for ropes, allowing for efficient and controlled vertical movement of personnel and equipment.

horizontal force The force exerted on an anchor point when the load is pulled or pushed horizontally away from it.

mechanical Prusik A mechanical type of friction hitch used to attach a loop of cord around a rope.

monopod A single-leg support structure designed to provide vertical stability and load-bearing capacity as a portable anchor system device.

portable winch system A portable anchor system device that typically consists of a winch unit, rigging attachment, and anchor point.

self-retracting lanyard A vertical fall-protection lifeline that allows the user to move around easily and freely while the line stays consistently taut, effectively preventing falls.

static analysis for force application Calculation of the total weight or force exerted on a structure or anchor point when all forces are in equilibrium.

tie-back An additional safety measure for tripods in which the anchor point or anchoring system used in the rescue setup is secured with an additional line or anchor.

vertical force The force exerted on an anchor point when a load is suspended directly below it.

REFERENCES

American Society of Safety Engineers. 2020. *American National Standard Occupational and Educational Personal Eye and Face Protection Devices.* ANSI/ISEA Z87.1-2020. Arlington, VA: International Safety Equipment Association.

"Eye and Face Protection." 1994. 29 C.F.R. 1910.133. https://www.ecfr.gov/current/title-29/subtitle-B/chapter-XVII/part-1910/subpart-I/section-1910.133.

National Fire Protection Association. 2018. *NFPA 1992: Standard on Liquid Splash-Protective Ensembles and Clothing for Hazardous Materials Emergencies, 2018 Edition.* Quincy, MA: National Fire Protection Association.

National Fire Protection Association. 2019a. *NFPA 1981: Standard on Open-Circuit SCBA for Emergency Services, 2019 Edition.* Quincy, MA: National Fire Protection Association.

National Fire Protection Association. 2019b. *NFPA 1989: Standard on Breathing Air Quality for Emergency Services Respiratory Protection, 2019 Edition.* Quincy, MA: National Fire Protection Association.

National Fire Protection Association. 2021. *NFPA 1006: Standard on Technical Rescue Personnel Professional Qualifications, 2021 Edition.* Quincy, MA: National Fire Protection Association.

National Fire Protection Association. 2022. *NFPA 2500: Standard on Operations and Training for Technical Search and Rescue Incidents and Life Safety Rope and Equipment for Emergency Services, 2022 Edition.* In *NFPA National Fire Codes Online.* Quincy, MA: National Fire Protection Association.

"Occupational Noise Exposure." 1974. 29 C.F.R. 1910.95. https://www.ecfr.gov/current/title-29/subtitle-B/chapter-XVII/part-1910/subpart-G/section-1910.95.

"Permit-Required Confined Spaces." 1993. 29 C.F.R. 1910.146. https://www.ecfr.gov/current/title-29/part-1910/section-1910.146.

"Personal Protective Equipment General Requirements." 1974. 29 C.F.R. 1910.132. https://www.ecfr.gov/current/title-29/subtitle-B/chapter-XVII/part-1910/subpart-G/section-1910.95.

"Respiratory Protection." 1998. 29 C.F.R. 1910.134. https://www.ecfr.gov/current/title-29/subtitle-B/chapter-XVII/part-1910/subpart-I/section-1910.134.

CHAPTER QUESTIONS

1. Which of the following is NOT a common type of portable anchor system used in confined-space rescue?
   a. monopods
   b. tripods
   c. beam clamps
   d. scaffolding

2. What is the primary purpose of a change-of-direction pulley in confined-space rescue operations?
   a. to increase mechanical advantage in a rope system
   b. to redirect the direction of force in a rope system
   c. to measure the weight of the victim
   d. to provide additional ventilation

3. When calculating force application on portable anchors, which of the following analyses considers movement or acceleration?
   a. static analysis
   b. dynamic analysis
   c. environmental analysis
   d. structural analysis

4. What is the main advantage of using a self-retracting lanyard in confined-space rescue operations?
   a. It provides fall protection and enables rapid rescue.
   b. It creates a separate anchor system.
   c. It allows for hands-free communication.
   d. It provides air to a victim in a confined space.

CHAPTER FEATURE: ON SCENE

A RESCUE IN A SILO

An emergency incident occurred at a large grain storage facility in the Midwest. In the incident, a worker entered a grain silo to perform routine maintenance on the auger system when he suddenly became engulfed by the flowing grain.

The Emergency Response

Local fire departments, along with specialized confined-space rescue teams, responded to the scene immediately. Upon arrival, they established incident command and assessed the situation. It was determined that the victim was submerged in the grain up to his waist and was unable to free himself.

The Rescue Operations

The confined-space rescue team conducted a thorough assessment of the silo, identifying potential hazards such as unstable grain conditions, limited ventilation, and the risk of grain collapse. Atmospheric monitoring equipment was deployed to monitor oxygen levels and detect any hazardous gases within the silo.

Portable anchor systems, including beam clamps and davit arms, were secured to provide stable anchor points for the rescue operation. Rigging ropes and pulley systems were set up to facilitate the extraction process, ensuring the safety of both rescuers and the victim.

Rescuers communicated with the victim to assess his condition and provide reassurance. A rapid victim harness using webbing was constructed to extricate the victim and prevent further submersion in the grain.

The Extrication and Treatment

Using a combination of mechanical advantage systems and manual lifting techniques, rescuers gradually lifted the worker out of the grain. Rescuers maintained constant communication with the victim throughout the extraction process, providing updates on their progress and ensuring his cooperation.

Once he was safely extricated from the silo, the victim was assessed by medical personnel for injuries and potential respiratory issues. He was transported to a nearby medical facility for further evaluation and treatment.

The Outcome

Despite the challenging conditions posed by the confined-space environment, the coordinated efforts of the rescue teams resulted in a successful extrication of the victim from the grain silo. The victim sustained minor injuries and was discharged from the hospital later that day. The incident underscores the importance of proper training, equipment, and coordinated rescue efforts in confined-space rescue operations.

This case study highlights the challenges and complexities involved in confined-space rescue operations and emphasizes the importance of preparedness, training, and teamwork in ensuring successful outcomes.

1. What was the main hazard that caused the worker to become trapped in the grain silo?
   - **a.** mechanical failure of the auger system
   - **b.** engulfment resulting from flowing grain
   - **c.** toxic gas release within the silo
   - **d.** collapse of the silo structure
2. What was one of the primary hazards identified during the rescue operation?
   - **a.** electrical fires inside the silo
   - **b.** faulty rigging systems for rescue
   - **c.** unstable grain conditions and risk of grain collapse
   - **d.** excessive accumulation of water inside the silo
3. Which piece of equipment was used to safely anchor the rescue operation inside the silo?
   - **a.** fire hoses
   - **b.** hydraulic jacks
   - **c.** lifting trucks
   - **d.** portable anchor systems
4. What method did the rescuers use to extricate the victim from the grain?
   - **a.** They cut through the wall of the silo.
   - **b.** They used a rapid victim harness with mechanical advantage systems.
   - **c.** They pumped out the grain with vacuum equipment.
   - **d.** They lowered a ladder to allow the victim to safely climb out.

SECTION

3

Technician

You Are the Rescuer

Your confined-space rescue team has responded to a frantic call from a child who reports that she was playing with two younger children in an abandoned plastics manufacturing plant when the two suddenly disappeared. She followed the sound of their cries and discovered they were trapped in a 30,000-gallon storage tank.

After your team arrived, you discovered that the children had crawled into the tank, which is 12 feet (3.6 meters) tall, through an opening 18 inches (20 cm) wide near the top of the tank and landed hard on several feet of plastic pellets that lay at the bottom of the tank. One of the children has head injuries and a probable broken leg but is semiconscious. The other child is apparently uninjured, alert, and conscious, but he is terrified.

Your team has decided to make entry through a 30-inch (76 cm) manway near the bottom of the tank, so you and your fellow team members are using a reciprocating saw to cut the bolts off.

The incident commander (IC) has ordered that continuous monitoring of the atmosphere is to be conducted near the children, so atmospheric monitors are placed strategically by them. You are designated as the rescue attendant and stationed on a ladder to peer into the tank through the hole the children used to enter the tank. From your perch, you are charged with maintaining constant visual contact with both children and verbal communication with the uninjured child.

Suddenly, the buzz of the saw pauses, and you hear a piercing cry from the uninjured child. He screams that he has begun to sink into the pellets and that his feet are getting wet. You shift your focus to the atmospheric monitor and realize that the oxygen level is falling—it had been 20.2 percent and has now fallen to 20.0 percent.

You Are the Rescuer Questions

1. In this situation, what is the first thing you should do?
2. How have your strategic factors changed?
3. How do you expect your priorities may change in this situation?
4. What can you do to change the atmospheric conditions within the space?

CHAPTER

10

Technician

Medical Concerns for Confined-Space Technicians

KNOWLEDGE OBJECTIVES

After studying this chapter, you should be able to:

- Differentiate between a rescue and a recovery operation. (pp. 256–257)
- Discuss the proficiencies of short spine immobilization equipment. (**NFPA 1006: 7.3.4**, p. 264; pp. 265–266)
- Discuss the identified constraints of short spine immobilization equipment. (**NFPA 1006: 7.3.4**, p. 264; p. 266)
- Describe the methods used to secure and move victims onto an immobilizer. (**NFPA 1006: 7.3.4**, pp. 266–267)
- Discuss the effects of hazardous atmospheres on victims and rescuers. (**NFPA 1006: 7.3.5**, pp. 267–268)
- Discuss the protocols for medical and psychological evaluations of responders. (**NFPA 1006: 7.3.5**, pp. 257–264; p. 268; pp. 270–272)
- Describe the methods for patient care. (**NFPA 1006: 7.3.6**, pp. 257–264)
- Discuss the importance of having operational protocols related to communication and medical treatment. (**NFPA 1006: 7.3.6**, pp. 259–264)

SKILL OBJECTIVES

- Demonstrate the process of immobilizing a victim in a confined-space incident. (**NFPA 1006: 7.3.4**, pp. 265–266)
- Demonstrate the process of monitoring a confined space. (**NFPA 1006: 7.3.5**, pp. 268–269)
- Assess the scene for personal protective equipment (PPE) and equipment requirements. (**NFPA 1006: 7.3.6**, pp. 261–262)
- Ensure that operational protocols are instituted and reassessed as needed. (**NFPA 1006: 7.3.6**, p. 260)
- Critique the scene and choose appropriate medical protocols to determine treatment priorities. (**NFPA 1006: 7.3.6**, p. 262)

Introduction

Confined-space rescue operations demand a high level of skill, precision, and adherence to stringent safety protocols to ensure the well-being of both victims and rescuers. In particular, providing medical care to victims in confined-space rescue events is a challenging process. In some cases, it is impractical or impossible to provide care in the space, and rescuers must determine whether the best course of action is to extricate the victim before administering advanced medical treatment. Teams must be aware of and trained in methodologies for providing effective patient care in confined spaces, the selection and use of appropriate personal protective equipment (PPE) for medical responders, and the operational protocols that facilitate efficient communication and medical treatment.

Operational protocols on communication, medical treatment, coordination, and safety procedures help ensure the safety and effectiveness of confined-space rescue teams and victims.

Confined-space incidents often require rescuers to immobilize victims with suspected spinal injuries using short spine boards, long spine boards, or Kendrick extrication devices (KEDs). Knowing which device is most appropriate for which situation requires experience and training.

Hazardous atmospheres pose significant risks to both victims and rescuers. Continuous monitoring is critical to maintaining safety in confined-space operations, where conditions can change rapidly.

Rescue technicians face unique challenges that go beyond the physical demands of their profession. Among the most harrowing experiences is dealing with loss of life, particularly in confined-space incidents. The psychological impact of such events can be profound, affecting mental health and overall well-being.

Leadership and organizational support are critical in fostering a culture of mental health awareness and support within rescue teams. Leaders should promote an environment where mental health is prioritized and seeking help is encouraged. **Post-traumatic stress disorder (PTSD)**—a mental health condition triggered by experiencing or witnessing a traumatic event, characterized by symptoms such as flashbacks, severe anxiety, and uncontrollable thoughts about the event—is a significant concern for emergency responders, who frequently encounter traumatic and high-stress situations.

Rescue Versus Recovery

Confined-space emergencies necessarily take place in locations with limited entry and exit points, poor ventilation, and potentially hazardous atmospheres. These formidable challenges significantly affect the outcomes of rescue operations. Understanding the statistics of rescue versus recovery in these scenarios is crucial for developing effective response strategies, improving safety protocols, and enhancing training programs for emergency personnel.

Defining Rescue and Recovery

Rescue is the process of safely extricating a living person from a confined space. Recovery is the process of retrieving the body of a deceased person from a confined space. Whether an emergency is a rescue or recovery effort often hinges on a team's rapid response, effective communication, and thorough preparedness.

According to the U.S. Bureau of Labor Statistics (2020), approximately 129 deaths occur annually in confined spaces. Most of these incidents involve workers in industries such as manufacturing, construction, utilities, and agriculture. Studies indicate that survival rates significantly decrease after the first hour of entrapment. Rescue rates within the first hour can be as high as 85 percent, but survival rates drop below 30 percent if the response is delayed beyond this period. See **TABLE 10-1** for a comparison of two case studies: One is a rescue operation, and the other is a recovery operation.

Asphyxiation due to oxygen deficiency or toxic atmospheres is the leading cause of confined-space fatalities; physical trauma and hazardous substances contribute to the remaining fatalities. Data from the National Institute for Occupational Safety and Health (NIOSH) show that approximately 60 percent of confined-space incidents result in recovery operations rather than rescues (Centers for Disease Control and Prevention 1986).

Successful rescue operations often depend on the immediate availability of trained rescue personnel and proper equipment. The promptness of the initial emergency call and the arrival of rescue teams are critical factors. Delays in detection and response significantly reduce the likelihood of rescue. The proficiency and preparedness of the rescue team, including their familiarity with confined-space protocols, have a tremendous influence on the success of the operation. Regular drills and scenario-based training improve response times and the efficiency of rescue operations. Access to

TABLE 10-1 Case Studies: Rescue Versus Recovery

Incident	Response Time	Outcome
A maintenance worker was trapped in a sewer line after being overcome by methane gas.	The rescue team arrived within 15 minutes of the incident.	The worker was successfully extricated and revived onsite using cardiopulmonary resuscitation (CPR) and oxygen therapy; the quick response and immediate application of rescue protocols were key factors in the successful outcome.
Two workers were trapped in a chemical storage tank after a sudden release of toxic fumes.	The incident was reported late, and rescue teams arrived 45 minutes after the initial exposure.	Both workers were deceased upon arrival of the rescue team; the delay in response and the high concentration of toxic gases resulted in a recovery operation.

specialized rescue equipment, such as ventilators, gas detectors, and proper PPE, is essential for conducting safe and effective rescues.

Continuous monitoring of atmospheric conditions within confined spaces is vital, as shown in the "You Are the Rescuer" feature at the beginning of this chapter. High levels of toxic gases, such as hydrogen sulfide or carbon monoxide, or low oxygen levels can quickly incapacitate victims.

Recommendations for Improving Rescue Rates

Institute regular and comprehensive training programs for rescue teams that emphasize confined-space entry procedures, atmospheric monitoring, and emergency medical care. Implementation of preincident planning and rapid response systems, including automated alarm systems and immediate dispatch protocols, can help ensure the timely arrival of rescue personnel.

State-of-the-art monitoring equipment used to continuously assess atmospheric conditions within confined spaces—such as remote monitoring systems, wireless gas detectors, and wearable sensors—can provide early warnings of hazardous conditions. Robust communication systems between the trapped individuals and the rescue team can help the team quickly assess the situation and plan the rescue operation—and quickly adapt the response when conditions change, as they did in "You Are the Rescuer."

The distinction between rescue and recovery operations in confined-space emergencies is stark, with survival often dependent on the rapidity and efficacy of the response. Although statistics indicate a high rate of recovery operations as a result of delayed responses and hazardous conditions, there is significant potential to improve rescue outcomes through enhanced training, rapid response initiatives, and advanced monitoring technologies. By understanding and addressing the factors that influence these outcomes, emergency response teams can better prepare for confined-space incidents, ultimately saving more lives and reducing the frequency of fatal recoveries.

Conducting Medical Care During a Confined-Space Rescue

Confined-space rescues often present significant obstacles to providing immediate medical care because of restricted access, limited mobility, and hazardous conditions. Administering medical care during a confined-space rescue—while following your local medical protocols—is complex and challenging, and sometimes, providing care within the space is impractical or impossible. Rescuers must make a critical decision in many cases and determine whether the best course of action is to extricate the victim before administering advanced medical treatment.

Confined spaces are areas that are not designed for continuous human occupancy, have limited or restricted means for entry or exit, and are large enough for a person to enter and perform tasks. They typically have narrow openings and restricted pathways, making it difficult for rescuers to reach and move within the space.

These spaces often contain hazardous atmospheres, including oxygen-deficient or oxygen-enriched air, toxic gases, and flammable vapors. Structural hazards, such as unstable surfaces or the risk of engulfment, add to the complexity. The restricted environment can hinder the ability of rescuers to maneuver, use equipment, and provide medical care effectively. Effective communication is crucial during rescues, but confined spaces can impede verbal communication and the use of radios or other devices.

The first step in any confined-space rescue is a thorough assessment of the situation. This includes identifying hazards, evaluating the condition of the victim, and determining the best approach for rescue and medical care. Initial medical care often involves the following steps:

- **Scene survey:** Determine whether the victim is entrapped, energized, or contaminated with a hazardous material.
- **Primary survey:** Conduct a primary survey to assess the victim's immediate needs. This includes "the ABCs"—checking the airway (A), breathing (B), and circulation (C). If the victim is unresponsive, rescuers must quickly determine the presence of life-threatening conditions.
- **Stabilization:** If possible, provide basic stabilization within the confined space. This may include controlling bleeding, providing oxygen, or immobilizing fractures.
- **Hazard mitigation:** Address any immediate hazards that pose a risk to both the victim and the rescuers. This could involve ventilating the space to improve air quality or shoring up unstable structures, which takes time. A best practice is to provide the victim with their own air supply.
- **Communication and coordination:** Establish effective communication with the rescue team and any medical personnel outside the confined space. Coordination is key to ensuring a unified approach to rescue and medical care.

Rapid Extrication of an Injured Victim Before Providing Medical Care

In some scenarios, providing medical care within a confined space is either impractical or impossible. These situations require quick decision-making and a focus on the safest and most effective way to extricate the victim. Factors that may necessitate immediate extrication in the best interest of the victim's care and well-being include the following:

- **Severe environmental hazards:** If the confined space contains life-threatening hazards such as toxic gases or structural instability, the priority should be to remove the victim as quickly and safely as possible.
- **Limited access:** In spaces where there is insufficient room to maneuver or provide care, attempting medical interventions can endanger both the victim and rescuers.
- **Critical medical conditions:** Victims with critical medical conditions, such as severe trauma, cardiac arrest, or respiratory failure, may require advanced medical interventions that are not feasible within the confined space.
- **Time sensitivity:** In cases where time is of the essence, the focus should be on rapid extrication to a safe environment where comprehensive medical care can be administered.

Extrication of an injured victim from a confined space requires meticulous planning and execution. The primary goal is to remove the victim safely and efficiently, minimizing further harm. To effectively extricate a victim, you must first develop a detailed plan that includes the method of extrication, necessary equipment, and the roles and responsibilities of each team member. The plan must consider factors such as the size and weight of the victim, the configuration of the confined space, and potential obstacles.

Your team must have the appropriate equipment for extrication, such as harnesses, ropes, stretchers, and mechanical lifting devices. This equipment must be tested and ready to go, and all rescuers must be trained in its use.

The victim must be moved with great care to prevent exacerbating the victim's injuries. If possible, the victim should receive basic stabilization before and during extrication. The victim's condition must be monitored continuously throughout the extrication process, and the team must be prepared to adjust the plan if the victim's condition changes or if unforeseen challenges arise.

Close coordination with medical personnel waiting outside the confined space is necessary because they must be ready to provide care immediately after extraction. The team must ensure they are briefed on the victim's condition and any interventions performed during extrication.

Administering Advanced Medical Treatment After Extrication

Once the victim is safely extricated from the confined space, the focus shifts to administering advanced medical treatment. This stage is critical for addressing any life-threatening conditions and stabilizing the victim for transport to a medical facility.

Depending on the victim's condition, **advanced life support (ALS) interventions** may be necessary. These include advanced airway management, intravenous access, administration of medications, and cardiac monitoring.

TABLE 10-2 Case Studies: Conducting Medical Care of Victims

Incident Type	Situation	Decision	Outcome
Toxic gas exposure	In an incident involving an unconscious victim trapped in a chemical storage tank, rescuers faced significant hazards that made providing care within the space impossible.	Rescuers performed rapid extrication of the victim before treatment.	Once outside, the victim received advanced medical treatment, including administration of antidotes and respiratory support, which were lifesaving.
Structural collapse	A collapsed trench trapped a worker; the confined space and unstable environment limited the ability to provide immediate medical care.	Rescuers focused on stabilizing the victim's position and extricating before treatment.	Advanced trauma care was administered post-extrication, addressing multiple fractures and internal injuries.
Limited-access rescue	A narrow underground tunnel severely restricted movement and access to the victim.	Rescuers prioritized rapid extrication using specialized equipment before treatment.	Once the victim was removed, paramedics initiated advanced cardiac life support, ultimately resuscitating the victim, who had suffered cardiac arrest.

A thorough secondary assessment can identify and address any additional injuries or medical conditions. This includes a head-to-toe examination and obtaining a detailed medical history of the victim, if possible. Continuous monitoring of the victim's vital signs and condition will allow the medical team to provide ongoing care and interventions as needed while preparing for transport. The receiving medical facility will need a detailed report on the victim's condition, the interventions performed, and any ongoing concerns to ensure a smooth transition of care upon arrival.

TABLE 10-2 presents case studies of three incidents where victims required extrication before the application of medical care.

Recognizing whether providing care within a confined space is impractical or impossible is crucial for ensuring the safety of both the victim and the rescuers. Rapid and safe extrication followed by advanced medical treatment is often the most effective approach in such scenarios. By understanding the complexities of confined-space rescues and adopting best practices, rescue teams can enhance their ability to save lives and protect their own well-being during these demanding operations.

Operational Protocols for Victim Care and Medical Treatment

Operational protocols play a pivotal role in ensuring the safety and effectiveness of confined-space rescue operations. These protocols include communication, medical treatment, coordination, and safety procedures that must be followed meticulously.

For communication, the team must establish clear lines of communication among all team members using radios or other reliable devices and use standardized communication protocols to convey critical information efficiently. To avoid confusion, you should use plain language instead of departmental or emergency response codes. This ensures clear and effective communication within the Incident Command System (ICS).

Regarding medical treatment, the team must follow established medical protocols for assessment, stabilization, and treatment and ensure all responders are trained in confined-space-specific medical procedures.

Coordination and safety operational protocols include clearly assigning roles and responsibilities to avoid confusion during operations, conducting regular safety briefings and debriefings to update the team on any changes or hazards, and placing the priority on responders' safety when working in a confined-space setting.

Operational protocols for victim care and medical treatment are designed to ensure consistent, high-quality care. They are developed by healthcare professionals based on best practices, clinical guidelines, and regulatory standards. Once created, these protocols are implemented across healthcare settings to guide treatment decisions and patient management.

The implementation process involves training staff, integrating the protocols into daily practice, and ensuring that all relevant personnel are familiar with them. The protocols are then periodically reassessed by reviewing recent medical research, analyzing victim outcomes, and gathering feedback from healthcare providers. This ongoing reassessment ensures that the protocols remain effective, up to date, and aligned with current medical standards, with adjustments made as needed to enhance patient care and safety.

Instituting and Reassessing Operational Protocols

These operational protocols must be dynamic and adaptable to evolving conditions within a confined-space emergency. Continuous reassessment and improvement are essential to maintaining high standards of safety and efficiency.

If the space's environment and conditions are continuously monitored throughout the operation, the team will be ready to adjust protocols as necessary based on real-time data and feedback. Debriefing sessions conducted after each operation allow team members to review their performance and identify areas for improvement. The feedback these sessions provide can then be incorporated into the protocols to enhance the quality of future responses.

Responders must be regularly trained on these updated protocols, as well as any new equipment that has been introduced. Drills allow team members to practice and refine their operational procedures.

A Medical Protocol Example: A Trapped Victim with Back Pain

A worker has fallen inside a confined space, such as a storage tank or underground utility vault, and is experiencing significant back pain. The victim is conscious and responsive but unable to move because of her pain. Your objective is to safely assess, stabilize, and extricate the victim while minimizing the risk of further injury, particularly to her spine. To do so, follow this step-by-step medical protocol:

Step 1: Initial assessment and scene safety. Ensure the scene is safe for entry by using appropriate PPE and atmospheric monitoring (checking for oxygen levels, toxic gases, etc.). Consider the routes of access and egress, as well as the configuration of the space, when determining how to remove the victim. Approach the victim cautiously, ensuring not to exacerbate any potential spinal injury. Brief the rescue team on the situation, outlining roles and responsibilities.

Step 2: Primary assessment (ABCs). Ensure the victim's airway is clear. Assess her breathing quality and rate. Place a supplemental air supply on the victim. Check the victim's pulse and assess her skin color, temperature, and moisture level. Control any visible bleeding.

Step 3: Secondary assessment. Evaluate the victim's level of consciousness using the **AVPU scale** (A = Alert, V = responds to Voice, P = responds to Pain, and U = Unresponsive). Ask her specifics about the nature of her fall and the symptoms she is suffering from (e.g., location and intensity of pain; numbness, tingling, or weakness in extremities). Perform a head-to-toe assessment focused on areas of pain and possible injury.

Step 4: Stabilization and immobilization. If spinal injury is suspected, manually stabilize her cervical spine immediately. Place a helmet and safety glasses on the victim as needed, and apply a cervical collar, if available. Prepare a short spine board or long spine board for immobilization. Depending on space constraints, use an immobilization device designed for technical rescue for durability and added handholds within the confined space. Secure the victim to the board. Ensure that her head, neck, torso, and legs are immobilized. Use straps and padding to prevent movement and provide comfort.

Step 5: Monitoring and support. Continuously monitor the victim's vital signs (e.g., pulse, respiration, blood pressure, oxygen saturation). Administer pain relief as per medical protocols and the rescuer's scope of practice. This may include oral pain medication if the victim can swallow and compromising the airway is not a risk. Keep the victim calm by explaining each step of the procedure and reassuring her that help is on the way.

Step 6: Communication. Maintain communication with the incident command, providing updates on the victim's condition and the progress of the rescue operation. Emergency medical services (EMS) systems typically include multiple levels of medical oversight, categorized as direct and indirect control. Consult with medical control if available, especially for pain management and any complications.

Step 7: Extrication. Determine the safest and most efficient route for extrication, considering the confined-space constraints. Carefully maneuver the victim out of the confined space. If necessary, use rope systems or mechanical advantage with progress capture devices (PCDs), which hold a rope in place to prevent a load from sliding back down when the pulling force is released. Keep monitoring the victim's condition during the extrication process. Be prepared to adjust her immobilization as needed.

Step 8: Post-extrication care. Once the victim is in a safer environment, conduct a more thorough secondary survey to identify any other injuries. Prepare her for transport to a medical facility. Ensure all immobilization devices remain secure during transport. Provide a detailed handover to the receiving medical team, including the mechanism of injury, the victim's condition, the treatment provided, and any changes during extrication.

Following is a summary of how the extrication and treatment of this victim could play out:

- **Initial contact:** Rescuers enter the confined space with full PPE and monitoring equipment. They find the victim lying on her back. She reports severe pain in her lower back and states that she is unable to move.
- **Assessment:** The victim's airway is clear. Her breathing is labored but adequate, and her pulse is strong but rapid. She has no visible bleeding and no other apparent injuries.
- **Stabilization:** Rescuers apply the cervical collar to the victim and use a spinal immobilization board to stabilize her spine in the confined space. The victim is constantly reassured and informed as each step is taken.
- **Extrication:** The victim is secured to the spinal immobilization device and carefully moved to the top of the space using a rope system. As the extrication process proceeds, the rescuers maintain communication with incident command and perform continuous monitoring of the victim's condition. She is safely extricated from the confined space and handed over to EMS.
- **Post-extrication:** Once the victim has been extricated, a full secondary assessment is conducted. The victim is prepared for transport to the hospital. A detailed handover assessment is provided to the EMS team.

Patient Care and Safety Protocols

Confined-space emergencies present unique challenges that require specialized training and protocols to ensure the safety and well-being of both victims and responders. Teams must be aware of and trained in methodologies for providing effective patient care in confined spaces, the selection and use of appropriate PPE for medical responders, and the operational protocols that facilitate efficient communication and medical treatment. It is also important to note that prior to entry, equipment needs and requirements should be anticipated and prepared for deployment—with the caveat that some equipment requirements may not be revealed until the entry team makes contact with the victim.

Effective patient care requires a systematic approach tailored to the environment of the confined space in question. Each team member should be thoroughly trained in *all* rescue disciplines prior to any emergency. Medical responders must be proficient in both basic and advanced medical interventions, as well as familiar with the constraints of confined spaces. Ideally, medical specialists will be part of the team, but specific levels of medical response training are necessary for all rescue team members.

The initial assessment of the victim must be a thorough primary survey (e.g., following the ABC [airway, breathing, circulation] protocol) to identify life-threatening conditions. A rapid secondary survey can be performed to assess additional injuries or medical conditions.

Next, the victim must be stabilized. Immobilize suspected spinal injuries using appropriate equipment, such as short spine boards, and administer oxygen and other critical treatments as needed.

For extrication and transport, plan and execute a safe and efficient extrication strategy, and ensure that the victim's condition is continuously monitored, reassessed, and documented throughout transport.

Assessing the Scene for Hazard-Specific PPE and Equipment Requirements for Medical Responders

Before any confined-space entry, it is essential to thoroughly assess the scene to determine the necessary PPE and equipment. This assessment ensures that medical responders are adequately protected and equipped to handle the emergency. Training preparation before an incident provides a higher chance for a positive outcome during an emergency response. This includes both physical and mental preparation for dealing with difficult circumstances.

The scene-assessment process consists of a survey for hazard identification, selection of appropriate PPE and equipment, and an evaluation of the team's operational readiness. The hazard-identification process involves locating and identifying potential hazards such as toxic gases, unstable structures, and poor environmental conditions. Proper selection and use of PPE are crucial for protecting medical responders from the various hazards present in confined spaces, including toxic atmospheres, physical obstructions, and biological hazards.

Types of hazard-specific PPE include respiratory protection (e.g., a self-contained breathing apparatus [SCBA], a supplied-air respirator [SAR], or air-purifying respirators [APRs], which are equipped with filters to protect against specific contaminants); protective clothing (e.g., chemical-resistant suits, abrasion-resistant coveralls, or flame-resistant clothing, the latter of which is essential in environments that contain flammable gases or materials); and additional gear, such as helmets and gloves. The PPE chosen should fit the identified hazards within the space, and all equipment used, including medical kits and extrication tools, must be readily available and vetted to ensure it is in good working condition.

The evaluation of the team's operational readiness includes taking time to verify that all responders are familiar with the equipment and PPE to be used and conducting pre-entry checks and drills to ensure their readiness.

Critiquing the Scene and Choosing Appropriate Medical Protocols

Critical evaluation of the scene and selection of appropriate medical protocols involves analyzing the specific conditions of the confined space and tailoring medical interventions accordingly.

The first step is to perform a comprehensive evaluation that assesses the overall environment, including the space's structural integrity and potential hazards. This evaluation also includes an evaluation of the condition of the victims and any immediate medical needs and a determination of the response's equipment requirements and logistical needs.

Next, the team must choose the appropriate medical protocols that can address the victims' identified needs and conditions. Priority should be given to interventions that stabilize victims and prepare them for safe extrication. Further treatment priorities should be based on the severity of injuries and the overall situation. A triage system is necessary if multiple victims are involved.

Triage of Multiple Patients

Triage, a critical process in emergency medical situations, is particularly important in confined-space rescues where multiple patients may be affected by hazardous conditions. The objective of triage is to prioritize victims' care based on the severity of their conditions, ensuring that those in the most critical need receive the most immediate attention. This process becomes more complex in confined-space scenarios because of the additional challenges of limited access, potential environmental hazards, and the physical constraints of the space.

Primary triage consists of a rapid assessment and categorization of victim needs. The rapid assessment involves a quick evaluation of each patient's condition using visual and auditory cues and a basic vital signs assessment, if possible. Victims are then assigned to the following categories (**FIGURE 10-1**) based on their condition and needs:

- **Immediate (red):** Patients with life-threatening injuries or conditions that require immediate intervention to survive (e.g., severe bleeding, airway obstruction).
- **Delayed (yellow):** Patients who need urgent medical attention but whose condition is stable enough to wait a short period (e.g., fractures, controlled bleeding).

Triage Chart		
Color	**Meaning**	**Description**
Red	Immediate	Patient has life-threatening injuries or conditions that require immediate intervention to survive.
Yellow	Delayed	Patient needs urgent medical attention, but their condition is stable enough that they can wait for a short period.
Green	Minor	Patient has minor injuries and can wait for treatment.
Black	Expectant	Patient has little or no chance of survival, given the available resources, and is not the immediate focus of rescue efforts.

FIGURE 10-1 A color-coded triage chart.

Scenario: A chemical spill in an industrial facility's underground storage area with five workers trapped inside.	
Triage Level	**Victims, Their Conditions, and Their Medical Needs**
Immediate (Red)	• Worker A: Unconscious, severe respiratory distress, and cyanosis. Requires immediate airway management and extraction. • Worker B: Severe bleeding from a leg wound, showing signs of shock. Needs urgent hemorrhage control and rapid evacuation.
Delayed (Yellow)	• Worker C: Conscious with a broken leg and moderate pain. Stable vitals but requires medical attention for fracture management. • Worker D: Minor head injury, alert but confused. Needs observation and possible imaging but can wait for treatment.
Minor (Green)	Worker E: Minor abrasions and cuts, fully conscious and ambulatory. Can assist with self-extraction if safe to do so.
Expectant (Black)	No workers in this category for this scenario, but the team remains prepared to make this difficult decision if necessary.
Actions Taken	
Conducted rapid assessment and triage tagging.	
Prioritized the immediate extraction of Workers A and B.	
Stabilized Worker B's bleeding and provided oxygen support to Worker A.	
Communicated ongoing conditions and needs to external medical teams.	
Reassessed Workers C, D, and E continuously, adjusting care and extraction plans as needed.	

FIGURE 10-2 A case study example of triaged victims.

- **Minor (green):** Patients with minor injuries who can wait for treatment (e.g., minor cuts, abrasions).
- **Expectant (black):** Patients with little or no chance of survival given the available resources, who are not the immediate focus of rescue efforts (e.g., massive trauma, no pulse or respiration).

Secondary triage is a reevaluation process. After victims have undergone the initial categorization process and have been removed to a safer environment, a more detailed assessment can be made to confirm or update their triage status. They must be continuously monitored because the conditions of those who have been classified as lower priority levels could deteriorate, and their triage category may change as well. **FIGURE 10-2** shows a case study of triaged victims in a confined-space incident.

The triage process in a confined-space emergency involves an initial scene and victim assessment, the categorization and tagging process, prioritization for extraction, stabilization, and continuous reassessment.

- **Scene assessment:** Evaluate the overall safety of the environment, including the presence of hazardous atmospheres, structural stability, and potential secondary hazards. Ensure the safety of rescuers by using appropriate PPE and monitoring equipment.
- **Initial patient assessment:** Approach each patient quickly but carefully. Check for responsiveness, airway patency, breathing, and circulation. Perform a quick head-to-toe examination to identify obvious injuries.
- **Categorization and tagging:** Use triage tags or markers to identify each patient's triage category visibly. Document key findings and vital signs on the tags if time permits.
- **Prioritization for extraction:** Coordinate with rescue teams to prioritize the extraction of patients based on their triage status. Ensure that those tagged as immediate are the first to be moved out of the confined space.
- **Stabilization:** Provide basic lifesaving interventions where possible within the confined space (e.g., controlling severe bleeding, opening the airway). Administer any necessary emergency medical care that can be performed onsite.
- **Continuous reassessment:** As conditions change or new information becomes available, reassess each patient's status. Update triage categories and priorities based on ongoing assessments.

Confined spaces often restrict movement, making it difficult to reach all patients quickly. Limited access points can slow down the extraction process and the delivery of medical care. Hazardous atmospheres (e.g., toxic gases, low oxygen levels) can compromise both patient and rescuer safety. Monitoring equipment must be used continuously to ensure a safe working environment.

Effective communication among rescue and medical teams is crucial but can be hampered by the confined environment. Use of radios, hand signals, and clear communication protocols is essential.

The availability of medical supplies and equipment may be limited within a confined space. Rescuers must be prepared to make critical decisions with limited resources.

Effective triage in confined-space incidents is essential for maximizing patient survival and ensuring efficient use of rescue resources. By adhering to structured triage protocols, continuously reassessing patient conditions, and addressing the unique challenges of confined spaces, rescue teams can provide the best possible care under difficult circumstances. Training and preparedness are key to successfully managing these complex and high-stress situations. Always be sure to check the protocols of the authority having jurisdiction (AHJ) in your area.

Immobilization Methods

Comparing Short and Long Spine Boards

In EMS and rescue operations, short and long spine boards are essential tools for immobilizing and transporting victims with suspected spinal injuries. Each type of spine board has its own unique applications, advantages, and limitations.

A long spine board (see Figure 6-26 in Chapter 6) is a full-length, rigid board used for immobilizing and transporting victims with suspected spinal injuries. It provides support from the head to the pelvis and is commonly used in a variety of rescue and medical scenarios. Long spine boards provide full-body support, immobilizing the head, neck, torso, and pelvis, and thus are ideal for victims with suspected injuries to any part of the spine. They are suitable for a wide range of emergency situations, including those requiring transportation over longer distances. Their longer length and multiple securing points offer better stability during movement and transportation and a reduced risk of further injury during extraction and transport.

The disadvantages of long spine boards include their bulkiness, their application time, and their relatively heavy weight and reduced portability. Larger and more cumbersome than short spine boards, they can be difficult to maneuver in confined spaces. They are less practical in situations with limited space, such as inside vehicles or narrow corridors. Long spine boards require more time and effort to properly secure a victim because of their size and the number of straps. They are heavier and more challenging to carry than short boards, especially in difficult terrain or when rapid deployment is needed. They place an increased physical demand on rescuers, potentially slowing down the rescue process.

A short spine board is a compact, rigid board designed for immobilizing the thoracic and cervical spine. It is primarily used in confined-space rescues and extrications from vehicles. Its size allows it to be more maneuverable, making it ideal for use in confined spaces like vehicle interiors, tight hallways, or small rooms. It is lightweight, is easy to position and carry, and causes less physical strain on rescuers. Thanks to its compact size and smaller number of securing points, a short spine board can be applied more quickly than a long spine board. As a result, the stabilization process is fast, which is crucial in emergency situations.

Short spine boards do have disadvantages—first and foremost, because they are designed to immobilize only the upper spine and torso, they are not suitable for victims with suspected injuries to the lower spine or pelvis. They do not provide the full-body support necessary for certain injuries. As a result, additional equipment may be required to fully immobilize the victim, increasing complexity and time. They are best suited for specific scenarios, such as vehicle extrications, and are less versatile than long spine boards.

Proficiencies in Short Spine Immobilization Equipment

Short spine immobilization equipment is specifically designed to stabilize the cervical and thoracic spine of a victim suspected of having spinal injuries. This equipment typically includes a short spine board, straps, head immobilizers, and supporting harnesses. The adjustable straps secure the victim to the board, preventing movement during extraction. Head immobilizers (see Figure 6-23 in Chapter 6) are cushioned blocks that are placed on either side of the victim's head to prevent movement and maintain spinal alignment. Additional padding can be used to fill gaps and ensure the victim is immobilized comfortably and securely. To better understand the process of immobilizing a victim in a confined space, follow the steps in **SKILL DRILL 10-1**.

SKILL DRILL 10-1

Immobilizing a Victim in a Confined Space

Step 1: Stabilize the Victim's Head and Neck

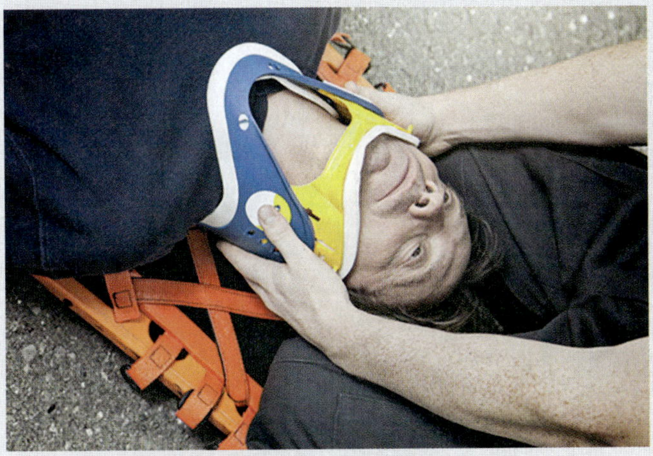

Manually stabilize the victim's head and neck.

Step 2: Position the Board

Position a spine board behind the victim without causing additional movement.

Step 3: Secure the Head Immobilizers

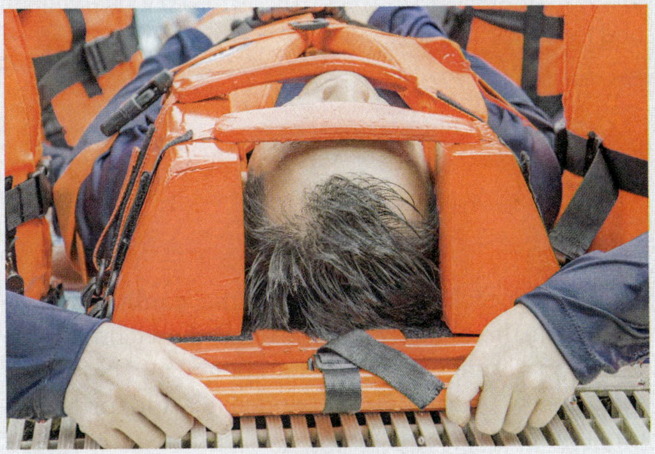

Apply the head immobilizers and secure them with straps, ensuring the victim's head is immobilized.

Step 4: Secure the Torso and Legs

Secure the victim's torso and legs with the straps provided, ensuring that the straps are snug but not overly tight. Recheck all straps and adjust as necessary to maintain proper alignment.

Continues.

SKILL DRILL 10-1 CONTINUED

Immobilizing a Victim in a Confined Space

Step 5: Coordinate the Move

The team should move the victim in a coordinated fashion, using a controlled and smooth motion to prevent any jarring movements.

Step 6: Monitor the Victim's Condition

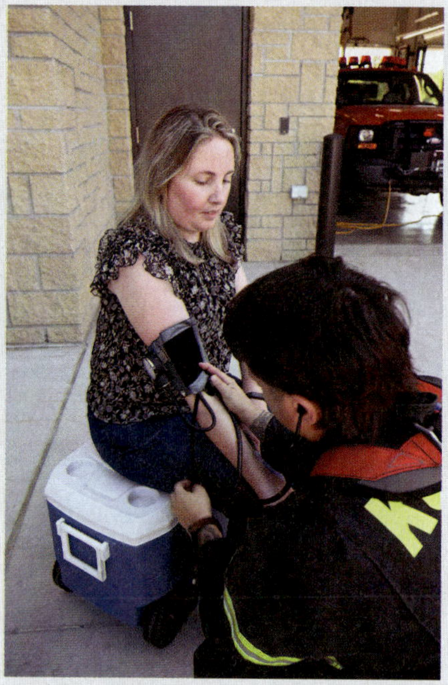

Closely monitor the victim's condition and adjust the approach as necessary based on real-time observations and feedback.

Steps 1, 4, and 5: © Courtesy of Illinois Fire Service Institute; **Step 2:** © Operation 2021/Alamy Stock Photo; **Step 3:** © narin phapnam/Shutterstock

Proficiency with short spine immobilization equipment ensures that rescuers can do the following:

- Quickly and effectively secure the victim to the immobilizer.
- Maintain spinal alignment throughout the rescue process.
- Adapt to the constraints and challenges posed by confined spaces.

Despite its many advantages, short spine immobilization equipment has several limitations and challenges. For example, confined spaces often have areas with restricted movement, making it difficult to position and secure the short spine board. Rescuers must work in close quarters, which can hinder their ability to properly immobilize and secure the victim.

The physical dimensions of the confined space may limit the ability to fully extend the victim's body and properly align the spine. Equipment and personnel may have to be carefully positioned to avoid further injury to the victim.

In emergencies, time is critical. The need to quickly stabilize and extract the victim can add stress, potentially compromising the immobilization process. Victims may be in awkward positions that complicate the application of the immobilization equipment. Additional injuries or medical conditions may require modified approaches to immobilization.

Properly securing and moving victims onto a short spine immobilizer involves several key steps:

Step 1: Initial assessment. Perform a primary survey to assess the victim's condition and check for life-threatening injuries. Stabilize the head and neck manually while preparing the immobilization equipment.

Step 2: Position the short spine board. Carefully slide the short spine board behind the victim, maintaining spinal alignment. Adjust the board so that it fits snugly against the victim's back.

Step 3: Apply the head immobilizers. Position the head immobilizers on either side of the victim's head. Secure the immobilizers with the provided straps, ensuring the head is stabilized and cannot move.

Step 4: Secure the straps and harnesses. Fasten the chest, waist, and leg straps snugly, but not so tight as to restrict breathing or circulation. Ensure that all straps are secured, and double-check for tightness and proper alignment.

Step 5: Move the victim. With the victim secured to the board, coordinate with the team to carefully move the victim out of the confined space. Use a smooth, controlled motion to avoid jarring or sudden movements that could exacerbate spinal injuries.

The Effects of Hazardous Atmospheres and Monitoring

Confined spaces can harbor various hazardous atmospheres, including toxic, flammable, or oxygen-deficient conditions. These hazards can significantly affect the health and safety of both victims and rescuers.

Sources of toxic atmospheres include chemical spills, decomposing organic matter, and industrial processes. Examples include carbon monoxide, hydrogen sulfide, and ammonia, which are present in the air. The physiological effects of toxic atmospheres can include respiratory distress, neurological impairment, and in severe cases, death.

The sources of flammable atmospheres can include leaking gases, volatile organic compounds, or combustible dusts (e.g., sawdust, sugar, or flour). Examples include methane, propane, grain dusts, and gasoline vapors. A flammable atmosphere bears the risk of explosion and fire, which can lead to burns, smoke inhalation, trauma, and death.

The causes of oxygen deficiency in a confined space can include displacement by other gases, chemical reactions, and biological processes. If the oxygen concentration reaches 19.5 percent or lower, the atmosphere is considered oxygen deficient. Symptoms of oxygen deficiency include dizziness, shortness of breath, loss of consciousness, and death from asphyxiation.

Psychologically speaking, any of these three types of atmospheric hazards can lead victims and rescuers to experience acute stress, which can include inducement of panic, confusion, and impaired decision-making. Long-term exposure to acute stress can lead to PTSD, anxiety, and depression in rescuers.

Monitoring Hazardous Atmospheres

Effective atmospheric monitoring is crucial to detect and mitigate hazards in confined spaces. Types of monitoring equipment used to ensure the safety of rescue operations include multi-gas detectors, single-gas detectors, photoionization detectors (PIDs), and infrared spectrometers.

- Multi-gas detectors simultaneously monitor multiple gases, such as oxygen, carbon monoxide, hydrogen sulfide, and flammable gases. They provide real-time readings and alarms for immediate hazard identification.
- Single-gas detectors measure the concentration of a specific gas. They are useful for targeted monitoring of known hazards.
- PIDs detect volatile organic compounds (VOCs) and toxic gases at low concentrations. They are the best choice for detecting hazardous substances not identified by standard gas detectors.
- Infrared spectrometers identify and quantify gases based on their infrared absorption characteristics. They are suitable for detecting a wide range of gases, including those in complex mixtures.

The operational requirements of these monitoring devices include proper calibration, careful maintenance of battery life, durability assessments, and proper training. Regular calibration is essential to ensure accurate readings. The meters used must be capable of monitoring the expected gases in the space, and the calibration gases must match the meter in use. Devices must be fully charged or determined to have adequate battery life for the duration of the operation. Spare batteries that are properly charged or backup devices should be kept on hand. The equipment should be rugged, intrinsically safe, and able to withstand harsh environmental conditions. Ideally, these devices should be waterproof and shock resistant.

All members of the team should be fully trained in the proper operation of the equipment in use. Technician-to-technician training in the field must be properly documented, and all records must remain current and well maintained. Because air monitoring is a hazardous materials skill set, team members must be properly trained to the hazardous materials operations level.

The Importance of Continuous Monitoring

Continuous monitoring is critical to maintaining safety in confined-space operations, where conditions can change rapidly, and the real-time data it yields are essential to informed decision-making. To better understand the air-monitoring process, follow the steps in **SKILL DRILL 10-2**.

These real-time data provide immediate information on atmospheric conditions and allow for quick response to emerging hazards. Audible and visual alarms on the monitoring equipment alert rescuers to dangerous conditions. This equipment also allows users to preset different thresholds for specific gases to set off alarms, which helps ensure that a warning is given while action is still possible.

Continuous monitoring also allows teams to record data over time, which assists in post-operation analyses. These recorded data can help identify trends and improve future safety protocols.

Physical and Mental Health of Responders

Rescue technicians face unique challenges that go beyond the physical demands of their profession. Among the most harrowing experiences is dealing with loss of life, particularly in confined-space incidents. The psychological impact of such events can be profound, affecting mental health and overall well-being.

Ensuring the health and fitness of responders is paramount before, during, and after confined-space operations. Proper evaluations mitigate risks and enhance operational effectiveness. There are protocols for pre-operation, during-operation, and post-operation evaluations.

A pre-operation evaluation consists of a medical check, where physical fitness, respiratory health, and any preexisting conditions are assessed, and a psychological assessment, which evaluates mental readiness and stress levels.

SKILL DRILL 10-2

The Air-Monitoring Process

Step 1: Determine the Proper Equipment

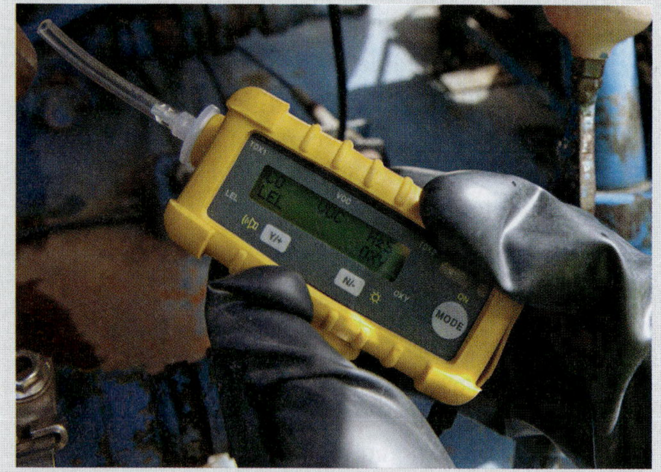

The right air-monitoring equipment must be selected for the operation.

Step 2: Perform Monitoring from Outside the Space

Perform air monitoring from outside the confined space to determine the safety of the space before entry.

SKILL DRILL 10-2 CONTINUED

The Air-Monitoring Process

Step 3: Continuously Monitor Throughout Entry Operations

Maintain continuous air monitoring while entry operations are being conducted.

Step 4: Continuously Monitor Throughout Rescue Operations

Maintain continuous air monitoring while rescue operations/victim packaging are being conducted.

Step 5: Interpret Data

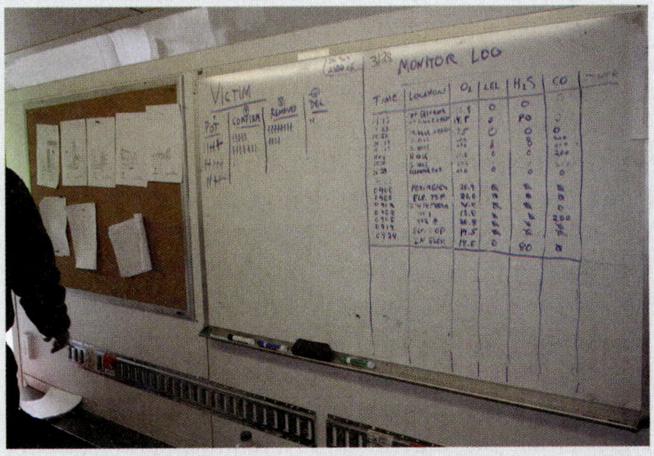

Interpret the air-monitoring data to identify hazardous conditions.

Step 1: © Jones & Bartlett Learning. Photographed by Glen E. Ellman; **Steps 2 to 4:** © King Ropes Access/Shutterstock; **Step 5:** Courtesy of Illinois Fire Service Institute.

During the operation, evaluation consists of continuous health monitoring of vital signs, hydration levels, and signs of fatigue. Responders are paired (the buddy system) to monitor each other's health and well-being.

The post-operation evaluation's medical assessment checks responders for exposure-related symptoms, injuries, or health issues. Psychological support is provided during debriefing sessions, and responders should have access to counseling if needed.

Understanding the Psychological Impact

Confined-space incidents are inherently dangerous and often involve complex rescue operations. When these situations result in loss of life, the emotional toll on rescue technicians can be significant. Understanding the psychological impact is the first step toward addressing mental health needs.

Immediately following a traumatic incident, rescue technicians may experience acute stress reactions. These can include shock, disbelief, anxiety, and hyperarousal. Although these reactions are normal, they can be overwhelming and distressing. Prolonged exposure to traumatic events can lead to PTSD, which is characterized by symptoms such as intrusive memories, flashbacks, nightmares, and severe anxiety. PTSD can significantly impair a technician's ability to function both personally and professionally. Repeated exposure to traumatic situations can result in compassion fatigue, where individuals become desensitized to suffering and feel emotionally drained. This can lead to burnout, a state of physical, emotional, and mental exhaustion.

Deaths of victims, especially in confined-space incidents where rescue efforts are intensely personal and physically close, can lead to profound feelings of grief and loss. Rescue technicians may struggle with feelings of guilt, helplessness, and sadness.

Recognizing the Signs of Mental Health Struggles

Early recognition of mental health struggles is crucial for timely intervention and support. Rescue technicians and their colleagues should be aware of the signs that indicate a need for help.

Emotional signs include persistent sadness, anxiety, irritability, and emotional numbness. Technicians may also experience mood swings or heightened emotional sensitivity. Behavioral signs include changes in behavior, such as withdrawal from social interactions, increased use of alcohol or substances, and difficulty concentrating, which can indicate mental health issues. Other signs include changes in eating or sleeping patterns.

Physical symptoms such as headaches, fatigue, gastrointestinal issues, and a general decline in physical health can be linked to mental health struggles. Cognitive signs technicians may experience include intrusive thoughts, difficulty making decisions, and a sense of hopelessness. Recurrent memories or flashbacks of the traumatic event are also common.

Strategies for Coping and Building Resilience

Building resilience and adopting effective coping strategies are essential for rescue technicians to manage the psychological impact of confined-space incidents. First, engaging in open conversations with colleagues who understand the unique challenges of the profession can provide emotional support and validation. Peer support groups can facilitate shared experiences and coping strategies.

Seeking help from mental health professionals, such as counselors or psychologists, can provide a safe space to process emotions and develop coping mechanisms. Therapeutic approaches such as cognitive–behavioral therapy (CBT) can be particularly effective.

Techniques such as mindfulness, meditation, and deep-breathing exercises can help reduce stress and promote relaxation. Regular physical activity and hobbies can also serve as healthy outlets for stress relief. Setting boundaries between work and personal life is crucial for maintaining mental health. Taking regular breaks and vacations and ensuring time for rest and recuperation can prevent burnout.

Participating in critical incident stress debriefing (CISD) sessions after traumatic incidents can help rescue technicians process their experiences and emotions. These structured debriefings allow for discussion of the event and emotional reactions in a supportive environment.

Ongoing education and training in mental health awareness can equip rescue technicians with the knowledge and skills to recognize and address mental health issues. Training should include recognizing signs of distress in themselves and others and knowing when and how to seek help.

The Role of Leadership and Organizational Support

Leadership and organizational support are critical in fostering a culture of mental health awareness and support within rescue teams. Leaders should promote an environment where mental health is prioritized and seeking help is encouraged. This includes creating policies that support mental health, such as providing access to mental health resources and support services.

Leaders should receive training on how to recognize signs of mental health struggles in their team members and how to provide appropriate support. This includes knowing how to initiate conversations about mental health and referring individuals to professional help when needed.

Organizations should implement wellness programs that include mental health components. These programs can offer regular mental health check-ins, workshops on stress management, and access to counseling services.

Organizations should also encourage rescue technicians to maintain a healthy work–life balance. This includes respecting time off, providing flexible scheduling when possible, and promoting activities that support physical and mental well-being.

TABLE 10-3 presents case studies of individuals and organizations involved in confined-space rescue facing mental health issues.

TABLE 10-3 Case Studies: Individuals and Organizations Facing Mental Health Issues

Case Study 1	In a fire department, a rescue technician experienced severe emotional distress after a confined-space rescue operation resulted in the loss of a victim. Through the department's peer support program, the technician found solace in sharing his experience with colleagues who had faced similar situations. The peer support group provided a platform for mutual understanding and coping, significantly aiding his recovery process.
Case Study 2	A rescue technician developed symptoms of post-traumatic stress disorder (PTSD) following a traumatic incident. Despite initial reluctance, she sought help from a mental health professional. Through regular counseling sessions, she learned to manage her symptoms and develop healthier coping mechanisms. Her experience underscores the importance of professional mental health support for rescue technicians.
Case Study 3	A fire department implemented a comprehensive wellness program following several high-stress incidents. The program included mandatory debriefings, access to counseling services, and workshops on stress management. This proactive approach resulted in a noticeable improvement in the overall mental health and morale of the rescue team.

PTSD in Responders

PTSD is a significant concern for emergency responders, who frequently encounter traumatic and high-stress situations, such as confined-space recovery operations. These environments, characterized by their hazardous conditions, limited access, and the often-grim task of recovering deceased individuals, can profoundly affect the mental health of those involved. Studies indicate that PTSD affects approximately 20 to 30 percent of emergency responders, with higher rates observed in those exposed to particularly traumatic incidents.

The nature of confined-space recoveries often involves dealing with severely injured or deceased individuals, which can be deeply disturbing. The visual and olfactory stimuli rescuers encounter, such as decomposing bodies or noxious gases, can be particularly traumatic. Many rescuers are troubled by a sense of helplessness and guilt when they find themselves unable to rescue a trapped victim alive. Witnessing the suffering or death of colleagues during rescue operations can exacerbate these feelings.

Working under a constant threat to their personal safety, including exposure to toxic atmospheres, risk of explosions, or structural collapse, adds to the stress experienced during these operations. Repeated exposure to confined-space incidents over time can lead to cumulative stress, increasing the likelihood of developing PTSD.

Symptoms of PTSD include distressing and intrusive memories (sometimes in the form of flashbacks); avoidance of places, activities, or people who remind the responder of the traumatic event; negative changes in thinking or mood; memory problems, including an inability to remember important aspects of the traumatic event; difficulty in maintaining close relationships; and changes in physical and emotional reactions.

These changes can include being easily startled or frightened; always being on guard for danger; self-destructive behaviors, such as drinking too much or driving too fast; trouble sleeping and concentrating; and irritability, angry outbursts, or aggressive behavior.

TABLE 10-4 Case Studies: Interventions in Responders

Situation	Intervention	Outcome	Result
A responder experienced a traumatic recovery operation involving multiple fatalities in a collapsed sewer system.	Immediate critical incident stress debriefing (CISD) was provided, followed by regular counseling sessions.	The responder reported significant improvement in coping with the trauma and continued to perform effectively in their role.	Successful early intervention
A responder with a history of multiple traumatic confined-space recoveries began experiencing post-traumatic stress disorder (PTSD) symptoms after several years.	The responder initially avoided seeking help, leading to worsening symptoms.	Although the responder faced a challenging recovery, ongoing support and therapy facilitated gradual improvement.	Cumulative stress and delayed onset of PTSD

Effective management of PTSD in responders requires early intervention, ongoing mental health and organizational support, and frequent training on the psychological impact of their work and self-care strategies.

Providing immediate psychological support following a traumatic incident can mitigate the development of PTSD. CISD allows responders to discuss their experiences in a structured and supportive environment shortly after the event, typically after 24 hours but within a week. Access to mental health professionals who specialize in trauma and PTSD is crucial. Regular counseling sessions and peer support groups can provide a platform for responders to share their experiences and coping strategies.

A supportive organizational culture that prioritizes mental health is necessary to help responders deal with the effects of PTSD. Policies that encourage taking time off for mental health and provide resources for coping with trauma can help (see Appendix G, An Example of Organizational Policy on Mental Health).

Comprehensive training programs that prepare responders for the potential psychological impact of their work can build resilience. Including mental health education in regular training ensures that responders recognize PTSD symptoms in themselves and their colleagues.

Encouraging responders to engage in self-care activities, such as physical exercise, hobbies, and spending time with loved ones, can help them reduce and manage stress. Teaching stress management techniques, such as mindfulness and relaxation exercises, can improve their overall well-being.

TABLE 10-4 presents case studies of two responders at risk for PTSD.

CHAPTER FEATURE: AFTER-ACTION

IN SUMMARY

- Confined-space rescue operations demand a high level of skill, precision, and adherence to stringent medical and safety protocols to ensure the well-being of both victims and rescuers
- Providing medical care to victims in confined-space rescue events is a challenging process. In some cases, it is impractical or impossible to provide care in the space, and rescuers must determine whether the best course of action is to extricate the victim before administering advanced medical treatment.
- Teams must be aware of and trained in methodologies for providing effective patient care in confined spaces, the selection and use of appropriate PPE for medical responders, and the operational protocols that facilitate efficient communication and medical treatment.
- Operational protocols on communication, medical treatment, coordination, and safety procedures help ensure the safety and effectiveness of confined-space rescue teams and victims.

- Confined-space incidents often require rescuers to immobilize victims with suspected spinal injuries using short spine boards, long spine boards, or KEDs. Knowing which device is most appropriate for which situation requires experience and training.
- Hazardous atmospheres pose significant risks to both victims and rescuers. Continuous monitoring is critical to maintaining safety in confined-space operations, where conditions can change rapidly.
- Rescue technicians face unique challenges that go beyond the physical demands of their profession. The psychological impact can be profound, affecting mental health and overall well-being.
- Leadership and organizational support are critical in fostering a culture of mental health awareness and support within rescue teams. Leaders should promote an environment where mental health is prioritized and seeking help is encouraged.
- PTSD is a significant concern for emergency responders, who frequently encounter traumatic and high-stress situations.

KEY TERMS

advanced life support (ALS) interventions Medical interventions such as advanced airway management, intravenous access, administration of medications, and cardiac monitoring.

AVPU scale A scale used to measure victim condition, with A = Alert, V = responds to Voice, P = responds to Pain, and U = Unresponsive.

post-traumatic stress disorder (PTSD) A mental health condition triggered by experiencing or witnessing a traumatic event that is characterized by symptoms such as flashbacks, severe anxiety, and uncontrollable thoughts about the event.

REFERENCES

Centers for Disease Control and Prevention. 1986. "Preventing Occupational Fatalities in Confined Spaces." https://www.cdc.gov/niosh/docs/86-110/default.html#print.

National Fire Protection Association. 2021. *NFPA 1006: Standard on Technical Rescue Personnel Professional Qualifications, 2021 Edition.* Quincy, MA: National Fire Protection Association.

U.S. Bureau of Labor Statistics. 2020. "Injuries, Illnesses, and Fatalities: Fatal Occupational Injuries Involving Confined Spaces." https://www.bls.gov/iif/factsheets/fatal-occupational-injuries-confined-spaces-2011-19.htm.

CHAPTER QUESTIONS

1. ALS interventions include which of the following?

- **a.** first-aid measures like bandaging wounds and applying splints
- **b.** medical measures like advanced airway management and cardiac monitoring
- **c.** a head-to-toe examination and procurement of the victim's medical history
- **d.** immediate psychological support following a traumatic incident

2. Which of the following is required for extrication of an injured victim from a confined space?

- **a.** thorough training and good weather
- **b.** thorough permits and entry approval
- **c.** meticulous planning and execution
- **d.** meticulous experts and permissions

3. Early recognition of mental health struggles is crucial for which of the following?

- **a.** after-action reviews
- **b.** timely intervention
- **c.** data collection
- **d.** team cohesiveness

4. Effective management of PTSD in responders requires

- **a.** organizational support.
- **b.** a change in duties.
- **c.** medical insurance.
- **d.** a request for help.

CHAPTER FEATURE: ON SCENE

A MEDICAL EMERGENCY IN A CONFINED SPACE

A worker was performing routine maintenance and cleaning inside a large industrial storage tank at a chemical processing plant when he slipped and fell. Immediately after the fall, the worker began experiencing severe back pain and difficulty breathing.

The tank, which was primarily used for storing chemical solvents, posed significant hazards to the worker and the rescuers who were called to his aid, including potential toxic atmospheres and limited access.

The Emergency Response

The incident occurred at 10:15 a.m., and the call came in to the responders—the plant's emergency response team (ERT) and its onsite medical team, as well as the local fire department's confined-space rescue team—at 10:20 a.m. The teams arrived on the scene within 10 minutes of the call.

Upon arrival, the fire department's confined-space rescue team conducted a rapid assessment of the scene. The tank's atmosphere was immediately monitored using portable gas detectors, revealing oxygen deficiency and the presence of VOCs. The tank was ventilated to improve air quality, and all rescuers donned appropriate PPE, including SCBAs.

The Rescue Operations

Two rescuers entered the tank while maintaining continuous communication with incident command outside. They found the victim lying on his back; he was conscious but in severe pain. He reported acute lower back pain and compromised breathing—likely because of the combination of his pain and the oxygen-deficient environment.

The rescuers checked the victim's vital signs and found that his pulse was rapid, his breathing was labored, and his skin was pale and clammy. A cervical collar was applied to prevent potential spinal movement. A spinal immobilization device was used to stabilize his spine. Oxygen was administered via a portable oxygen cylinder to improve his respiration, and continuous monitoring of vital signs was initiated.

During the triage process, the victim was categorized as "Immediate" (red) because of his severe pain, respiratory distress, and potential spinal injury. Ongoing updates were provided to incident command and the medical team outside.

The Extrication and Treatment

The team evaluated the safest and most efficient route for extraction, considering the tank's narrow entry point. A high-angle rescue system with ropes and pulleys was set up to facilitate safe lifting. The victim was carefully secured to a spinal immobilization device, ensuring minimal movement. Using the rope system, rescuers slowly and steadily extricated the victim from the tank, ensuring his continuous stability. Continuous coordination was maintained with the medical team to prepare for immediate care upon extraction.

A detailed secondary assessment was conducted outside the confined space. No other major injuries were identified, but the victim's lower back pain and respiratory distress persisted.

The victim was transferred to the onsite medical team for further stabilization. Detailed information about the fall, initial findings, and care provided was communicated to the paramedics. The victim was transported to a nearby hospital for further evaluation and treatment, with continuous monitoring of his condition performed throughout his transit.

The Outcome

The victim was diagnosed with a lumbar vertebral fracture and respiratory distress from his exposure to the space's oxygen-deprived environment. He received appropriate medical treatment, including pain management, spinal stabilization, and respiratory support, and his condition stabilized after a few days in the hospital. He underwent physical therapy for his back injury and was eventually discharged with a follow-up plan.

1. What triggered the medical emergency inside the storage tank?
   - **a.** A worker was exposed to chemicals.
   - **b.** A worker slipped and fell.
   - **c.** There was a fire inside the tank.
   - **d.** There was a structural collapse inside the tank.

2. What immediate hazards did rescuers face when entering the tank?
   a. slippery floors and sharp equipment
   b. poor lighting and high temperature
   c. oxygen deficiency and VOCs
   d. high pressure from gas leaks
3. Which devices were used to stabilize the worker's spine during the rescue?
   a. a splint and an LSP Half Back
   b. a cervical collar and a spinal immobilization device
   c. a short spine board and a Yates Spec Pak
   d. a stretcher and a long spine board
4. How was the worker safely extricated from the confined space?
   a. Rescuers lowered a ladder and helped him climb out.
   b. Rescuers used a crane to lift him out.
   c. Rescuers used a high-angle rescue system.
   d. Rescuers guided him through an emergency exit.

CHAPTER

11

Technician

Methods of Entry, Respirators, and Other Equipment

KNOWLEDGE OBJECTIVES

After studying this chapter, you should be able to:

- Describe the methods of entry used in confined-space incidents. (**NFPA 1006: 7.3.5**, pp. 278–279)
- Describe the methods used to conserve air, manage the air line, and communicate in a confined space. (**NFPA 1006: 7.3.3**, pp. 280–283)
- List the types and operational requirements of supplying respirators. (**NFPA 1006: 7.3.5**, pp. 279–280)
- Determine the need for utilizing a supplied-air respirator (SAR). (**NFPA 1006: 7.3.3**, p. 285)
- Discuss the proficiencies of an SAR. (**NFPA 1006: 7.3.3**, p. 280; p. 284)
- Discuss the identified constraints of an SAR. (**NFPA 1006: 7.3.3**, p. 280; p. 284)
- Describe hazard-specific personal protective equipment (PPE). (**NFPA 1006: 7.3.6**, pp. 279–280)

SKILL OBJECTIVES

- Ensure victim and rescuer communications are established. (**NFPA 1006: 7.3.5**, pp. 281–283)
- Execute a self-rescue. (**NFPA 1006: 7.3.1**, pp. 294–295)
- Manage air lines and communicate with outside sources while wearing an SAR. (**NFPA 1006: 7.3.3**, pp. 281–283)
- Demonstrate air supply management while wearing an SAR. (**NFPA 1006: 7.3.3**, p. 286)

Introduction

Methods of entry to confined spaces for rescue operations include direct, vertical, horizontal, and technical entry, and preconstructed confined-space entry kits can be very useful in all these types of rescue operations. The kits facilitate quick, efficient, and safe responses to confined-space emergencies. Proper care, maintenance, and setup of the kits are essential to maximizing their effectiveness and reliability.

Once responders are inside a confined space, rope systems are often necessary. These include main line, belay line, two-tensioned, and mechanical advantage systems. Using these systems to ascend and descend within a confined space requires many other types of equipment, including progress capture devices (PCDs), such as ascenders and Prusik knots.

Because many confined spaces harbor the potential for hazardous atmospheres, adequate respiratory protection is a key part of personal protective equipment (PPE) for rescuers. A self-contained breathing apparatus (SCBA; for short-duration entries) and a supplied-air respirator (SAR; for those of longer durations) are the primary methods of protection in use.

As outlined in NFPA 1006 7.3.3 (National Fire Protection Association [NFPA] 2021), the decision to use an SAR involves evaluating several factors: atmospheric conditions, duration of the operation, mobility requirements, and air supply management needs. Proper management of SAR air supplies requires keeping hoses clear of obstructions, avoiding sharp bends, and securing hoses to prevent entanglement.

Establishing and maintaining clear communication with outside sources is always critical in rescuer operations but even more so when rescuers must use an SAR or SCBA. As per NFPA 1006 7.3.5, the following methods are used (NFPA 2021):

- **Voice communication systems:** Hands-free devices that enable clear communication between rescuers and support teams.
- **Hand signals:** Preestablished signals for situations where verbal communication is impractical.
- **Tethered communication lines:** Ensuring a direct line of communication through tethered lines that can also serve as guide ropes.

In addition to preconstructed entry kits and an SAR, this chapter details other forms of critical confined-space rescue equipment for technicians, including rope systems, self-retracting lanyards, rescue winches, and hardline communication systems.

Methods of Entry

Methods of entry into confined spaces must be chosen based on the specific scenario, the nature of the confined space, and the risks involved. These methods include direct entry, vertical entry, horizontal entry, and technical entry.

Direct Entry

Direct entry is the simplest and often the safest method when the entry point allows easy access and egress without the need for specialized equipment. This method is typically used in situations where the confined space is at ground level and has a sufficiently large opening. As always, a thorough risk assessment is the first step, to ensure the atmosphere in the space is safe. All rescuers must be equipped with appropriate PPE, including helmets, gloves, and respiratory protection, if needed.

Once all necessary equipment and personnel are ready for the direct entry, atmospheric monitoring devices can be used to test for oxygen levels, flammable gases, and toxic substances. The space must be entered cautiously while maintaining communication with the outside team. The atmosphere and the condition of the rescuer must be continuously monitored throughout the process. The strategy for egress must be planned and practiced in order to ensure a quick and safe escape if conditions change.

The benefits of direct entry include the minimal equipment it requires and the relative speed of implementation (as long as conditions are favorable). On the other hand, direct entry is only suitable for confined spaces that are easily accessible, and it also carries a risk of unanticipated hazards inside the space.

Vertical Entry

Vertical entry involves accessing confined spaces from above using ladders, ropes, or other descent mechanisms. This method is often used in confined spaces such as manholes, tanks, or silos.

Vertical entries require fall protection systems to safeguard rescuers during their descent and ascent. Rescuers must identify and secure reliable anchor points for their ropes and other equipment, and a clear rescue plan must be in place for quickly retrieving a rescuer if needed.

To prepare for a vertical entry, rescuers must first set up ladders, preengineered systems, davit arms, rope systems, or tripods over the entry point and inspect all equipment for wear and proper functionality. The rescuer must be carefully lowered into the space, maintaining a controlled descent. Rescuer progress and the atmosphere within the space must be continuously monitored. When the rescuer is ready to leave the space, the ascent must be safe and controlled.

The benefits of vertical entry include its versatility—it is suitable for a greater range of space types—and the fact that it requires controlled entry and exit from the space. However, it is more complicated, so it requires more equipment and setup time, and it also involves more risk, specifically, risk of a fall if the equipment fails.

Horizontal Entry

Horizontal entry involves accessing confined spaces by crawling or walking through side openings. This method is often used for spaces such as ducts, tunnels, or pipes.

Horizontal entry can only be conducted if the space's opening or entrance is large enough for the safe passage of rescuers. Like the other types of entry, it requires thorough and continuous monitoring of the atmosphere of the space. Horizontal entries often have a restricted range of movement and likely will involve limitations on visibility within the space.

Initiating a horizontal entry requires following the same rules as direct entry: Ensure that all necessary equipment is available and in good condition; test the air quality initially and continuously throughout the operation; enter the space cautiously, using appropriate PPE and maintaining communication; move through the space slowly and carefully to avoid injury and maintain balance; and have a clear plan for egress, ensuring that the exit is unobstructed.

Like vertical entry, horizontal entry is suitable for a wider range of spaces. It also provides greater ease of movement because rescuers may crawl or walk into the space. However, movement may be restricted if the space is narrow, and as with vertical entry, limited visibility and maneuverability can present challenges.

Technical Entry

Technical entry involves the use of advanced techniques and specialized equipment, such as confined-space entry kits, tripods, winches, and retrieval systems. This method is necessary for more complex and hazardous confined spaces.

As with all other forms of entry, all team members must be trained and proficient in using the equipment necessary for technical entry—and they must be prepared for the fact that this type of entry requires more complex setup and operation procedures. Safety needs are significant, as well—redundant safety systems must be in place to manage the increased risk involved.

The first step in initiating a technical entry is to gather and inspect all necessary equipment, which likely includes tripods, winches, harnesses, and entry kits. The equipment must be set up according to the manufacturer's guidelines and must be tested to ensure all components are secure. Atmospheric testing and constant communication must be maintained throughout the process as the equipment is used to facilitate a controlled and safe entry into the space and monitor the team's progress. Upon completion of the rescue, winches and other retrieval systems are used to assist rescuers as they exit the space.

The technical entry method is well suited to high-risk entries because it provides additional safety measures for high-risk entries and permits access to complex and hazardous confined spaces. Downsides include the fact that it requires more time, resources, and specialized training and requires more setup time compared with simpler methods.

Respirators and Air Management

Types and Operational Requirements of Air-Supplying Respirators

Confined-space rescues require careful consideration of the respiratory protection provided to both rescuers and victims: an SCBA (see Figure 1-2 in Chapter 1) or an SAR (**FIGURE 11-1**). SCBAs provide a portable

FIGURE 11-1 An SAR.

supply of breathable air from a cylinder worn by the user, as previously covered in this book. They are widely used in confined-space rescues because of their mobility and reliability. SARs deliver breathable air through a hose connected to a remote source, providing a continuous air supply for extended operations. The operational requirements, benefits, and challenges of SCBAs and SARs are shown in **TABLE 11-1**.

Methods of Air Conservation, Air-Line Management, and Communication

In confined-space rescues, conserving air is paramount to ensuring the safety and operational effectiveness of rescuers and victims. Effective air conservation extends the duration of the air supply, providing crucial time for rescue operations.

TABLE 11-1 Operational Requirements, Benefits, and Challenges of Self-Contained Breathing Apparatuses (SCBAs) and Supplied-Air Respirators (SARs)

Operational Requirements		
	SCBA	**SAR**
Certification	Must meet NFPA 1981 standards (National Fire Protection Association [NFPA] 2019b) for SCBA units and NFPA 1852 (NFPA 2019a) for selection, care, and maintenance	Must comply with NFPA 1989 (NFPA 2019c) for respiratory protection and Occupational Safety and Health Administration (OSHA) 29 C.F.R. 1910.134 ("Respiratory Protection" 1998)
Training	Users must be trained in the proper use and maintenance of SCBAs as per the OSHA regulation 29 C.F.R. 1910.134 ("Respiratory Protection" 1998)	Comprehensive training in SAR use, hose management, and emergency procedures
Fit testing	Fit testing is required to ensure a proper seal as per OSHA guidelines	Must fit well to maintain a clean and continuous air supply from a safe, uncontaminated location
Maintenance	Regular inspection and maintenance are mandatory to ensure functionality	Proper hose management techniques are necessary to avoid entanglement and damage
Benefits and Challenges		
	SCBA	**SAR**
Benefits	▪ **Mobility:** Allows rescuers to move freely within the confined space ▪ **Independence:** Provides air independent of the surrounding environment ▪ **Reliability:** Generally reliable under various environmental conditions	▪ **Extended use:** Suitable for longer operations because the air supply is not limited by a cylinder ▪ **Lighter weight:** The user carries less weight compared with an SCBA ▪ **Reduced heat stress:** Less physical strain because the equipment is lighter
Challenges	▪ **Weight:** Can be heavy and cumbersome, potentially causing fatigue ▪ **Limited air supply:** The duration is limited by the capacity of the air cylinder ▪ **Cost:** High initial cost and ongoing maintenance expenses	▪ **Limited mobility:** Hoses can restrict movement and become entangled ▪ **Dependence on external source:** Requires a reliable external air source and long hoses ▪ **Setup complexity:** More complex setup and logistics compared with SCBAs

The primary techniques for conserving air are controlled breathing, the use of air-saving devices (e.g., **positive-pressure respirators**, which maintain a slight positive pressure in the mask, reducing the likelihood of inward leaks but using air more efficiently), and operational efficiency.

Rescuers should be trained in techniques such as controlled breathing to minimize air consumption. Regular practice can help rescuers maintain calm and control their breathing rate under stress.

The operational efficiency approach combines task planning, where tasks are planned as efficiently as possible to reduce the time that each individual spends in the confined space, and effective rotation of personnel to conserve air supply and limit exposure time.

Managing Air Lines

Proper management of SAR air lines is crucial for maintaining a continuous and unobstructed air supply, preventing entanglement, and ensuring the safety of personnel. Techniques for their management include routing and securing air lines and monitoring air-line integrity.

Routing and securing air lines involves planning the route of the air line before entering the confined space to avoid sharp edges, heat sources, and potential obstructions and using clips, straps, or other securing devices to keep the air line in place and prevent it from getting tangled or damaged.

Monitoring air-line integrity involves conducting regular visual and tactile inspections of the air line to ensure there are no kinks, abrasions, or leaks and establishing and practicing procedures for dealing with air-line failures, including the use of emergency air supplies.

To protect their integrity, air lines must be stored coiled to prevent kinks and damage. Follow manufacturer guidelines for maintenance and replacement schedules to ensure the air line remains in good condition. To better understand the process of managing SAR air lines and communication, follow the steps in **SKILL DRILL 11-1**.

SKILL DRILL 11-1

Managing SAR Air Lines and Communication

Step 1: Assemble the Components

Use essential SAR components—including masks, air lines, air supply sources, and regulators—to ensure safe and reliable air delivery during confined-space operations.

Step 2: Carefully Route the Air Lines

Carefully route SAR air lines to prevent kinks or damage, maintaining a consistent air supply for safe operations.

Continues.

SKILL DRILL 11-1 CONTINUED

Managing SAR Air Lines and Communication

Step 3: Uncoil the Air Lines to Prevent Tangles

Properly uncoil and lay out SAR air lines to prevent tangles and ensure efficient use during emergency response.

Step 4: Secure the Air Lines with Clips and Straps

Secure SAR air lines with clips and straps to minimize entanglement and reduce tripping hazards.

Step 5: Use Intrinsically Safe Radios for Reliable Communication

To ensure reliable communication in hazardous environments where safety is critical, use an intrinsically safe radio.

Step 6: Use Standardized Hand Signals

Use standardized hand signals to maintain clear communication when radios are impractical or malfunctioning.

The Importance of Effective Communication

Clear and effective communication is vital for the coordination and safety of confined-space rescue operations. It ensures that all team members are aware of the status of the operation, potential hazards, and necessary actions.

Verbal communication is managed with intrinsically safe radios, which are designed to operate safely in potentially explosive atmospheres, and clear communication protocols, including specific phrases and signals to convey critical information efficiently.

For situations in which verbal communication cannot take place, nonverbal communication methods, such as standardized hand signals, the OATH method (O for okay = one tug; A for advance = two tugs; T for take up slack = three tugs; H for help = four tugs) or the SUDRH method (S for stop = one short blast; U for up = two short blasts; D for down = three short blasts; R for rope free = four short blasts; and H for help = one long blast) (which are covered in Chapter 2), can be used if line of sight is possible (meaning that rescuers can see each other clearly).

Regular communication drills help ensure that all team members are familiar with the protocols and can effectively communicate under stress. Realistic scenarios should be used for training teams on communication strategies. To better understand the process of establishing victim and rescuer communications while wearing an SAR, follow the steps in **SKILL DRILL 11-2**.

SKILL DRILL 11-2

Establishing Victim and Rescuer Communications While Wearing an SAR

Step 1: Perform an Initial Assessment

Assess the situation, check the environment for hazards, and ensure all safety protocols are in place before attempting communication.

Step 2: Establish Communication

Attempt to establish initial contact, using a loud voice or communication device, to reassure the victim and confirm their condition.

Step 3: Use Communication Tools

Use a two-way radio or a hardwired communication system to ensure clear, uninterrupted communication, even in environments where wireless signals might be weak.

Step 4: Listen for Victim Response

Listen carefully for any response, noting the victim's voice, tone, and content to assess their physical and mental state.

Step 5: Provide Reassurance

Provide reassurance to the victim, explaining the rescue plan and encouraging them to stay calm. Clear, calm communication is crucial to maintaining the victim's trust and cooperation.

Step 6: Direct the Victim

If the victim is mobile, you may give specific instructions to help guide them to a safer location within the confined space, minimizing risks during the extraction.

Step 7: Monitor the Victim's Condition

Continuous monitoring of the victim's condition is essential. Take notes on the victim's condition and any changes, and be ready to adjust the rescue plan as needed.

Step 8: Establish Emergency Signal Procedures

Establish an emergency signal, such as pressing a button or using a pre-agreed-upon phrase, to indicate if the victim's condition deteriorates or if the situation becomes more dangerous.

Step 9: Communicate During Extraction

As the extraction begins, continue to communicate clearly with the victim, ensuring they understand each step and move safely toward the exit.

Step 10: Offer Final Reassurance

Once the victim is out of the confined space, provide final reassurance, confirming their safety and coordinating with medical personnel for immediate care.

Using SAR

SARs are critical equipment in confined-space rescue operations because they supply clean air from a remote source through a hose. This makes them suitable for long-duration tasks in hazardous environments where the air is contaminated or oxygen deficient. Unlike an SCBA, which has a limited air supply, SARs provide a continuous flow of breathable air.

Like all equipment, an SAR has proficiencies and constraints. Its proficiencies include its ability to do the following:

- Ensure a reliable and uncontaminated external air supply, such as a high-capacity air compressor or air cylinders
- Switch to backup air supplies if the primary source fails
- Adjust airflow to match the user's needs (only with some units)
- Be used for extended operations beyond the typical capacity of an SCBA

Its constraints include the following:

1. Limited mobility because of hose length and potential for entanglement
2. The impact of limited mobility on the user's safe movement through the confined space
3. Some units' inability to adjust airflow to match the user's needs
4. Communication barriers when an SAR is in place (most jurisdictions that use SARs have an integrated communication line along with the air line and tag line)
5. Reliance on an external air source
6. Hose integrity, which may be compromised if an explosive environment erupts into fire

To mitigate the first and second constraints, users must plan and secure the routing of hoses to minimize the risk of entanglement and conduct thorough training on hose management techniques to improve mobility.

As for the third constraint, SARs provide a continuous flow of air, which may not always be adjustable to the user's needs. Some SARs have fixed flow rates that cannot be adjusted based on the user's activity level or comfort. This can result in discomfort or inefficiency, especially in varying work conditions. Continuous airflow can cause overheating or excessive cooling, depending on the environment. Mitigating this concern is simple: Choose a unit that offers selectable flow rates.

The fourth constraint, communication barriers, includes noise from the air supply and the mask muffling the sound of voices, making verbal communication challenging. Full-face masks can limit visibility and make nonverbal communication difficult. These issues can be mitigated by using intrinsically safe communication devices integrated with SARs to facilitate clear communication or deploying hand-signal or rope-tug protocols for nonverbal communication.

The fifth constraint, an SAR's reliance on an external air source, can be mitigated through the use of redundant air supply systems to ensure continuous operation in case of primary system failure.

Finally, the sixth constraint—the possibility of compromised integrity of the SAR hose—can be addressed by constant monitoring of the hose itself.

SAR Use, Care, and Maintenance

An SAR is an essential part of any rescuer's equipment, and its proper use, care, and maintenance are crucial to ensure the effectiveness and reliability of these systems. SARs must be selected based on the specific hazards present in the confined space, following Occupational Safety and Health (OSHA) regulation 29 C.F.R. 1910.134 ("Respiratory Protection" 1998) and NFPA 1989 standards (NFPA 2019c). All users must be adequately trained in the proper use, limitations, and maintenance of SARs, as specified by OSHA regulations and NFPA guidelines. Fit testing should be conducted according to OSHA 29 C.F.R. 1910.134 ("Respiratory Protection" 1998) to confirm that the respirator seals correctly on the user's face.

An SAR's air supply hose must be of adequate length to allow for free movement but no longer than 300 feet (91 m) so that it does not become a tripping hazard or get entangled. Use clips or guides to secure the hose and prevent it from snagging or becoming tangled. Keep the hose away from sharp edges and hot surfaces. All users must be well trained in emergency procedures for an SAR, including how to disconnect the unit quickly and safely if necessary.

Before using an SAR, inspect it for any damage or defects. Check the face piece, hoses, and connections. Follow the manufacturer's instructions for donning the SAR. Be sure that the face piece fits securely and that all connections are tight. Perform a positive- and negative-pressure check to ensure the respirator is functioning correctly. SARs must be removed carefully to avoid contaminating the face piece or other components. Follow proper decontamination procedures if SARs are exposed to hazardous substances.

Follow a routine maintenance schedule based on the SAR manufacturer's guidelines and NFPA 1989 standards (NFPA 2019c). This includes replacing filters, checking hoses, and inspecting the air supply system. Maintain detailed records of all inspections, maintenance, and repairs as required by OSHA 29 C.F.R. 1910.134 ("Respiratory Protection" 1998) and NFPA 1989 (NFPA 2019c). Store SARs in a clean, dry area away from direct sunlight and contaminants. Ensure that hoses are coiled properly and not kinked. They must be stored in a state of readiness, ensuring that all components are accessible and in good working condition.

Determining the Need for SAR

Factors that influence the decision to use an SAR or SCBA include atmospheric conditions, the duration of the operation, mobility requirements, and air supply management. An SAR is the preferred choice for operations expected to exceed the capacity of an SCBA (typically 30–60 minutes). Always plan for a safety margin to accommodate unforeseen delays or complications.

Atmospheric hazards are the primary consideration when determining the need for an SAR. Key hazards include the following:

- **Toxic gases:** The presence of harmful gases such as carbon monoxide, hydrogen sulfide, or chlorine.
- **Oxygen deficiency:** Oxygen levels below 19.5 percent can impair cognitive and motor functions and can be life threatening.
- **Explosive atmospheres:** The presence of flammable gases or vapors that can ignite.

Multi-gas detectors are used to measure a space's oxygen levels, detect toxic gases, and assess the potential for explosive atmospheres—all determining factors in whether an SAR is necessary. If toxic gases exceed permissible exposure limits (PELs), if oxygen levels fall below 19.5 percent, or if the lower explosive limit (LEL) of a flammable gas exceeds safe thresholds, use of respiratory protective equipment or an air source is mandatory.

SARs are suitable for extended rescue operations because they provide a continuous supply of breathable air from an external source. All equipment, including communication devices and gas detectors, must have sufficient battery life available for the duration of the operation. Contingency plans, including backup air supplies and redundant systems, must be in place to address any potential failures of an SAR system.

NFPA 1989's Role in Maintaining Air Quality for SARs

NFPA 1989: Standard on Breathing Air Quality for Emergency Services Respiratory Protection (NFPA 2019c) sets the requirements for the quality of air supplied to respirators used by emergency services. This standard is crucial for ensuring that the air delivered to SARs is free from harmful contaminants and meets specific purity criteria.

The primary purpose of air quality testing is to protect the health and safety of personnel using SARs. Contaminated air can lead to serious health issues, including respiratory problems and toxic exposures. Regular air quality testing ensures compliance with NFPA 1989 (NFPA 2019c), which is essential for maintaining operational standards and avoiding legal and regulatory penalties.

NFPA 1989 (NFPA 2019c) specifies that the air supplied to respirators must be free from the following contaminants:

- **Carbon monoxide:** A poisonous gas that can cause serious health effects or death.
- **Carbon dioxide:** High levels can lead to respiratory issues.
- **Oil mist:** Can cause respiratory irritation and other health problems.
- **Water vapor:** Excessive moisture can degrade respirator performance and increase the risk of microbial contamination.
- **Particulate matter:** Dust and other particles can cause respiratory issues.

TIP

A best practice is to purge all SAR supply hoses before use because of the potential for buildup of carbon monoxide during storage. This purge should be performed for one minute per 100 feet (30 m) of supply hose.

NFPA 1989 (NFPA 2019c) sets specific limits for oxygen and each contaminant to ensure the air quality remains within safe limits. These limits are as follows:

- **Oxygen content:** 19.5 percent to 23.5 percent.
- **Carbon monoxide:** Not to exceed 10 parts per million (ppm).
- **Carbon dioxide:** Not to exceed 1000 ppm.
- **Oil mist:** Not to exceed five milligrams per cubic meter (mg/m^3).
- **Particulate matter:** Air must be free from noticeable particulates.

Regular testing intervals are specified to ensure ongoing compliance. Initial testing is performed before the air system is put into service. Periodic testing proceeds at intervals not exceeding three months. Post-maintenance testing is completed following any maintenance or repair work on the air supply system.

Various equipment and methods are used to test air quality, including gas analyzers, which are used to measure concentrations of carbon monoxide, carbon dioxide, and oxygen; impactors and filters, which detect oil mist and particulate matter; and dew point meters, which measure water vapor levels.

Air samples for testing are collected as close to the respirator connection point as possible. The sample volume must be sufficient to get accurate readings. The air sample must be stored in a clean and sealed container to prevent contamination during transport to an accredited laboratory that uses standardized methods to ensure the accuracy and reliability of the results.

High-efficiency filtration systems can significantly reduce contaminant levels in the air supply. These filters must be regularly checked and replaced as needed. Air dryers can reduce water vapor levels in the system, preventing moisture-related issues in the air supply.

Regular maintenance of the system is critical for preventing contaminant buildup. Proper maintenance includes replacing filters as directed by the manufacturer or when testing indicates it is necessary, cleaning and inspecting the compressor to ensure it is not introducing contaminants into the air supply, and checking seals and connections to prevent leaks and contamination ingress.

OSHA 29 C.F.R. 1910.134 Compliance for SARs

Ensuring compliance with OSHA 29 C.F.R. 1910.134 ("Respiratory Protection" 1998) is critical for the safe and effective use of an SAR. Also known as the *Respiratory Protection Standard,* this regulation sets forth requirements for the proper selection, use, and maintenance of respirators to protect workers from inhaling harmful contaminants. The regulation covers several aspects, including respirator selection, fit testing, medical evaluations, training, and maintenance. To better understand the process of air supply management while wearing an SAR, follow the steps in **SKILL DRILL 11-3**.

SKILL DRILL 11-3

Air Supply Management While Wearing an SAR

Step 1: Optimize the Airflow

Adjust the regulator for optimal airflow.

Step 2: Don the SAR Correctly

When donning the SAR, check for functionality, secure the face piece for an airtight seal, and ensure the air supply is connected and flowing before entry into the confined space.

Fit testing ensures that the SAR mask forms a proper seal on the user's face, preventing contaminants from bypassing the mask and entering the respiratory system. A proper fit is essential for the respirator to provide adequate protection, and it must be tested before the first use of the respirator. There are two main types of fit testing: qualitative and quantitative.

Qualitative fit testing (QLFT) relies on the user's sensory response to detect the presence of a test agent (**FIGURE 11-2**). Agents used in testing include isoamyl acetate (banana oil), saccharin, bitrex, and irritant smoke. This type of testing is suitable for testing negative-pressure respirators and can be used for half-mask respirators and SARs.

Quantitative fit testing (QNFT) uses an instrument to measure the amount of leakage into the respirator (**FIGURE 11-3**). QNFT provides a numerical measure of fit and can be used for any tight-fitting respirator.

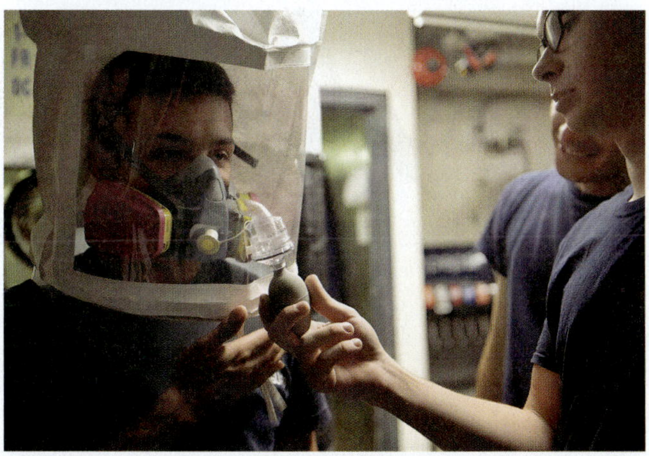

FIGURE 11-2 Qualitative fit testing (QLFT) on a respirator.

Courtesy of Petty Officer 3rd Class Nicole J. Groll/U.S. Coast Guard.

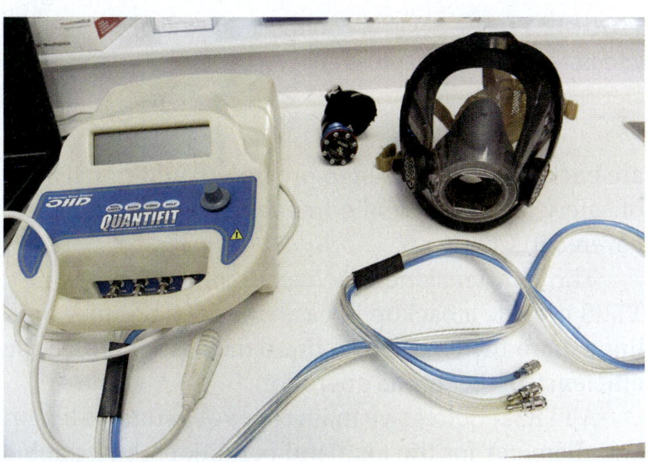

FIGURE 11-3 A QNFT instrument.

© Lori Van Buren/Albany Times Union/Hearst Newspapers/Getty Images

Devices used in QNFT include Portacount and controlled negative pressure (CNP).

Periodic fit testing must be completed at least annually or whenever there are changes that could affect the fit (e.g., facial hair, weight loss/gain, dental work). During the fit test, the user must check the seal by exhaling gently while blocking the exhalation valve to check for leaks (a positive-pressure check) and inhaling sharply while blocking the inhalation valve to check for inward leakage (a negative-pressure check).

Medical evaluations are crucial to determine whether a rescuer is physically capable of wearing a respirator. Wearing a respirator can place additional stress on the body, particularly the cardiovascular and respiratory systems. Medical evaluations, which include a medical questionnaire and a follow-up examination, help ensure that rescuers can safely perform their duties while wearing a respirator. These evaluations must take place prior to the first wearing of the respirator and then again every one to three years, depending on the rescuer's health status and exposure risks.

Other Equipment

Preconstructed Confined-Space Entry Kits

A preconstructed confined-space entry kit contains all the essential tools and gear required for safe entry, rescue, and egress from most confined spaces. A typical confined-space entry kit includes the following items:

- PPE, such as helmets, gloves, safety glasses/goggles, protective clothing, and respirators (an SAR or SCBA, depending on the environment)
- Entry and retrieval equipment, such as tripods, winches, harnesses, and rope systems
- Atmospheric monitoring equipment, such as gas detectors and calibration kits
- Communication devices, including radios
- Intrinsically safe lighting to provide illumination without the risk of sparking in a potentially explosive atmosphere
- Medical supplies, such as first aid kits and a stretcher or backboard

The advantages of a preconstructed confined-space entry kit include the undeniable benefit of having all necessary equipment in a single, organized kit and the confidence that the equipment meets required safety standards and is compatible. It can be set up quickly, which is critical in emergency situations, and time spent gathering and checking individual items is minimized. Such kits confer a greater level of safety because they promote consistent safety practices and procedures and help ensure that the same high-quality, well-maintained equipment is used for every operation.

Care and maintenance of preconstructed confined-space entry kits require regular visual inspections for signs of wear, damage, or contamination and regular testing of the operational functionality of items such as winches, gas detectors, and communication devices. Equipment should be maintained according to manufacturer guidelines and relevant standards. Rope systems and PPE, such helmets, gloves, and clothing, must be cleaned after each use with mild soap and water, and manufacturer recommendations must be followed for cleaning respirators. Equipment must be wiped down to remove dirt and contaminants. Detailed records of inspections, cleaning, and maintenance activities must be retained for accountability and traceability reasons.

Kits should be stored in a clean, dry environment, away from direct sunlight and contaminants. Keep the kit organized to ensure that all items are readily accessible and in good working condition.

To use the kit, first determine the specific hazards and requirements of the operation, and then check that all items in the kit are present, undamaged, and functional. The entry and retrieval system is the first item to set up, so position the tripod over the entry point and secure it according to the manufacturer's instructions. Attach the winch to the tripod; verify that it operates smoothly; and arrange the ropes, harnesses, and carabiners. Use the kit's gas detectors to perform an initial test of the atmosphere and set them so that they continuously monitor air quality during the operation. Ensure all team members properly don their PPE, including helmets, gloves, and respirators. Distribute radios and test them to make sure they are operating correctly.

Rope Systems

Rope retrieval systems and methods are crucial for safely extracting individuals from hazardous environments, and NFPA 1006 (NFPA 2021) covers these haul systems in both its Operations and Technician sections. All confined-space technical rescue personnel must hold training qualifications to rope rescue operations or technician levels. Common systems and methods used in these operations include a main line system, a belay line system, and a **two-tensioned rope system (TTRS)**; a TTRS offers enhanced safety and redundancy with two ropes that are both tensioned and actively managed, providing a backup in case one fails (**FIGURE 11-4**).

Main line systems are primary ropes used to raise or lower a rescuer and victim, and belay line systems are backup systems used to provide an additional layer of safety in case the main line fails. These systems are covered in Chapter 6.

FIGURE 11-4 A TTRS.

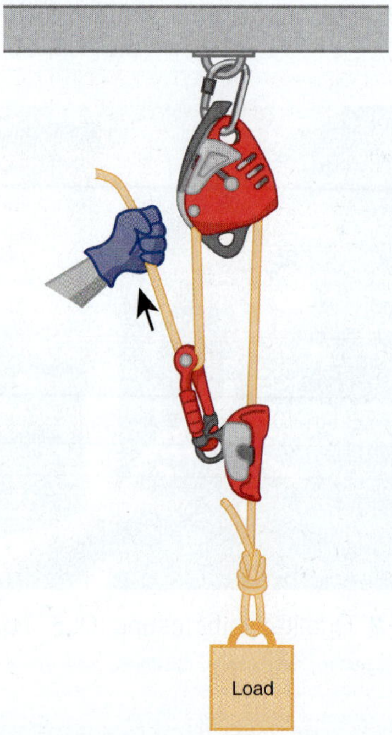

FIGURE 11-5 A descent control device (DCD) with an integrated progress capture pulley.

Used with permission from Petzl.

The two-tensioned, actively managed ropes in a TTRS provide a backup in case one fails. Because both lines share the load, this redundancy can improve the efficiency of the rescue operation.

A TTRS requires two main ropes of a suitable length and diameter for the load and conditions; two double pulleys or a combination of pulleys; tri-locking carabiners for secure connections; two solid and reliable anchor points; two PCDs (**FIGURE 11-5**), which include ascenders, Prusik knots, or mechanical devices; webbing or slings to create anchor points and attach pulleys; a rescue load (which could be a litter containing a victim, rescuer, or equipment); and edge protection to guard against abrasion on the ropes.

To set up a TTRS, follow these steps:

- **Step 1: Set up the anchors:** Choose solid anchor points that are secure and can withstand the load. Ideally, they should be positioned to provide the best angles for load distribution. Use webbing or slings to create anchor points, if necessary, and attach them securely to the anchors using tri-locking carabiners.
- **Step 2: Rig the main lines:** Attach the pulleys to each anchor and use tri-locking carabiners to secure the pulleys to the anchor points. Pass the first rope through the pulley at anchor 1. Secure the end of the first rope to the load using a figure-eight follow-through knot or a similar secure knot.
- **Step 3: Rig the second line:** Pass the second rope through the pulley at anchor 2. Secure the end of the second rope to the load using a figure-eight follow-through knot or a similar secure knot.
- **Step 4: Create mechanical advantage (FIGURE 11-6):** Attach a single pulley to the load and run the first rope through it. Then, back up to a second pulley at anchor 1. Attach a single pulley to the load and run the second rope through it, then back up to a second pulley at anchor 2.

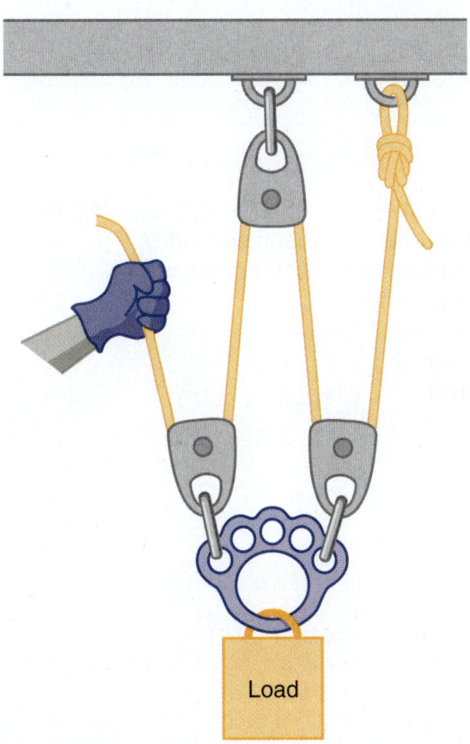

FIGURE 11-6 A 4:1 mechanical advantage rope system.

- **Step 5: Set up PCDs:** Use webbing or slings to attach a PCD to anchor 1 or near it. Run the first rope through the PCD to capture progress and prevent the load from slipping back. Use webbing or slings to attach the PCD to anchor 2 or near it. Run the second rope through the PCD to capture progress and prevent the load from slipping back.
- **Step 6: Conduct final adjustments and safety checks:** Ensure all carabiners are locked and all knots and attachments are secure. Test the system without load by pulling on both ropes to ensure the pulleys move freely and the PCDs engage properly. Carefully attach the load to the system, ensuring everything remains secure. Adjust as necessary to maintain equal tension in both ropes.

To haul with a TTRS, pull both ropes simultaneously through their respective PCDs to lift the load. Always make sure both ropes are kept tensioned equally to distribute the load evenly between the two ropes. When using a TTRS for lowering, control the descent by managing both PCDs simultaneously, ensuring smooth and safe lowering of the load.

Progress Capture Devices (PCDs)

PCDs are critical components of rope systems used to control the movement of the load (e.g., a rescuer, victim, or equipment). The primary function of a PCD is to allow rope to move in one direction (typically upward when hauling a load) while preventing it from slipping back in the opposite direction (downward), thus capturing the progress made during the rescue operation. This ensures that once the load has been lifted or moved, it cannot slide back down if the hauling force is released. See **TABLE 11-2** for details on different types of PCDs.

To set up a PCD in a rope system, first ensure the PCD is suitable for the rope's diameter and the expected load. Secure the PCD to the rope according to the manufacturer's instructions. Anchor the device by attaching the PCD to a solid anchor point using a tri-locking carabiner and webbing or a sling. Before loading the system, ensure the PCD is functioning correctly by pulling the rope through it and checking that it locks when tension is applied.

Practical applications for PCDs include vertical hauling systems (when lifting a load from a vertical shaft, a PCD prevents the load from slipping back), TTRS (both ropes in a TTRS use PCDs to maintain load tension and safety redundancy), and mechanical advantage systems (PCDs are essential in **Z-rig systems**—3:1 mechanical advantage rope systems that allow rescuers

TABLE 11-2 Progress Capture Device (PCD) Types

Device Type	Description	Use	Advantages	Disadvantages (if any)
Mechanical ascender  © shoricelu/iStock/Getty Images Plus/Getty Images	Device that uses camming mechanisms to grip the rope; examples include Petzl® ASCENSION, CMI Ultrascender, and Kong Lift™	Attached to the rope, they allow upward movement and automatically lock to prevent downward movement	Easy to use, reliable, and can handle significant loads	None
Prusik knot  © Hubert Knoblauch/Shutterstock	A friction hitch made with a loop of Prusik cord wrapped around the main rope	Tightens and grips the main rope when weight is applied, allowing movement in one direction	Simple to create with standard rope and cord, versatile	Requires skill to tie correctly and can be slower to engage and disengage compared with mechanical devices
Gibbs ascender 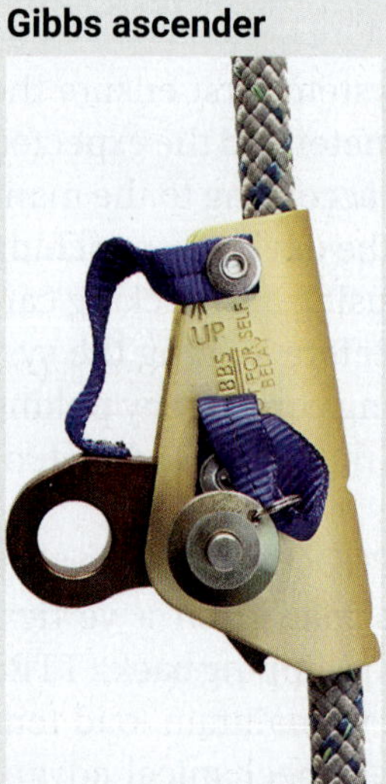 Used with permission from Gibbs Products.	A type of mechanical rope grab that uses a spring-loaded cam	Slides easily up the rope but locks onto the rope when downward pressure is applied	Durable and handles muddy or icy ropes well	None

Device Type	Description	Use	Advantages	Disadvantages (if any)
Rescuecender  © BangJals/Shutterstock	A cam-loaded device that is commonly used in rescue systems	Allows rope movement in one direction and locks when force is applied in the opposite direction	Robust and easy to attach and detach from the rope	None
Rope grab  © King Ropes Access/Shutterstock	A device that automatically locks onto the rope when loaded	Frequently used in fall arrest systems, they can also be used in progress capture setups	Effective and simple to use	Can damage lifeline if not used properly

to lift or lower a load with one-third of the force that would otherwise be needed—and other mechanical advantage systems to capture progress during hauling operations).

Always inspect PCDs before use for wear, damage, or malfunction, and verify that the PCD is compatible with the rope being used and the load expected. Ensure rescuers are trained in the correct use and limitations of the PCDs being employed.

Tag Lines and Guiding Lines

Tag lines are used to control the position and orientation of a litter or load. **Guiding lines** help prevent a load from swinging or spinning during the operation. Regular training and drills are crucial for rescuers to be proficient with these systems, and using proper PPE, such as helmets, harnesses, gloves, and eye protection, is essential to use these systems safely.

Mechanical Advantage Systems

Mechanical advantage systems use pulleys to reduce the amount of force needed to lift or lower a load. The following discussion provides more detail on 3:1, 4:1, and 5:1 mechanical advantage systems, which are covered in Chapter 6. The difference between an even simple system (4:1, 2:1) and an odd simple system (3:1, 5:1) is where the rope terminates—it is either tied to the load or tied back at the anchor.

A 3:1 mechanical advantage—or Z-rig—rope system is a configuration that allows rescuers to lift or lower a load with one-third of the force that would otherwise be needed (**FIGURE 11-7**).

The necessary equipment for such a system resembles that of the TTRS: one single pulley, a rope of suitable length and diameter for the load and conditions, locking carabiners for secure connections, a solid and reliable anchor point, a Prusik cord to use as a PCD or

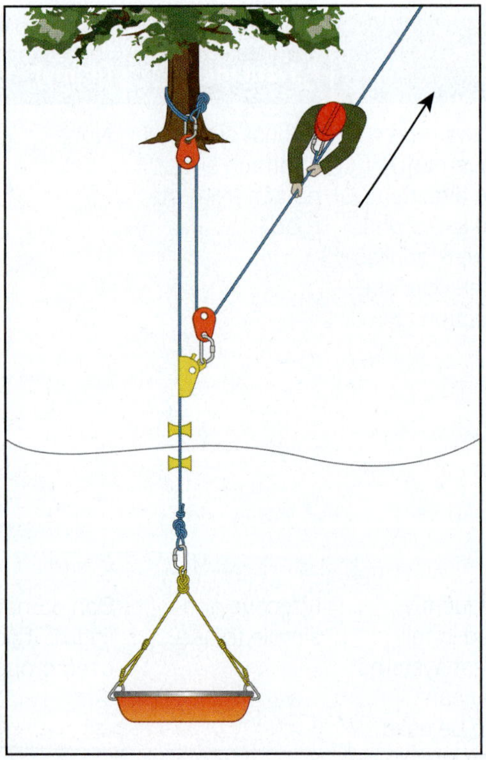

FIGURE 11-7 A 3:1 mechanical advantage (or Z-rig) haul system.

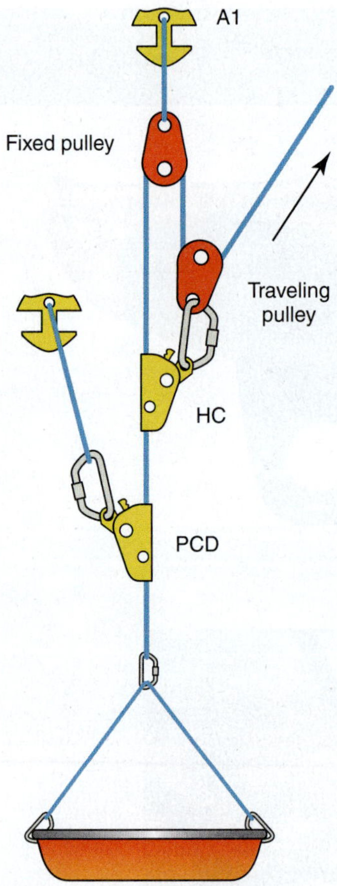

FIGURE 11-8 Constructing a Z-rig system.

a mechanical ascender, webbing or slings to create anchor points and attach pulleys, a rescue load (which could be a litter containing a victim, rescuer, or equipment), and edge protection to guard against abrasion on the ropes.

To set up a Z-rig system, follow these steps (**FIGURE 11-8**):

- **Step 1: Set up the anchor:** Choose a solid anchor point that is secure and can withstand the load. Attach the double pulley using a tri-locking carabiner to secure it to the anchor point.
- **Step 2: Rig the main line:** Secure the end of the rope to the load using a figure-eight follow-through knot or a similar secure knot.
- **Step 3: Create the mechanical advantage:** After attaching the end of the rope to the load, run the rope up and through one sheave of the double pulley at the anchor. Use a tri-locking carabiner to secure the single pulley to the load. After passing it through the double pulley, run the rope down to the single pulley attached to the load and pass it through the pulley. Run the rope back up to the double pulley and pass it through the remaining sheave.
- **Step 4: Set up the PCD:** Use webbing or slings to attach the PCD to or near the anchor point. The rope coming out of the double pulley should run through the PCD to capture progress and prevent the load from slipping back.
- **Step 5: Conduct final adjustments and safety checks:** Ensure all carabiners are locked and all knots and attachments are secure. Test the system without a load by pulling on the rope to ensure the pulleys move freely and the PCD engages properly. Carefully attach the load to the system, ensuring everything remains secure.

To haul with a Z-rig system, pull the rope through the PCD to lift the load. The 3:1 mechanical advantage means that for every three feet (1 m) of rope pulled, the load will rise one foot (30 cm), reducing the force needed by a factor of three. When using a Z-rig for lowering, control the descent by managing the PCD, ensuring smooth and safe lowering of the load.

A 4:1 mechanical advantage rope system (see Figure 11-6) allows rescuers to lift or lower a load with a quarter of the force that would otherwise be needed, as covered in Chapter 6. This system requires the same equipment as a 3:1 system, except that it also requires a Prusik cord, and the PCD would be an ascender or Prusik knot.

The steps to build the 4:1 system are the same as those for the 3:1 system, except with these differences:

- In **Step 2: Rig the main line**, the rope should be passed through both sheaves of the double pulley, and another locking carabiner should be used to attach the single pulley to the load.
- In **Step 3: Create the mechanical advantage**, after passing the rope through the double pulley, run it down to the single pulley attached to the load and then back up. After the rope goes through the single pulley, run it back up and pass it through the remaining sheave of the double pulley.

To haul with a 4:1 system, pull the rope through the PCD to lift the load. The 4:1 mechanical advantage means that for every four feet (1.2 m) of rope pulled, the load will rise one foot (30 cm), reducing the force needed by a factor of four.

TIP

Remember these key points when using a mechanical advantage system or TTRS:

- Anchor points must be strong and reliable.
- Pulleys must be correctly aligned to minimize friction.
- Keep the ropes free of twists and tangles.
- Always double-check all connections, carabiners, and knots.
- Using PCDs is essential for safety because they prevent the load from slipping back.
- For a TTRS, equal tension is critical to maintaining balance and ensuring both ropes share the load evenly.
- Retrieval systems used in confined-space rescue should be rigged so that the entrants are immediately retrievable while being lowered.

A 5:1 mechanical advantage rope system for confined-space rescue involves creating a setup that allows rescuers to lift or lower a load with one-fifth of the force that would otherwise be needed (**FIGURE 11-9**). Unlike the 3:1 and 4:1 systems, this type of system requires two double pulleys and one single pulley.

The steps to build a 5:1 system are as follows:

- **Step 1: Set up the anchor:** Attach the first double pulley to the secure anchor point using a tri-locking carabiner.

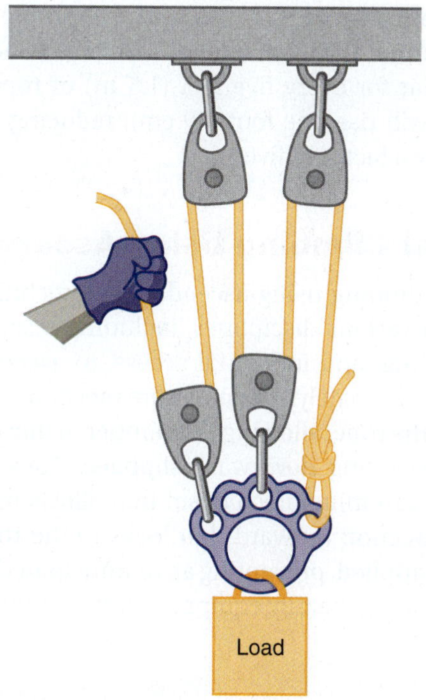

FIGURE 11-9 A 5:1 mechanical advantage rope system.

- **Step 2: Rig the main line:** Pass the rope through one sheave of the first double pulley. Use another tri-locking carabiner to attach the second double pulley to the load.
- **Step 3: Create the mechanical advantage:** After passing through the first double pulley, run the rope down to the second double pulley attached to the load and pass it through one of its sheaves. Run the rope back up and pass it through the remaining sheave of the first double pulley. After passing through the first double pulley, run the rope down to the second double pulley and pass it through the remaining sheave. Use another tri-locking carabiner to secure the single pulley to the anchor point or near it.
- **Step 4: Set up the PCD:** Use webbing or slings to attach the PCD to or near the anchor point. The rope coming out of the single pulley should run through the PCD to capture progress and prevent the load from slipping back.
- **Step 5: Conduct final adjustments and safety checks:** Ensure all carabiners are locked and all knots and attachments are secure. Test the system without a load by pulling on the rope to ensure the pulleys move freely and the PCD engages properly. Carefully attach the load to the system, ensuring everything remains secure.

To haul with a 5:1 system, pull the rope through the PCD to lift the load. The 5:1 mechanical advantage means that for every five feet (1.5 m) of rope pulled, the load will rise one foot (30 cm), reducing the force needed by a factor of five.

Vertical Climbing Using Ascenders

Vertical climbing using ascenders is a technique employed in various disciplines, including rescue operations, caving, and industrial access, to ascend a rope efficiently and safely. Ascenders are mechanical devices that grip the rope, allowing the climber to move upward while preventing downward slippage. They typically feature a camming mechanism that allows movement in one direction (upward) but locks to the rope when weight is applied, preventing an unanticipated descent. They are also generally equipped with a handle for easy grip and control during ascent. Ascenders are primarily designed for ascent, but they can be used for descent in some circumstances.

To better understand the process of executing a self-rescue, follow the steps in **SKILL DRILL 11-4**.

Ascenders help climbers overcome obstacles by managing transitions over edges or around them by adjusting the ascenders' position and maintaining tension on the rope. They also allow for rest periods because platforms or foot loops can be used to take breaks and relieve muscle fatigue during extended climbs.

Types of ascenders include hand ascenders, which are used with one hand to climb and often paired with a foot loop or another ascender; chest ascenders, which are worn on a harness to keep the body upright and assist in smooth ascent; and foot ascenders, which are attached to the foot to facilitate upward movement and reduce strain on the arms.

SKILL DRILL 11-4

Executing a Self-Rescue

Step 1: Prepare PPE and Secure a Harness

Don the appropriate PPE and secure a harness, ensuring safety and preparedness for a self-rescue operation.

Step 2: Set Up a Fall Arrest System

Set up a fall arrest system, a critical step in preventing uncontrolled falls during vertical movement.

SKILL DRILL 11-4 CONTINUED

Executing a Self-Rescue

Step 3: Use an Ascender for Upward Movement

Use an ascender, like the one pictured here, in a vertical confined space to maintain controlled and safe upward movement.

Step 4: Use a DCD for Lowering

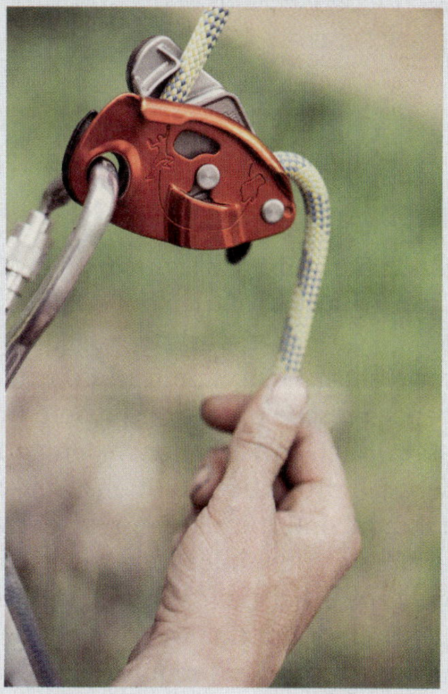

Use a descent control device (DCD), like the one pictured here, to safely lower yourself in a vertical confined space during self-rescue.

Most ascenders are made from high-strength materials, such as aluminum or steel, to withstand significant loads and harsh conditions. They are designed to be lightweight yet durable, ensuring ease of use and long-term reliability.

Ascenders are utilized in different types of rescue operations but are particularly essential in vertical confined-space rescues, where rescuers may need to climb up and down ropes. They are also used in self-rescue scenarios to ascend out of a space with a blocked primary exit.

Ascenders are used in conjunction with ropes and harnesses. A full-body or sit harness provides support and distributes weight evenly, enhancing safety and comfort. A static or low-stretch rope is preferred for vertical climbing because it offers stability and minimizes bounce.

If using hand ascenders to grip the rope, pull upward with the arms while pushing down with the legs. If using foot ascenders, use leg strength to drive upward movement. Chest ascenders assist in maintaining an upright position, reducing the need for constant manual adjustments.

To prepare for a climb using ascenders, first check all equipment for damage or wear, ensuring everything is in good working condition. Securely attach the ascenders to the rope and harness, ensuring proper alignment and locking mechanisms are engaged. Coordinate hand and foot movements to ensure smooth and continuous upward progress, and keep the body close to the rope to maintain balance and reduce strain on the arms and legs. Regularly check the engagement of ascenders and ensure no slack develops in the system.

TIP

Rope-climbing ascenders, specialized devices used to facilitate upward movement on a rope, are commonly employed in rescue operations, caving, climbing, and industrial settings. Designed to allow a user to ascend a rope efficiently and safely, they provide both grip and support.

Self-Retracting Lanyards

Self-retracting lanyards (see Figure 9-10 in Chapter 9) are vital components of personal fall arrest systems used in various industries to ensure worker safety at heights. Their application extends into confined-space rescue operations, where the unique challenges of such environments demand reliable and efficient fall protection solutions. Self-retracting lanyards should only be used for fall arrest when working around an opening; if they are used by a rescuer, it will arrest their fall, but without a means of retrieval. OSHA requires that working at any height of more than five feet (1.5 m) requires the use of a retrieval system.

Self-retracting lanyards are designed to arrest a fall within a few inches, providing effective fall protection by reducing fall distance and impact forces on the body. Key components of a self-retracting lanyard include the following:

- **Housing:** Encases the retractable lifeline, usually made of durable materials, such as stainless steel or high-impact plastic.
- **Lifeline:** The retractable cord or cable, typically made of high-strength materials, such as steel cable or webbing.
- **Energy absorber:** Reduces the impact force on the user during a fall.
- **Retracting mechanism:** Allows the lifeline to extend and retract smoothly, maintaining constant tension on the lifeline.
- **Connector hook:** Attaches to the user's harness and the anchor point.

Self-retracting lanyards are predominantly used in both industrial work and rescue operations, with each context presenting unique requirements and benefits. In industrial settings, self-retracting lanyards are used to protect workers operating at heights, such as on scaffolding, ladders, or platforms. They are integral in preventing falls and ensuring compliance with safety regulations. In confined-space rescue, self-retracting lanyards are used to protect rescuers as they enter and exit hazardous environments. The quick arrest capabilities and controlled descent features make self-retracting lanyards particularly valuable in these scenarios. See **TABLE 11-3** for the benefits and challenges of using self-retracting lanyards in confined-space rescue.

Best Practices for Using Self-Retracting Lanyards in Confined-Space Rescue

To maximize the benefits of self-retracting lanyards in confined-space rescue operations, the following best practices should be observed:

- Ensure that all personnel are adequately trained in the use and maintenance of self-retracting lanyards. Training should cover proper donning of harnesses, correct attachment to anchor points, and procedures for regular inspection and maintenance.
- Implement a routine inspection schedule to check for wear, damage, and proper function of the self-retracting lanyard. Inspections should be documented, and any issues should be addressed immediately.
- Secure self-retracting lanyards to appropriate anchor points that meet or exceed the strength requirements specified by safety regulations. The anchor points should be positioned to minimize fall distance and avoid obstructions.
- Ensure that self-retracting lanyards are compatible with other components of the fall protection systems the team uses, including harnesses and anchor points. Incompatibility can compromise the effectiveness of the system and increase risk.
- Develop and rehearse emergency procedures for rescuing individuals who may become suspended in a confined space after a fall. Quick response is essential to prevent suspension trauma and other injuries.
- Take into account the specific conditions of the confined space, such as the presence of chemicals, moisture, or abrasive surfaces, which may affect the performance and durability of the self-retracting lanyard.

TABLE 11-3 Benefits and Challenges of Self-Retracting Lanyards in Confined-Space Rescue

Benefits	Challenges
▪ **Quick arrest capability:** Arrests falls within inches (compared with traditional lanyards), significantly reducing the risk of injury in confined spaces, where falls can be particularly hazardous because of limited space and potential obstacles ▪ **Automatic retraction:** Retracting mechanism keeps the lifeline taut, reducing slack and the risk of tripping or entanglement, which is essential in confined (and often cluttered) spaces ▪ **Mobility:** Allows users to move more freely compared with traditional fall arrest systems; lifeline extends and retracts as the user moves, providing continuous protection without hindering movement ▪ **Versatility:** Can be used in various configurations and environments, making it suitable for a wide range of rescue scenarios; can be mounted overhead or used with portable anchor points, providing flexibility in deployment ▪ **Reduced impact forces:** Built-in energy absorber reduces the impact forces exerted on the body during a fall, minimizing the risk of injury, which is particularly important in rescue scenarios where the rescuer's safety is paramount	▪ **Cost:** Tends to be more expensive than traditional lanyards and other fall protection systems; can be a limiting factor for organizations with tight budgets ▪ **Maintenance:** Requires regular inspection and maintenance to ensure reliability and functionality; the retracting mechanism and other components can be susceptible to wear and damage, necessitating careful upkeep ▪ **Weight and bulk:** Can be heavier and bulkier than standard lanyards, potentially adding to the physical burden on rescuers operating in confined spaces ▪ **Complexity:** Mechanically complex, so they can be more prone to mechanical failure if not properly maintained; rescuers must be trained to understand and operate these devices effectively ▪ **Limited length:** Length is finite, which can be a limitation in some confined-space scenarios where extended reach is required; may necessitate additional planning and equipment to ensure effective coverage

Rescue Winches

Rescue winches are used to lift or lower personnel and equipment safely into or out of confined spaces (**FIGURE 11-10**). These winches can be either manual or automatic and are utilized by industrial teams for routine operations and emergency response teams for rescue missions. Winches must be capable of retrieving individuals from confined spaces without exacerbating the risk to their safety. These devices should be part of a broader fall protection and retrieval system, which may include tripods, davit arms, and other anchoring mechanisms designed to provide a safe and effective means of rescue.

FIGURE 11-10 A rescue winch.

The NFPA provides several standards that relate to the use of winches and other retrieval equipment, including NFPA 1006 (NFPA 2021), NFPA 2500 (NFPA 2022), and NFPA 1983 (NFPA 2017). OSHA provides detailed regulations related to their use under 29 C.F.R. 1910.146 ("Permit-Required Confined Spaces" 1993). Key points include the following:

- **Retrieval systems:** OSHA mandates the use of retrieval equipment, such as body harnesses with retrieval lines, for nonentry rescue. For vertical spaces over five feet (1.5 m) deep, a mechanical device, such as a winch, is required to aid in retrieval.
- **Equipment standards:** The rescue equipment must not introduce additional hazards and should be suitable for the specific confined-space configuration.

- **Training and maintenance:** Personnel must be trained in using the equipment, and regular maintenance and inspections are required to ensure the equipment's functionality and safety.

OSHA mandates that nonentry rescue is the preferred method for confined-space retrieval, requiring each entrant to use retrieval equipment, such as a body harness with a retrieval line. The other end of the line must be attached to a mechanical device or a fixed point outside the confined space. A mechanical haul system, such as a winch, must be available for vertical spaces more than five feet (1.5 m) deep to facilitate the retrieval of personnel.

The equipment used for rescue, including winches, must not introduce additional hazards. This means that winches should be reliable and suitable for the specific confined-space configuration. For instance, equipment prone to tangling or malfunctioning in certain spaces should be avoided.

Personnel involved in confined-space rescues must be adequately trained in using the equipment, including winches. Regular maintenance and inspections of the winch systems are necessary to ensure they remain functional and safe. This includes checking for wear and tear, ensuring the mechanical components are working correctly, and verifying that the system is ready for immediate use.

The outer end of the retrieval line must be affixed to a fixed point or a lifting device. This setup allows for a prompt rescue initiation as soon as the rescuer becomes aware of the need. In vertical permit spaces, the use of a mechanical retrieval device is mandatory. This requirement enhances safety by facilitating efficient and secure rescues in such environments.

Although power winches and overhead cranes can aid in rescue operations, they also pose significant risks to workers. If the use of these devices could potentially harm employees, they should not be utilized. For rescues at depths exceeding 50 feet (15 m), it is recommended to use powered retrieval winches. These winches should be equipped with torque limiters set at approximately 450 pounds (204 kg) to prevent injury to an incapacitated individual during the rescue process. See **TABLE 11-4** for the benefits and challenges of using winches in confined-space rescue.

Hardline Communication Systems

Hardline communication systems (**FIGURE 11-11**) use physical cables to transmit voice and data signals between rescuers inside the confined space and those outside. Unlike wireless systems, which rely on radio signals, these physical cables ensure consistent and interference-free communication. The main components of a hardline communication system include the following:

- **Communication units:** These devices, often resembling traditional telephones or intercoms, are placed both inside the confined space and at strategic points outside.
- **Cabling:** The durable, high-quality cables that connect the communication units are resistant to physical damage and environmental hazards, such as chemicals and moisture.
- **Central control unit:** The central control unit that manages the communication network is typically located outside the confined space and monitored by a dedicated operator.
- **Power supply:** Hardline systems require a stable power supply, often provided by batteries or portable generators, to ensure uninterrupted operation.

See **TABLE 11-5** for the benefits and challenges of hardline communication systems in confined-space rescue.

To maximize the effectiveness of hardline communication systems in confined-space rescues, certain best practices should be followed:

- **Preplanning:** Conduct thorough preplanning to identify the best locations for communication units and the most efficient cable routes. This planning should take into account potential hazards and obstacles within the confined space.
- **Training:** Ensure that all team members are trained in the use of hardline communication systems, including setup, operation, and troubleshooting. Familiarity with the equipment will enhance efficiency and response times during actual rescues.
- **Regular maintenance:** Implement a routine maintenance schedule to inspect and test all components of the hardline communication system. Regular checks will help identify and address issues before they become critical during a rescue.
- **Backup systems:** Have backup communication systems in place, such as portable radios, to provide redundancy in case the hardline system fails. This ensures continuous communication throughout the operation.

TABLE 11-4 Benefits and Challenges of Rescue Winches in Confined-Space Rescue

Benefits	Challenges
▪ **Quick arrest capability:** Arrests falls within inches, reducing injury risk; this quick response is crucial in confined spaces with potential obstacles ▪ **Controlled entry and exit:** Provides controlled, stable lowering and lifting of personnel or equipment, reducing the risk of falls or uncontrolled movements in confined spaces ▪ **Automatic retraction:** Mechanism keeps the lifeline taut, reducing slack and the risk of tripping or entanglement ▪ **Increased operator safety:** The mechanical advantage reduces physical strain on rescuers, making operations safer and less physically demanding ▪ **Mobility:** Allows for more freedom of movement, providing continuous protection without hindering movement ▪ **Enhanced rescue capability:** Allows for fast and efficient rescue in emergencies, enabling rescuers to lift a person out of a confined space with precision and control ▪ **Versatility:** Can be used in various configurations and environments, making winches suitable for a wide range of rescue scenarios ▪ **Compatibility with fall protection:** Many winch systems are compatible with fall protection equipment, combining descent, ascent, and fall arrest capabilities in one system ▪ **Reduced impact forces:** Built-in energy absorbers minimize the risk of injury during a fall, which is vital in rescue operations ▪ **Durability in tough conditions:** Built to withstand rugged environments, making them suitable for confined spaces that may contain harsh elements or challenging conditions	▪ **Cost:** More expensive than traditional fall protection systems, potentially limiting accessibility ▪ **Maintenance:** Requires regular inspection and maintenance because the retracting mechanism and other components can wear or get damaged ▪ **Weight and bulk:** Can be heavier and bulkier than standard lanyards, adding to the physical burden on rescuers ▪ **Complexity:** Its complexity increases the likelihood of mechanical failure if not properly maintained ▪ **Limited length:** The finite length can be a constraint in certain confined-space scenarios, necessitating extra planning and equipment

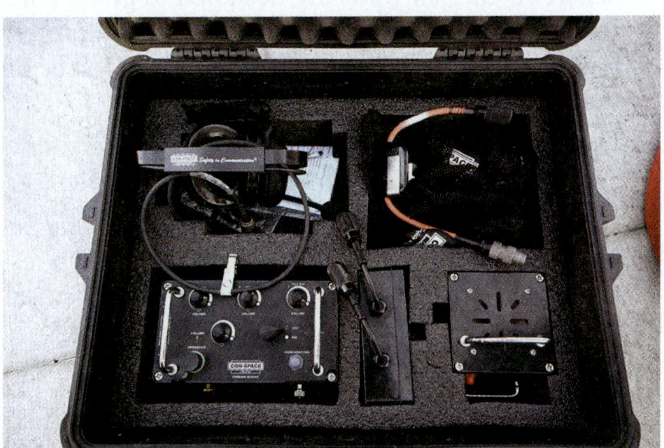

FIGURE 11-11 A hardline communication system.

- **Efficient setup and takedown:** Develop efficient procedures for the setup and takedown of the hardline system. Practice these procedures regularly to minimize delays during actual rescues.
- **Cable management:** Use cable management techniques to prevent entanglement and tripping hazards. Secure cables along walls or overhead when possible to keep pathways clear.
- **Power management:** Ensure that backup power sources, such as spare batteries or portable generators, are readily available. Monitor power usage to prevent unexpected outages during operations.

TABLE 11-5 Benefits and Challenges of Hardline Communication Systems in Confined-Space Rescue

Benefits	Challenges
▪ **Reliability:** Provides a stable and reliable communication channel, unaffected by environmental factors that can disrupt wireless signals, such as metal structures or electromagnetic interference ▪ **Clear communication:** Delivers high-quality, clear audio, which is crucial in noisy environments or when precise instructions are needed ▪ **Resistance to interference:** Not susceptible to interference from other electronic devices ▪ **Security:** Wired connections are inherently more secure than wireless ones, reducing the risk of eavesdropping or signal interception ▪ **Durability:** Cables and units are designed to withstand harsh conditions, making them suitable for rugged environments	▪ **Limited mobility:** Use of physical cables restricts the mobility of rescuers, potentially hindering movement; cables can also become entangled ▪ **Setup time:** Requires time to lay out and connect the cables, which can delay the start of rescue operations ▪ **Maintenance and repair:** Requires regular maintenance to ensure the integrity of cables and units; damaged cables need to be repaired or replaced, which can be challenging ▪ **Weight and bulk:** The equipment associated with hardline systems, including cables and communication units, adds to the weight and bulk that rescuers must carry, potentially affecting their efficiency and stamina ▪ **Dependency on power supply:** Requires a reliable power source; in situations where power supply is compromised, such as during extended operations or in remote locations, maintaining consistent power can be problematic

CHAPTER FEATURE: AFTER-ACTION

IN SUMMARY

- Methods of entry into confined spaces must be chosen based on the specific scenario, the nature of the confined space, and the risks involved. These methods include direct, vertical, horizontal, and technical entry.
- The primary techniques used to conserve air when working in a confined space are controlled breathing, the use of air-saving devices, and operational efficiency.
- SARs are suitable for long-duration tasks in hazardous environments where the air is contaminated, oxygen deficient, or potentially explosive. Unlike an SCBA, which has a limited air supply, SARs provide a continuous flow of breathable air.
- A typical confined-space entry kit includes the following items: PPE, respirators, entry equipment, retrieval equipment, atmospheric monitoring equipment, communication devices, intrinsically safe lighting, and medical supplies.
- Rope retrieval systems and methods are crucial for safely extracting individuals from hazardous environments. These include a main line, a belay line, a TTRS, and mechanical advantage systems.
- Self-retracting lanyards are designed to arrest a fall within a few inches, providing effective fall protection by reducing fall distance and impact forces on the body.
- Rescue winches are used to lift or lower personnel and equipment safely into or out of confined spaces.
- Hardline communication systems use physical cables to transmit voice and data signals between rescuers inside the confined space and those outside.

KEY TERMS

guiding line A rope line designed to help prevent a load from swinging or spinning during the operation.

positive-pressure respirator An air-saving device that maintains a slight positive pressure in the mask, reducing the likelihood of inward leaks but using air more efficiently.

two-tensioned rope system (TTRS) A system used in confined-space rescues to provide enhanced safety and redundancy by providing two ropes that are both tensioned and actively managed, providing a backup in case one fails.

Z-rig system A 3:1 mechanical advantage rope system that allows rescuers to lift or lower a load with one-third of the force that would otherwise be needed.

REFERENCES

National Fire Protection Association. 2017. *NFPA 1983: Standard on Life Safety Rope and Equipment for Emergency Services, 2017 Edition.* Quincy, MA: National Fire Protection Association.

National Fire Protection Association. 2019a. *NFPA 1852: Standard on Selection, Care, and Maintenance of Open-Circuit Self-Contained Breathing Apparatus (SCBA), 2019 Edition.* Quincy, MA: National Fire Protection Association.

National Fire Protection Association. 2019b. *NFPA 1981: Standard on Open-Circuit SCBA for Emergency Services, 2019 Edition.* Quincy, MA: National Fire Protection Association.

National Fire Protection Association. 2019c. *NFPA 1989: Standard on Breathing Air Quality for Emergency Services Respiratory Protection, 2019 Edition.* Quincy, MA: National Fire Protection Association.

National Fire Protection Association. 2021. *NFPA 1006: Standard on Technical Rescue Personnel Professional Qualifications, 2021 Edition.* Quincy, MA: National Fire Protection Association.

National Fire Protection Association. 2022. *NFPA 2500: Standard for Operations and Training for Technical Search and Rescue Incidents and Life Safety Rope and Equipment for Emergency Services, 2022 Edition.* Quincy, MA: National Fire Protection Association.

"Permit-Required Confined Spaces." 1993. 29 C.F.R. 1910.146. https://www.ecfr.gov/current/title-29/part-1910/section-1910.146.

"Respiratory Protection." 1998. 29 C.F.R. 1910.134. https://www.ecfr.gov/current/title-29/subtitle-B/chapter-XVII/part-1910/subpart-I/section-1910.134.

CHAPTER QUESTIONS

1. The methods for entry into a confined space include direct entry, technical entry, vertical entry, and
   - **a.** below-grade entry.
   - **b.** side entry.
   - **c.** horizontal entry.
   - **d.** solo entry.
2. Which of the following tasks is part of routing and securing air lines before entering a confined space?
   - **a.** keeping the air line clean
   - **b.** planning the air-line route
   - **c.** attaching line hangers
   - **d.** removing all debris
3. What is the maximum allowable limit for carbon dioxide to ensure air quality remains within safe limits, as set by specific oxygen and contaminant guidelines?
   - **a.** 50 ppm
   - **b.** 100 ppm
   - **c.** 500 ppm
   - **d.** 1000 ppm
4. The rope line designed to help prevent a load from swinging or spinning during an operation is known as a
   - **a.** guiding line.
   - **b.** tag line.
   - **c.** haul line.
   - **d.** belay line.

CHAPTER FEATURE: ON SCENE

A TRAPPED WORKER IN AN IMMEDIATELY DANGEROUS TO LIFE OR HEALTH (IDLH) ENVIRONMENT

One hot summer afternoon, a worker became trapped inside a storage tank at a chemical processing plant. The worker was performing routine maintenance inside the tank, which was 20 feet (6 m) high and 10 feet (3 m) in diameter, when an unexpected release of gas—specifically, methyl isocyanate, a highly toxic and volatile chemical, created an immediately dangerous to life or health (IDLH) environment. The worker, who had 15 years of industrial maintenance experience, was also a certified confined-space entry and rescue technician.

The Emergency Response

The emergency call was received within three minutes of the incident by the plant's internal emergency response team. Two minutes later, the team arrived on the scene; local fire department personnel and a specialized HazMat team arrived 10 minutes later.

An incident commander (IC) was appointed, along with officers responsible for safety, operations, and communications. A dedicated accountability officer was assigned to track all personnel and ensure adherence to safety protocols.

The responders identified the presence of the gas and determined the environment was IDLH. Continuous air monitoring was established to assess gas concentrations. As gas levels fluctuated, rescue strategies were adjusted to minimize responder exposure. Additional hazards included limited visibility, potential for secondary releases, and structural integrity concerns. Given the IDLH environment, a risk–benefit analysis was performed, and it was decided to prioritize responder safety while attempting to rescue the trapped worker.

The Rescue Operations

The primary entry team consisted of three highly trained confined-space rescue technicians equipped with SCBAs and chemical-resistant suits. Clear roles and responsibilities were assigned based on the responders' expertise and training. A secondary entry team was positioned as backup, ready to assist or relieve the primary team if needed. Throughout the operation, the teams maintained constant radio contact with incident command. The equipment selected by the team included gas detectors, portable ventilation fans, and rope rescue systems.

All responders employed a tag-in/tag-out system to ensure accurate tracking as they entered and exited the confined space. Radio-frequency identification (RFID) tags were used to monitor their real-time location inside the tank.

The Extrication and Treatment

The primary team located the victim, secured him with a rescue harness, and used a rope and pulley system to extract him through the side manway. The victim received first aid and oxygen therapy onsite before being transported to the hospital. There, he underwent a thorough medical evaluation and treatment for potential chemical exposure and respiratory distress.

The Outcome

A debriefing session was conducted immediately after the successful rescue to review actions, identify challenges, and gather feedback. A detailed report was compiled that documented the incident, response actions, and lessons learned. The lessons learned included needs for the following:

- **Improved tracking:** The team recommended that enhanced electronic tracking systems were needed for better real-time monitoring of responder locations.
- **Regular drills:** The response revealed that the team required more frequent and varied confined-space rescue drills to ensure readiness for different scenarios.
- **Improved tactical decisions:** The team determined that making improvements in risk–benefit analysis procedures would lead to better tactical decision-making.
- **Advanced training:** Responders felt a need for increased training in dynamic risk assessment and decision-making under pressure.
- **Enhanced PPE:** Following this experience in an IDLH environment, responders determined that upgrades to their PPE would be necessary to ensure maximum protection.

1. What caused the IDLH environment inside the storage tank?
   a. an unexpected release of methyl isocyanate gas
   b. an unexpected rise in temperature inside the tank
   c. a chemical fire
   d. a structural collapse
2. What was one of the hazards faced by the rescue teams during the operation?
   a. limited visibility inside the tank
   b. high levels of water inside the tank
   c. accumulation of debris inside the tank
   d. aggressive wildlife inside the tank
3. What did responders use to track personnel entering and exiting the confined space?
   a. a manual sign-in sheet
   b. security cameras
   c. a tag-in/tag-out system
   d. fingerprint scanning
4. What was one of the main lessons that the team learned from the incident?
   a. Fewer rescue drills would help the team avoid overtraining.
   b. Upgrades to responders' PPE would ensure maximum protection.
   c. Reducing the number of responders would ensure better communication.
   d. Improving communication systems would help the team avoid confusion.

CHAPTER

12

Technician

Preplanning, Controlling, Searching, and Other Topics

KNOWLEDGE OBJECTIVES

After studying this chapter, you should be able to:

- Describe the process for establishing search parameters. (**NFPA 1006: 7.3.1**, p. 309)
- Discuss the methods used for establishing a victim profile. (**NFPA 1006: 7.3.1**, p. 314)
- Discuss search result information and how it should be communicated. (**NFPA 1006: 7.3.1**, pp. 314–315)
- Describe the techniques used to ensure responder assignments match their level of expertise. (**NFPA 1006: 7.3.1**, p. 315)
- Discuss the importance of accountability and risk–benefit analysis. (**NFPA 1006: 7.3.1**, pp. 315–316)
- Explain the methods used to preplan a confined-space incident. (**NFPA 1006: 7.3.2**, pp. 306–308)
- Discuss the types of hazards associated with confined spaces and what should be documented on a preplan. (**NFPA 1006: 7.3.2**, pp. 306–308)
- Describe the importance of controlling the entry and egress routes of a confined space. (**NFPA 1006: 7.3.6**, pp. 308–309)

SKILL OBJECTIVES

- Demonstrate the process of working within a confined space. (**NFPA 1006: 7.3.1**, pp. 316–318)
- Establish communication with outside sources. (**NFPA 1006: 7.3.1**, pp. 310–312)
- Illustrate the process of preplanning a confined-space incident. (**NFPA 1006: 7.3.2**, pp. 306–308)
- Discover entry and exit points. (**NFPA 1006: 7.3.5**, pp. 308–309)

Introduction

In the realm of technical rescue, the importance of thorough preparation and precise execution cannot be overstated. Documenting hazards and preplanning for confined-space incidents are critical components of ensuring the safety and efficiency of rescue operations. By thoroughly identifying and assessing potential hazards, implementing control measures, and adhering to National Fire Protection Association (NFPA) standards and Occupational Safety and Health (OSHA) regulations, organizations can create a robust preplan that mitigates risks and enhances readiness for confined-space emergencies.

Establishing search parameters is a foundational step in any search and rescue operation. This process involves a systematic assessment to define the scope and boundaries of the search area, the development of a scene assessment, and an evaluation of the operation to determine which type of search strategy—grid, zone, point to point, or sequential—is appropriate. Key considerations include scene assessment, victim profiling, and implementation of search strategies. Victim profiles are developed by collecting firsthand accounts from co-workers, family, or bystanders who have witnessed the incident or know the victim; examining logs, security footage, and communication records; and getting details on any medical conditions that may affect the victim's behavior or location.

Clear and consistent communication is vital during search operations. All findings, whether significant or not, should be documented and relayed to the incident command. This ensures a shared awareness of the situation and that a detailed log of the incident will be available for analysis and accountability checking after the operation is complete.

Effective rescue operations depend on matching each responder's skill level and training with the appropriate tasks. Developing clear job descriptions and inventories of the current skills and qualifications of all team members allows a team to make sure each task is performed by a qualified individual.

Accountability ensures that all actions are coordinated, avoiding redundant efforts and ensuring that all areas of a confined space are covered, and accurate tracking of personnel and equipment helps optimize the allocation and utilization of resources. Performing a risk–benefit analysis on an operation gives insight into whether the potential benefits of a rescue may be outweighed by the potential risks to rescuers.

Preplanning a Confined-Space Incident

Preplanning a confined-space incident involves a systematic approach to anticipating and preparing for potential emergencies, ensuring the safety and efficiency of rescue operations. This process includes assessing hazards, developing detailed rescue plans, and coordinating resources.

Preplanning is important because it helps reduce the risk of injury or fatality among rescuers and helps them act more efficiently and effectively in an emergency. Clear, predefined procedures facilitate quick and coordinated responses, minimizing confusion and delays during an emergency. Preplanning also ensures that the necessary resources, including personnel and equipment, are available and properly allocated.

To assess hazards, first catalog all confined spaces within the facility, including their locations, sizes, and configurations. Assess each confined space for its potential hazards, such as toxic atmospheres, flammable gases, engulfment risks, and mechanical hazards. Record the identified hazards and ensure they are communicated to all relevant personnel.

A detailed rescue plan must include procedures for safe entry, including permits, isolation of hazards, and atmospheric testing. Determine appropriate rescue techniques, such as rope rescue, retrieval systems, or breathing apparatus, based on the confined space's characteristics. Specify the equipment needed for safe entry and rescue, including personal protective equipment (PPE), communication devices, and emergency medical supplies.

Resource coordination requires training all team members in confined-space entry and rescue procedures, as well as the use of specialized equipment; establishing and maintaining a list of emergency contacts, including local fire departments, medical facilities, and specialized rescue teams; and conducting regular mock drills to test the rescue plan, refine procedures, and ensure team readiness.

Documenting Hazards in a Preplan

Confined spaces pose unique and significant hazards that must be thoroughly understood and documented to ensure the safety of personnel during entry and the rescue operation. Potential atmospheric hazards, such as oxygen deficiency or enrichment or toxic or flammable vapors, and physical hazards, such as engulfment risks, mechanical equipment, live wires, electrical panels, poor-visibility conditions, or structural problems, should be documented on preplans so that rescuers have the information at their fingertips when an emergency occurs. Regular review and updates of the preplan are essential to maintain its effectiveness and reflect any changes in the operational environment. To better understand the process of preplanning a confined-space incident, follow the steps in **SKILL DRILL 12-1**.

SKILL DRILL 12-1

Preplanning a Confined-Space Incident

Step 1: Conduct a Hazard Analysis

Conduct a hazard analysis to assess potential dangers and hazards within the confined space.

Step 2: Make a Resource List

Compile a resource list, outlining necessary personnel, equipment, and support services for the confined-space operation.

Step 3: Document with Incident Command System (ICS) Forms

Utilize ICS forms (ICS 201 and ICS 208) to document the confined-space preincident plan and ensure proper communication and coordination.

Step 3: Courtesy of Illinois Fire Service Institute.

An effective hazard assessment documentation for a preplan includes the exact location and physical characteristics of the confined space; all possible entry and exit points, including emergency exits; the types of atmospheric hazards present, required monitoring equipment, and acceptable levels for safe entry; the specific physical hazards, control measures in place, and emergency procedures; any environmental conditions that could affect safety and the measures to mitigate these effects; and the necessary PPE for safe entry and work in the space. The ventilation requirements and systems in place to ensure safe atmospheric conditions should also be noted, as well as procedures for isolating hazardous energy sources (e.g., lockout/tagout).

OSHA Regulations on Shipyard and Shipboard Confined Spaces

Confined-space operations in shipyards and onboard ships present unique challenges and hazards. OSHA has established specific regulations to ensure the safety and health of workers in these environments. These regulations are outlined primarily in 29 C.F.R. 1915 ("Confined and Enclosed Spaces and Other Dangerous Atmospheres in Shipyard Employment" 1994), which pertains to occupational safety and health standards for shipyard employment.

Key Requirements for Confined and Enclosed Spaces in Shipyards and Onboard Ships

Prior to starting work in a confined or enclosed space on a ship or in a shipyard, atmospheric testing for oxygen levels, flammable gases, and toxic substances must be performed before entry; if initial tests indicate potential hazards are present, continuous monitoring is required. Mechanical ventilation must be used to maintain safe atmospheric conditions, and controls must be put in place to eliminate or mitigate identified hazards (e.g., isolation of energy sources or use of PPE).

A daily entry placard must be issued by a competent person certifying the space as safe for entry and displayed at the confined or enclosed space. There must also be a "competent person certification" issued by a competent person, ensuring all safety precautions are in place.

Workers who may be assigned to perform tasks in confined or enclosed spaces on ships and in shipyards must receive comprehensive training on confined-space hazards, entry procedures, and emergency response. Effective communication systems must be in place to ensure constant contact between workers inside the confined or enclosed space and those outside.

Advanced technology, such as real-time gas-monitoring systems, remote-controlled ventilation, and enhanced PPE, can significantly improve safety. Continuous investment in safety technology is crucial for adapting to the evolving challenges in shipyard and shipboard environments.

In the realm of emergency preparedness, detailed rescue plans that include the availability of rescue equipment and trained rescue personnel must be developed, with regular emergency drills performed to prepare for emergencies in accordance with OSHA and the United States Coast Guard (USCG).

Discovering and Controlling Entry and Egress Routes

Discovering

Identifying the best entry and exit points for various types of confined spaces (e.g., manholes, hatches, vents) involves the development of a firm understanding of the space's characteristics, an evaluation of each potential entry and exit point's structural integrity and accessibility, and identification and mitigation of the potential hazards associated with each entry and exit point (e.g., toxic gases, physical obstructions).

Controlling

NFPA 1006 (NFPA 2021) outlines essential competencies for rescuers to ensure safety and effectiveness during emergency operations, and Section 7.3.6 emphasizes the importance of controlling entry and egress routes in confined-space incidents. Controlling these routes ensures that rescuers can access and evacuate the confined space efficiently while minimizing risks such as entrapment, exposure to hazardous substances, and environmental dangers. Clear routes help rescuers avoid prolonged exposure to extreme temperatures and toxic atmospheres.

Uncontrolled entry and egress into and out of a confined space can exacerbate the space's structural weaknesses, leading to collapses, and debris and equipment left in pathways can hinder movement and cause injuries. Uncontrolled entry and egress can also lead to confusion, causing rescuers to become disoriented and increasing the time required to perform rescues.

Best practices for controlling entry and egress routes include the following:

- **Site assessment:** Conduct thorough assessments of potential confined spaces to identify and map entry and egress points.
- **Resource allocation:** Ensure that resources, including ladders, ropes, and nonentry retrieval systems, are available and appropriately positioned.
- **Designated entrances and exits:** Clearly mark and communicate designated entry and egress points to all personnel.
- **Continuous monitoring:** Regularly monitor entry and egress routes for changes in conditions or new hazards.
- **Briefings:** Outline the designated routes, potential hazards, and communication protocols to all team members.
- **Continuous updates:** Maintain open lines of communication during the operation to provide updates on conditions and instructions.
- **Safety officers:** Assign safety officers to oversee entry and egress points, ensuring compliance with established protocols.
- **Coordination:** Ensure seamless coordination between different teams, such as fire services, HazMat teams, and medical personnel.
- **Incident Command System (ICS):** Utilize ICS to manage resources and communication effectively.
- **Debriefing sessions:** Conduct debriefings to review the effectiveness of entry and egress control and identify areas for improvement.
- **Documentation:** Document all actions taken, observations made, and lessons learned to enhance future operations.

Establishing Search Parameters

The establishment of search parameters is the cornerstone of effective search and rescue operations. NFPA 1006 7.3.1 (NFPA 2021) outlines a methodical approach to defining the scope and boundaries of the search area.

The first step in establishing search parameters is conducting a comprehensive scene assessment. This involves a meticulous examination of the search area to identify potential hazards, evaluate structural integrity, and understand the environmental conditions that may affect the rescue operation.

Next is the identification of potential physical, atmospheric, or environmental hazards. Identify any immediate dangers, such as unstable structures, sharp objects, or debris, that could pose risks to both victims and rescuers. Assess the presence of toxic gases, oxygen deficiency, or other atmospheric conditions that may require specialized equipment or protocols. Consider factors such as temperature, weather conditions, and time of day, which can affect both the search process and the well-being of victims and rescuers. The terrain within the search area must also be included—in particular, any difficult or hazardous features, such as cliffs, bodies of water, or dense vegetation. Visibility conditions may be hampered by such factors as smoke, fog, or darkness, which may necessitate additional lighting or equipment.

Following the hazard assessment, the structural integrity of the space must be evaluated. Assess the stability of structures within the search area, especially in cases of collapse or potential collapse. Identify and evaluate potential entry and exit points that are safe and accessible for rescue operations. Determine if any spaces or isolated areas within the search zone require special attention or equipment.

Victim profiles are also essential to establishing search parameters because accurate victim profiles are crucial for directing search efforts. Detailed victim profiles include information on the missing person's last known location, physical description, medical history, and likely behavior patterns.

Search Strategies

Implementing effective search strategies is essential for covering the search area thoroughly and efficiently. Various search patterns can be utilized based on the specific circumstances and characteristics of the search area.

The most common search strategies are grid, zone, point-to-point, and sequential searches, and all four share the same post-search procedures. After a search concludes, teams debrief with central command to discuss their findings and any issues encountered. This helps improve future search operations and address any immediate concerns. All findings are documented in detail, including the areas searched, evidence found, and any environmental conditions noted. The search operation is evaluated to identify any areas for improvement. This includes assessing the effectiveness of the search and the overall coordination.

Grid Searches

Grid searches offer a structured approach to covering the search area, ensuring no section is overlooked. They involve dividing the search area into a grid and systematically assigning teams to search each section. Grid searches are particularly effective in confined spaces where visibility is limited and the risk of missing a victim or crucial clue is high.

As a first step, the search area is divided into manageable grid sections. This can be done using physical markers or by mapping out the area on paper or digitally. Each grid section is assigned to a specific search team. This ensures that all areas are covered without confusion or overlap. Searchers move methodically through their assigned sections, covering every inch of the space to ensure no clues or victims are missed.

Searchers move systematically within their grid sections, often using a sweep technique to cover the area thoroughly. This involves moving from one side of the grid to the other in a controlled and deliberate manner. Every nook and cranny within the grid section is examined. This includes looking under debris, inside containers, and in any possible hiding spots. Any findings, no matter how small, are documented and reported. This includes physical evidence, environmental conditions, and any potential hazards.

A central command unit oversees the entire search operation. This unit is responsible for assigning grid sections, coordinating teams, and compiling search data. Teams maintain constant communication with central command and with each other. This is typically done using radios or other reliable communication devices. Regular reports to central command of the teams' progress and findings allow for real-time adjustments to the search strategy and ensure that all sections are covered without overlap or gaps.

Clear communication and coordination are vital to success and ensuring that the entire search area is covered systematically and efficiently. Part of this communication includes information gathered from safety checks as teams check in at regular intervals to confirm their status and ensure that all members are safe and accounted for. To better understand the process of establishing communication with outside sources, follow the steps in **SKILL DRILL 12-2**.

SKILL DRILL 12-2

Establishing Communication with Outside Sources

Pre-Rescue Planning and Setup

Step 1: Conduct Initial Briefing

The team leader should conduct an initial briefing, reviewing the rescue plan, assigning roles, and establishing communication protocols with the team.

Step 2: Check Equipment

The team should ensure that all communication devices are tested and functioning correctly before entering the confined space.

SKILL DRILL 12-2 CONTINUED

Establishing Communication with Outside Sources

Step 3: Set Up Communication Devices

The team should set up communication equipment at the entry point, including external antennas or repeaters, to maintain a clear line of communication during the rescue.

Establishing Initial Communication

Step 4: Establish Communication from the Entry Point

A team member should establish communication with the command center from the entry point, ensuring clear contact before the rescue begins.

Step 5: Set Up Command Center Communication

The command center staff should maintain communication with the rescue team, monitoring the operation and coordinating support.

Continues.

SKILL DRILL 12-2 CONTINUED

Establishing Communication with Outside Sources

During the Rescue Operation

Step 6: Communicate with the External Team

Rescue personnel inside the confined space should use communication devices to maintain contact with the external team and report ongoing conditions.

Step 7: Relay Information to the Command Center

Key information, such as victim status and operation progress, should be relayed back to the command center by team members inside the confined space.

Emergency Communication Protocols

Step 8: Deploy a Backup Communication System

A backup communication system should be deployed to ensure continuous communication in case of primary system failure.

Step 9: Signal for Evacuation

A team member should signal for an emergency evacuation after detecting a hazardous condition inside the confined space.

Step 7: Courtesy of Illinois Fire Service Institute.

When a team prepares for a grid search, the team members start by mapping out the area and clearly defining the grid sections. This can involve physical marking of the space or using digital tools to create a grid layout. Search teams are briefed on the grid layout, their assigned sections, and the specific procedures they need to follow. This includes safety protocols and communication procedures. Each team is equipped with the necessary tools and equipment, such as flashlights, radios, first-aid kits, and PPE.

As a grid search is executed, teams move systematically through their assigned sections, covering every part of their grid. This often involves using a search pattern, such as zigzag or parallel lines, to ensure complete coverage. They thoroughly examine their sections, checking for any signs of victims or clues. This includes looking under, over, and inside all potential hiding places and obstacles. Throughout the process, they maintain continuous communication with central command and the other teams, providing updates on their progress and any findings.

Zone Searches

Zone searches are effective for methodically covering large and complex search areas. This approach involves dividing the search area into distinct zones based on natural or artificial boundaries and assigning specific teams to each zone to ensure focused and manageable coverage. This method allows for a structured and organized approach, ensuring thorough coverage of each designated zone.

As with a grid search, the first step in a zone search is to divide the search area. This division of zones can be based on a variety of factors, including natural boundaries, such as rivers, cliffs, and forests, that naturally segment the area; artificial boundaries, such as roads, walls, and buildings, that create distinct sections; or functional boundaries, such as different parts of a facility, separate rooms in a building, or compartments in a ship.

Once the search area has been divided into zones, specific teams are assigned to each zone. This targeted assignment has several benefits. Teams can focus on their own smaller, more manageable areas, increasing the thoroughness and efficiency of the search. The teams can be selected and assigned based on their skills and experience, ensuring that the most qualified personnel are working in the areas where they are needed most. Clearly defined zones and team assignments reduce the risk of overlap or gaps in the search coverage.

Confined-space rescue operations often require the use of specialized techniques and equipment to address the unique challenges presented by specific zones. These techniques must be adapted to the characteristics of the area being searched. For a high-angle rescue, such as one taking place on a cliff, on a building exterior, or in a deep pit, a team will require ropes, harnesses, pulleys, and anchor systems, along with special training in high-angle rescue techniques. Teams responding to a rescue taking place in a flooded compartment, such as a basement, tunnel, or submerged area, require dive gear, water pumps, and flotation devices, as well as training in water rescue techniques.

Before implementing a zone search, a detailed preplan must be developed that identifies potential hazards, maps out the zones, and preassigns teams based on their expertise. All team members must be briefed on the overall plan, the specific zones, and the roles of each team. This ensures everyone is aware of their responsibilities and the overall strategy.

As a zone search progresses, teams systematically search their assigned zones, using the appropriate techniques and equipment for the specific challenges they encounter. They maintain clear and constant communication among themselves and central command and follow strict safety protocols, including use of PPE and regular check-ins.

Point-to-Point Searches

Point-to-point searches involve identifying and investigating specific locations within a confined space where victims are likely to be found. Searchers move from one point of interest to the next, ensuring each location is thoroughly investigated before proceeding, and they adapt the search pattern as new information becomes available. This method is particularly effective in scenarios where initial information from victim profiles or the scene assessment suggests probable victim locations.

Victim profiles are essential to point-to-point searches. A comprehensive victim profile includes the victim's last known location, physical description, medical history, and behavior patterns. This information helps predict potential locations where the victim might be found. The scene assessment can also identify potential points of interest. This process involves examining the layout of the confined space, identifying hazards, and noting any environmental factors that may affect the search. Findings from the victim profiles and scene assessment are combined to identify specific points of interest. These could include areas where the victim was last seen, locations with high visibility, spots near exits, areas under debris, or places offering shelter or relative safety. Secondary locations, such as adjoining rooms or secondary chambers, may be included as plausible locations based on victim profiles.

Sequential Searches

Sequential searches, a subset of point-to-point searches, involve moving from one point of interest to the next in a systematic manner. This method ensures that each location is thoroughly investigated before moving on, reducing the risk of overlooking the victim. Each point of interest is meticulously examined using appropriate techniques and equipment. This includes visual inspection, physical probing, and the use of search tools such as cameras or thermal imaging devices.

Conducting a sequential search begins with a briefing to outline the search plan, including identified points of interest and team assignments, and implement safety protocols. As the search begins, once a point of interest has been thoroughly investigated and documented, searchers move systematically to the next location. This sequential approach ensures that no area is left unchecked and that search efforts are organized and methodical. Each investigated location is clearly marked off to avoid confusion among the team and ensure that all points are covered. Effective communication and coordination among teams and central command facilitates smooth transitions between search points.

If a victim is attached to a retrieval system or airline equipment within the line of sight, a hasty search that follows the equipment to the victim can be initiated for the most efficient rescue.

Methods for Establishing Victim Profiles

Detailed victim profiles that feature comprehensive information about the missing person(s) to predict their possible locations and behaviors are crucial for directing search efforts and increasing the likelihood of a successful rescue. Chapter 7 contains much more detail on victim profiles.

Among the data gathered in a victim profile are their last known location (LKL) based on eyewitness accounts, electronic tracking, or other reliable sources; a detailed physical description, including age, gender, height, weight, clothing, and any distinguishing features; and relevant medical information, including any conditions that may affect the victim's behavior or survival time, such as diabetes, heart conditions, or disabilities.

Victim profiles are reliant on thorough interviews with anyone who might have seen or interacted with the victim prior to their disappearance, as well as security footage, phone records, and social media activity to piece together the victim's movements and actions.

An assessment of the victim's likely behavior patterns based on past actions, psychological state, and environmental influences is particularly useful.

Consider the victim's psychological state, which can significantly influence their actions and decision-making processes, and how the victim might interact with their environment, such as seeking shelter, moving toward or away from certain features, or following natural paths.

Communicating Search Results

Effective communication and dissemination of search result information ensures that all team members are informed, coordinated, and able to respond promptly to evolving situations. During a confined-space emergency, the timely and accurate communication of search results can mean the difference between life and death. Effective communication of search results enhances coordination efforts, improves decision-making capacity for the command center, increases safety through rapid identification of hazards, and boosts efficiency by reducing redundancies and ensuring resources are used in an effective way.

Accurate and thorough information collection is the first step in ensuring effective communication. This involves the following:

- **Documenting findings:** Each search team must meticulously document their findings at each point of interest.
- **Recording conditions:** Note the environmental conditions, including air quality, visibility, and any hazards encountered.
- **Cataloging evidence:** Any clues or evidence related to the victim's location should be carefully cataloged.

Tools for information collection include standardized checklists, digital devices like tablets or smartphones, and photos or videos that can provide valuable context reports might miss.

Reporting Protocols

Reporting protocols for search results involve both an initial and a detailed report, regular updates, and a final summary. After each search segment, the initial report to the command center should include the locations searched, any findings and observations, the environmental conditions, and any immediate hazards. After the conclusion of the search, the detailed report should include comprehensive findings, detailed descriptions of any evidence, and recommendations for further actions or areas to search.

Teams should provide regular updates to the command center, even if no new findings are reported, to keep the command center informed of the team's progress and any changes in conditions.

At the end of the search operation, a final summary should be submitted. This should include an overview of all areas searched, the consolidated findings, a summary of the environmental conditions and hazards, and recommendations for next steps.

Communication Tools

Effective communication requires the use of reliable tools and technologies chosen based on the specific needs and conditions of the operation. Two-way radios are essential for real-time communication between search teams and the command center. They use dedicated channels to prevent interference and ensure clear communication.

Intercom systems are useful in large or complex confined spaces, allowing for direct communication with multiple teams. Public address systems can be used to broadcast important messages to all personnel simultaneously. (Note: Consider Health Insurance Portability and Accountability Act [HIPAA] laws when using open communications.) Smartphones and tablets can be used for real-time data entry and communication, and specialized apps can facilitate the sharing of real-time updates, maps, and other critical information.

Signal codes, which can be either visual or audible, allow for quick communication of important information, such as hazards or the discovery of a victim. Flashlights and other light signals can be used to communicate in low-visibility conditions.

Best Practices for Communication

Best practices for communication include the following:

- **Pre-search:** Detailed briefings outlining the communication plan, reporting protocols, and safety measures, as well as role assignments for each team member, including communication duties.
- **During the search:** Regular check-ins for teams to report their status and findings and a review of emergency signals and protocols with the team to ensure they are all familiar with their meanings.
- **Post-search:** Conduct thorough debriefings after the search to review the operation, discuss findings, and identify areas for improvement. Ensure all information is documented and archived for future reference and training.

Responder Assignments

Operation safety, efficiency, and success are dependent on ensuring that responders are assigned to tasks that align with their expertise and training. All responders must hold the necessary certifications outlined in NFPA 1006 (NFPA 2021) and must complete initial training programs that cover the basics of confined-space rescue, including hazard recognition, entry procedures, and emergency response.

As a first step in ensuring responder assignments are appropriate, clear job descriptions must be developed for each role within the team. These must detail the required skills and qualifications for each role. An inventory of the current skills and qualifications of all team members can be developed in conjunction with the job descriptions. Once these are prepared, responders can be matched to roles based on their assessed competencies, ensuring each task is performed by a qualified individual.

At the same time, task assignments must be flexible enough to adapt to changing conditions and the emergence of new information during an operation. Cross-training in multiple skills is the best way to provide this flexibility.

Some organizations use incident management software to manage assignments, track the progress of responders, and ensure that tasks are allocated based on each team member's level of expertise.

Accountability and Risk–Benefit Analysis

Confined-space emergencies present significant risks to both victims and responders, so ensuring accountability and conducting thorough risk–benefit analyses are necessary to maximize safety and operational efficiency.

The term *accountability* in this context means ensuring that every member of the rescue team is aware of their responsibilities, actions, and location at all times. This not only enhances the safety of responders but also contributes to the overall efficiency of the rescue operation.

Knowing the exact location and status of each team member at all times helps prevent accidents and ensures rapid response in the case of an emergency.

Having clear definitions for roles and responsibilities prevents confusion and overlapping duties, thus reducing the risk of errors.

Accountability ensures that all actions are coordinated, avoiding redundant efforts and ensuring that all areas of the confined space are covered, and accurate tracking of personnel and equipment helps optimize the allocation and utilization of resources.

The two most common types of accountability systems are personnel accountability systems (PASs) and the ICS. With PASs, each responder is assigned a tag that they must check in and out when entering or exiting the confined space. Radio-frequency identification (RFID) tags or global positioning system (GPS) devices make it possible to monitor the location of responders in real time. The ICS establishes a clear command hierarchy, ensuring that all actions are directed and monitored by a designated incident commander (IC). Accountability officers are appointed to track the status and location of all personnel during the operation.

A risk–benefit analysis is a systematic approach to evaluating the potential risks and benefits of a course of action. In confined-space emergencies, these are used to balance the urgency of rescuing victims with the safety of responders. These analyses provide a structured framework for making informed decisions about whether to proceed with a rescue operation and help in prioritizing actions based on the potential benefits and the associated risks.

The analysis features a risk–benefit matrix that is used to assist with decision-making. The matrix is a grid that organizes the operation in the following format:

- **Low risk/high benefit:** Proceed with the operation, ensuring all safety measures are in place.
- **High risk/low benefit:** Consider alternative strategies or abort the operation to avoid unnecessary exposure to hazards.
- **Moderate risk/moderate benefit:** Implement additional safety measures and proceed with caution.

As an incident progresses, a risk–benefit analysis permits a team to continuously assess the risks as the situation evolves, considering new information and changing conditions, and weigh the potential benefits of rescue actions against the risks to both the victim and the responders.

Working Safely Within a Confined Space

SKILL DRILL 12-3 provides a high-level summary of the steps that need to be taken in order to work safely within a confined space.

SKILL DRILL 12-3

Working Within a Confined Space

Step 1: Conduct a Risk Assessment

Perform a comprehensive risk assessment, identifying potential hazards in the confined-space environment.

Step 2: Conduct Atmospheric Testing

Use approved equipment to conduct atmospheric testing, ensuring the air inside the confined space is safe for entry.

SKILL DRILL 12-3 CONTINUED

Working Within a Confined Space

Step 3: Review the Confined-Space Entry Permit

Carefully review the confined-space entry permit to ensure compliance with all safety protocols.

Step 4: Perform Correct Entry Procedures

Follow established procedures for safe confined-space entry, verifying all safety measures are in place.

Step 5: Maintain Effective Communication

Stay in constant communication with incident command and team members to ensure coordination during the operation.

Step 6: Use Communication Devices

Use communication devices for effective coordination and signaling in case of emergencies.

Continues.

SKILL DRILL 12-3 CONTINUED

Working Within a Confined Space

Step 7: Perform Self-Rescue Techniques

Perform self-rescue techniques that demonstrate the ability to safely exit the confined space in an emergency situation.

Step 8: Execute a Confined-Space Victim Rescue

Perform a confined-space victim rescue using proper techniques and specialized equipment for safe extraction.

Step 9: Decontaminate Equipment and Personnel

Follow decontamination procedures to clean equipment and personnel after the confined-space entry.

Step 10: Complete Required Documentation

Ensure that all necessary documentation, such as incident reports and equipment logs, is completed accurately.

Step 1: © Red_Shadow/Shutterstock; **Step 8:** Courtesy of Illinois Fire Service Institute.

CHAPTER FEATURE: AFTER-ACTION

IN SUMMARY

- Potential hazards must be thoroughly understood and documented in a preplan to ensure the safety of personnel during entry as well as the rescue operation.
- Identifying the best entry and exit points for various types of confined spaces requires an understanding of the space's characteristics, as well as an evaluation of the structural integrity, accessibility, and potential hazards associated with each entry and exit point.
- Establishment of search parameters requires a comprehensive scene assessment; the identification of potential physical, atmospheric, or environmental hazards; evaluation of the structural integrity of the space; and completion of victim profiles.
- The most common search strategies are grid, zone, point-to-point, and sequential searches.
- Effective communication of search results enhances coordination efforts, improves decision-making capacity for the command center, increases safety through rapid identification of hazards, and boosts efficiency by reducing redundancies and ensuring resources are used in an effective way.
- Matching each responder's skill level and training with the appropriate tasks first requires the development of clear job descriptions and inventories of the current skills and qualifications of all team members.
- Accountability ensures that all actions are coordinated, avoiding redundant efforts and ensuring that all areas of a confined space are covered, and accurate tracking of personnel and equipment helps optimize the allocation and utilization of resources.
- Performing a risk–benefit analysis on an operation gives insight into whether the potential benefits of a rescue may be outweighed by the potential risks to rescuers.

REFERENCES

"Confined and Enclosed Spaces and Other Dangerous Atmospheres in Shipyard Employment." 1994. 29 C.F.R. 1915 Subpart B. https://www.ecfr.gov/current/title-29/subtitle-B/chapter-XVII/part-1915/subpart-B.

National Fire Protection Association. 2021. *NFPA 1006: Standard on Technical Rescue Personnel Professional Qualifications, 2021 Edition.* Quincy, MA: National Fire Protection Association.

CHAPTER QUESTIONS

1. The systematic approach used to anticipate and prepare for potential emergencies during a confined-space incident is known as

a. purging.
b. preplanning.
c. incident control.
d. incident command.

2. To ensure compliance with established protocols, who should be assigned to oversee the entry and egress points of a confined space?

a. safety officers
b. attendants
c. backup teams
d. medical officers

3. Which of the following processes ensures that no section of a search area is overlooked?

a. a point-to-point search
b. a scene size-up
c. a preincident briefing
d. a grid search

4. As a first step in ensuring responder assignments are appropriate, which of the following must be developed for each role within the team?

a. strategy plans
b. equipment assignments
c. job descriptions
d. performance criteria

CHAPTER FEATURE: ON SCENE

THE BOILER ROOM INCIDENT

On a summer afternoon, a paper mill located in a rural area was the scene of a terrible accident. In the mill's boiler room, inside a confined space measuring 15 feet (4.5 m) deep by 8 feet (2.4 m) wide, a maintenance worker became unresponsive as a result of exposure to high levels of carbon monoxide. The emergency quickly escalated once the local fire department responded to the call and found that they would face significant challenges due to inadequate training and improper equipment.

The Emergency Response

At 2:15 p.m., the emergency response team at the paper mill encountered the unresponsive maintenance worker inside the boiler room. The team quickly contacted the local fire department to report the emergency, and fire personnel arrived on the scene within 10 minutes.

The fire personnel quickly assessed the situation but did not have the specialized confined-space rescue training—or the necessary equipment, such as gas detectors, ventilation devices, or proper rescue equipment—they needed to attempt the rescue.

The Rescue Operations

Even though the team lacked the necessary equipment and training, one of the firefighters attempted to enter the confined space without a self-contained breathing apparatus (SCBA) because the entry hatch was too narrow for it to fit through. He was quickly overcome by the carbon monoxide, became disoriented, and required immediate assistance. Fortunately, he was still close enough to the hatch that another firefighter was able to reach inside and pull him back.

Realizing the severity of the situation and their inability to safely conduct the rescue, fire personnel requested mutual aid from a neighboring city that had a specialized confined-space rescue team. The specialized confined-space rescue team arrived onsite at 3:20 p.m. and immediately assessed the space's environment using gas detectors. They confirmed the presence of high carbon monoxide levels and initiated ventilation to reduce the concentration of the gas.

The Extrication and Treatment

The initial rescue attempt of the firefighter entering the space ended up being a self-rescue with the help of another firefighter without an SCBA at the opening hatch. The specialized team focused on extricating the original victim using proper confined-space entry techniques and equipment. The victim was extricated from the boiler room within 15 minutes of the team's arrival.

The Outcome

The fire personnel who initially responded lacked proper training and also had none of the rescue equipment, gas monitors, and confined-space retrieval systems necessary to perform a successful rescue. They were never able to establish contact in any form with the victim inside the space, other than visible verification that he was there. Their lack of training directly led to the improper and dangerous rescue attempt that nearly cost the life of one of their own.

Unfortunately, despite the extrication and immediate lifesaving efforts, the original victim—the maintenance worker—succumbed to carbon monoxide poisoning and could not be revived. The delay caused by the necessary call for mutual aid could have been a factor in the death of the worker.

The firefighter who entered the confined space without proper equipment was treated for carbon monoxide exposure and made a full recovery after receiving oxygen therapy and other supportive care. The second firefighter stationed at the opening hatch who assisted the first firefighter also required oxygen therapy and supportive care before recovering.

This tragic incident highlights the critical importance of proper training and equipment in confined-space rescue operations. The loss of the maintenance worker emphasizes the need for preparedness and the swift, informed action of emergency responders. The recovery of the firefighter serves as a reminder of the risks involved when entering hazardous environments without proper protection. This story is proof that comprehensive training and better-equipped first responders are necessary factors for a successful confined-space rescue.

1. What was the primary reason why the maintenance worker became unresponsive inside the boiler room?
   - **a.** overheating from high temperatures
   - **b.** lack of ventilation in the boiler room
   - **c.** a physical injury from a fall
   - **d.** exposure to high levels of carbon monoxide

2. What significant challenge did the local fire department face when responding to the emergency?
   - **a.** They were unable to locate the boiler room.
   - **b.** They arrived too late to attempt the rescue.
   - **c.** They lacked the necessary specialized training and equipment.
   - **d.** They experienced a power outage that delayed their response.
3. Why did one of the firefighters attempt to enter the confined space with no SCBA?
   - **a.** The SCBA malfunctioned.
   - **b.** The entry hatch was too narrow for the SCBA to fit through.
   - **c.** The firefighter forgot the SCBA.
   - **d.** The firefighter was not aware of the carbon monoxide levels.
4. How was the trapped firefighter rescued after being overcome by carbon monoxide inside the confined space?
   - **a.** He was extricated by the rest of the emergency response team.
   - **b.** He was pulled out by another firefighter stationed at the entry hatch.
   - **c.** He was lifted out with a crane from the local fire department.
   - **d.** He was extricated by the mutual aid team from a neighboring city.
5. What important lesson does this incident highlight for emergency responders?
   - **a.** the necessity of comprehensive training and proper equipment in confined-space rescue operations
   - **b.** the importance of arriving quickly to any emergency scene
   - **c.** the importance of using properly calibrated gas detectors in confined-space rescue operations
   - **d.** the necessity of backup power systems at any emergency scene

Appendix A

NFPA 1006 2021 Edition Correlation Grid for Confined-Space Rescue (Awareness, Operations, and Technician Level)

NFPA 1006: Standard for Technical Rescue Personnel Professional Qualifications, 2021 Edition		
Job Performance Requirement	**Corresponding Chapter(s)**	**Learning Objective(s)**
Confined-Space Rescue		
7.1: Awareness Level		
7.1.1	1, 2, 4	▪ Define the meaning of the term *confined space*. (pp. 4–5) ▪ Recognize immediate hazards. (pp. 12–17) ▪ Describe the process of access control at a scene. (pp. 17–20) ▪ List different types of technical references. (pp. 21–24) ▪ Predict the potential hazards associated with a confined-space incident. (pp. 12–17) ▪ Describe isolation procedures for a confined-space incident. (pp. 39–43) ▪ Explain how to initiate different isolation methods. (pp. 39–43) ▪ Demonstrate the process of securing and mitigating access to a confined-space incident. (pp. 43–48) ▪ Demonstrate the methods used to identify the required resources for a confined-space incident. (pp. 95–96)
7.1.2	1, 2, 3	▪ Define the terms *awareness*, *operations*, and *technician* in the context of confined-space rescue, and identify the roles appropriate to these personnel levels. (pp. 8–9) ▪ Describe local policies relating to confined-space access areas. (p. 21) ▪ Describe the process for selecting appropriate personal protective equipment (PPE). (pp. 20–21) ▪ Identify search parameters at a confined-space incident. (pp. 48–49; 50–52) ▪ Explain the process for establishing a victim profile. (pp. 53–55) ▪ Describe the process for interviewing bystanders. (pp. 55–56) ▪ Construct a plan for escape from the area of a confined-space incident. (pp. 63–65) ▪ Demonstrate the ability to operate in and exit adjacent areas to a confined space. (pp. 80–82)
7.1.3	2, 3	▪ Interpret and document information appropriately. (pp. 38–42) ▪ Describe the process for communicating with confined-space victims. (pp. 49; 53) ▪ Document all aspects of a confined-space incident, including victim transfer information. (pp. 38–42) ▪ Choose communication devices that can be utilized anywhere in a confined-space incident. (pp. 69–71)

Job Performance Requirement	Corresponding Chapter(s)	Learning Objective(s)
7.1.4	3	▪ Identify confined-space protocols related to safety, communication, and medical procedures. (pp. 62–82) ▪ Describe fall prevention techniques. (pp. 84–87) ▪ Explain the process used for retrieving a victim. (pp. 87–89) ▪ Implement all needed safety and communication protocols. (pp. 62–82) ▪ Don applicable PPE. (pp. 66–69) ▪ Demonstrate fall prevention methods. (pp. 84–87) ▪ Utilize nonentry retrieval systems. (pp. 87–89) ▪ Demonstrate methods for securing a portal for victim passage. (pp. 87–89)
7.1.5	1, 2, 4	▪ Utilize reference materials to ensure that all needed information is obtained and communicated. (p. 24) ▪ Utilize interview techniques. (pp. 55–56) ▪ Describe the criteria for a basic search. (pp. 94–95) ▪ List the elements of an incident action plan (IAP). (pp. 96–103) ▪ Describe how a size-up is incorporated into an IAP. (p. 103) ▪ Explain how to gain information from witnesses during the incident size-up. (pp. 103–104)
7.1.6	2, 3, 4	▪ List the steps for securing a confined-space incident. (pp. 43–48) ▪ Indicate common safety measures used in confined-space incidents. (pp. 62–82) ▪ Establish safety measures. (pp. 62–82) ▪ Explain how to determine the need for additional resources. (pp. 95–96) ▪ Discuss how operational protocols are determined by the authority having jurisdiction (AHJ). (pp. 104–107) ▪ Ascertain the needed resources for a confined-space incident. (pp. 95–96) ▪ Select appropriate planning forms and document hazards within the AHJ. (pp. 105–107)
7.1.7	1, 3, 4	▪ Explain how awareness level personnel support operations or technician level personnel. (pp. 8–9) ▪ Discuss the need for rehabilitation during confined-space incidents. (pp. 82–83) ▪ Communicate needed information while following an IAP while functioning in an Incident Command System (ICS). (pp. 98–100)
7.2: Operations Level		
7.2.1	6, 7	▪ Describe the process for establishing search parameters. (pp. 144, 145–146) ▪ Indicate the process for interviewing witnesses. (pp. 144, 145–147) ▪ Recognize and identify all visible victims. (p. 185) ▪ Explain the process for developing a victim profile. (p. 185) ▪ Demonstrate emergency escape procedures while operating in a confined-space entry portal. (p. 184)
7.2.2	5, 6	▪ Discuss the process for ensuring adequate resources. (p. 133) ▪ During a size-up, gather all pertinent information and relay it to the incident commander (IC). (pp. 134–135) ▪ Determine the type of rescue. (pp. 144, 147–148)

Job Performance Requirement	Corresponding Chapter(s)	Learning Objective(s)
7.2.3	8	▪ Discuss the process for monitoring the environment. (p. 205) ▪ Explain what documentation should be completed during air-monitoring operations. (pp. 214–216) ▪ During a size-up, gather all pertinent information and relay it to the IC. (pp. 224–225) ▪ Calibrate monitoring equipment. (pp. 213–214) ▪ Acquire and analyze air samples from a confined-space incident. (pp. 205–206)
7.2.4	5, 6, 7, 8, 9	▪ Describe the risk–benefit analysis process. (pp. 135–136) ▪ Identify a means of retrieval for rescuers. (pp. 162–166) ▪ Describe how to build victim removal systems. (pp. 152–153, 156, 160–161) ▪ Ensure rescue systems are operational. (p. 161) ▪ Recognize ingress and egress routes and the process for measuring them. (pp. 182–183) ▪ Describe how to identify space-specific hazards. (pp. 200–201) ▪ Explain immediate and ongoing monitoring needs of the space, as well as victim conditions. (pp. 218, 220–221) ▪ Identify immediate hazards. (pp. 200–201) ▪ Choose appropriate PPE and needed equipment. (pp. 246, 248–249)
7.2.5	5, 6, 8, 9	▪ Deploy scene control measures. (pp. 118–119) ▪ Describe the process for establishing a rescue area. (pp. 148–149) ▪ Identify and control energized and physical hazards. (pp. 201–204) ▪ Explain the process for managing exposures from hazardous materials. (pp. 218, 220–221) ▪ Explain the procedures for operating ventilation equipment. (pp. 216–219) ▪ Identify and mitigate hazards. (pp. 200–204) ▪ Acquire and analyze air samples from a confined-space incident. (pp. 205–206) ▪ Employ ventilation for a confined-space incident. (pp. 216–219) ▪ Demonstrate proper PPE selection. (pp. 246, 248–249)
7.2.6	5, 6, 8	▪ Develop communication methods so that information is adequately sent and received. (pp. 119–121) ▪ Demonstrate communication methods. (pp. 119–121) ▪ Explain the process for ensuring a clear and unobstructed space for rescuer retrieval systems. (pp. 163–166) ▪ Describe the capabilities and limitations of a self-contained breathing apparatus (SCBA). (pp. 190–194, 220–221) ▪ Discuss the need for the rescue entrant to communicate their air levels to outside rescuers. (p. 221) ▪ While using an SCBA, demonstrate conservative breathing techniques. (pp. 221–223)
7.2.7	5, 7	▪ Discuss the importance of constant communication with outside rescuers. (pp. 120–121) ▪ Describe the capabilities and limitations of an SCBA or a supplied-air respirator (SAR) for victims. (pp. 190–194) ▪ Determine the expected air usage for victims. (pp. 192–194) ▪ Describe how to secure a breathing apparatus to a packaged victim attached to a retrieval system. (pp. 190, 192) ▪ Explain the process for donning face pieces on victims with helmets or injuries. (p. 192) ▪ Establish an air supply for a victim. (pp. 190, 192) ▪ Maintain a patent air supply for a victim throughout the incident. (p. 192)

Job Performance Requirement	Corresponding Chapter(s)	Learning Objective(s)
7.2.8	6	▪ Discuss the hazards of removal of victims on immobilization devices. (pp. 169–174) ▪ Describe the process for immobilizing victims and securing them to retrieval devices. (pp. 166–170) ▪ Using two rescuers, demonstrate the process for immobilizing a victim and securing them to a retrieval device. (pp. 167–168)
7.2.9	5, 6, 7, 8	▪ Identify the process for ensuring rescuers have been cleared for confined-space operations. (pp. 124–125) ▪ Demonstrate pre-entry personnel evaluations. (pp. 124–125) ▪ Identify communication needs. (pp. 119–121) ▪ Demonstrate evacuation methods. (pp. 165–166) ▪ Discuss the different methods used for entry into confined spaces. (pp. 180–182) ▪ Ensure the point and route of entry are clear of hazards. (pp. 180–182) ▪ Describe the hazardous atmospheres associated with confined-space incidents. (pp. 205–207) ▪ Describe the types of operations of hazard-specific monitoring devices. (pp. 210–213)
7.2.10	5, 7, 8, 9	▪ Describe the operational protocols related to safety, communication, medical, and operations aspects. (pp. 119–124) ▪ Ensure that operational protocols are instituted. (pp. 123–124) ▪ Explain the process for controlling access and egress for confined spaces. (pp. 183–185) ▪ Describe the processes involved when rendering patient care during a confined-space incident. (pp. 185–186) ▪ Critique the scene and choose appropriate medical protocols to determine treatment priorities. (pp. 185–186) ▪ Discuss the importance of continued atmospheric monitoring. (pp. 207–210) ▪ Discuss the importance of portable lighting systems and the different methods used in setting them up. (pp. 245–246) ▪ Deploy scene lighting. (pp. 247–248) ▪ Assess the scene for PPE and equipment requirements. (pp. 246, 248–249)
7.2.11	6, 7, 9	▪ Explain the importance of spinal immobilization. (pp. 152–161) ▪ Discuss the processes used to package victims. (pp. 186–188) ▪ Describe the differences and similarities between packaging for confined-space events versus others. (p. 188) ▪ Describe the processes used to maneuver low-profile packaging devices. (p. 188) ▪ Demonstrate immobilization and patient treatment methods in horizontal spaces. (pp. 186–187) ▪ List the steps for limiting damage to confined-space equipment. (p. 249)

Job Performance Requirement	Corresponding Chapter(s)	Learning Objective(s)
7.2.12	9	▪ Discuss the proficiencies of different types of portable anchor systems. (pp. 230–236) ▪ Discuss the identified constraints of portable anchor systems. (pp. 230–236) ▪ Describe the process of assembling portable anchor systems. (p. 236) ▪ Explain the importance of stabilizing a portable anchor system to limit movement. (pp. 236–237) ▪ Explain the direction of force on a portable anchor. (p. 237) ▪ List the steps necessary to calculate force application on a portable anchor. (pp. 237, 240–241) ▪ Using a portable anchor device, demonstrate how to remove a vertically oriented litter. (pp. 238–240)
7.2.13	5, 6, 7, 8	▪ Identify the process for ensuring rescuers have been cleared for confined-space operations. (pp. 124–125) ▪ Demonstrate pre-entry personnel evaluations. (pp. 124–125) ▪ Identify communication needs. (pp. 119–121) ▪ Ensure victim and rescuer communications are established. (p. 152) ▪ Demonstrate evacuation methods. (pp. 165–166) ▪ Discuss the different methods used for entry into confined spaces. (pp. 180–182) ▪ Ensure the point and route of entry are clear of hazards. (pp. 180–182) ▪ Discover entry and exit points. (p. 183) ▪ Describe the hazardous atmospheres associated with confined-space incidents. (pp. 205–207) ▪ Describe the types of hazard-specific monitoring devices and their operation. (pp. 210–213) ▪ Demonstrate the process of monitoring a confined space. (pp. 205–210)
7.2.14	5, 7, 8, 9	▪ Describe the operational protocols related to safety, communication, medical, and operations aspects. (pp. 119–124) ▪ Ensure that operational protocols are instituted. (pp. 123–124) ▪ Ensure that operational protocols are instituted and reassessed as needed. (pp. 123–124) ▪ Explain the process for controlling access and egress for confined spaces. (pp. 183–185) ▪ Describe the processes involved when rendering patient care during a confined-space incident. (pp. 185–186) ▪ Critique the scene and choose appropriate medical protocols to determine treatment priorities. (pp. 185–186) ▪ Discuss the importance of continued atmospheric monitoring. (pp. 207–210) ▪ Discuss the importance of portable lighting systems and the different methods used in setting them up. (pp. 245–246) ▪ Deploy scene lighting. (pp. 247–248) ▪ Assess the scene for PPE and equipment requirements. (pp. 246, 248–249)

Job Performance Requirement	Corresponding Chapter(s)	Learning Objective(s)
7.2.15	6, 7, 9	▪ Explain the importance of spinal immobilization. (pp. 152–161) ▪ Discuss the processes used to package victims. (pp. 186–188) ▪ Describe the differences and similarities between packaging for confined-space events versus others. (p. 188) ▪ Describe the processes used to maneuver low-profile packaging devices. (p. 188) ▪ Demonstrate immobilization and patient treatment methods in vertical spaces. (pp. 186–187) ▪ List the steps for limiting damage to confined-space equipment. (p. 249)
7.2.16	6, 7	▪ Establish rescue systems and perform a rapid victim extraction. (pp. 153–155) ▪ Discuss the steps needed to use a rapid victim harness technique. (pp. 188–189)
7.2.17	5, 6, 7, 9	▪ List the multiple types of confined spaces. (pp. 115–117) ▪ Discuss having operational protocols related to medical issues, emergency medical services (EMS) providers, and decontamination. (pp. 121–124) ▪ Discuss the need for personnel and equipment and resources lists. (p. 134) ▪ Discuss the types of rescue and retrieval systems and the equipment needed to build them. (pp. 157–161) ▪ Utilizing appropriate PPE and retrieval systems, remove all victims from a confined-space incident. (pp. 172–174) ▪ Demonstrate equipment decontamination procedures. (p. 191) ▪ Describe the different types of confined-space PPE. (pp. 246, 248–249)
7.2.18	5	▪ List the functions established by the ICS. (p. 118) ▪ Describe the processes used for hazard identification, risk management, logistics, and resource management. (pp. 126–134) ▪ Discuss the need for personnel accountability systems. (pp. 115, 137–139) ▪ Discuss procedures and protocols involving personnel rehabilitation. (pp. 115, 126) ▪ Demonstrate the use of the accountability system. (pp. 138–139) ▪ Manage the termination of a confined-space incident. (pp. 115, 136–137)
7.3: Technician Level		
7.3.1	11, 12	▪ Execute a self-rescue. (pp. 294–295) ▪ Describe the process for establishing search parameters. (p. 309) ▪ Discuss the methods used for establishing a victim profile. (p. 314) ▪ Discuss search result information and how it should be communicated. (pp. 314–315) ▪ Describe the techniques used to ensure responder assignments match their level of expertise. (p. 315) ▪ Discuss the importance of accountability and risk–benefit analysis. (pp. 315–316) ▪ Demonstrate the process of working within a confined space. (pp. 316–318) ▪ Establish communication with outside sources. (pp. 310–312)
7.3.2	12	▪ Explain the methods used to preplan a confined-space incident. (pp. 306–308) ▪ Discuss the types of hazards associated with confined spaces and what should be documented on a preplan. (pp. 306–308) ▪ Illustrate the process of preplanning a confined-space incident. (pp. 306–308)

Job Performance Requirement	Corresponding Chapter(s)	Learning Objective(s)
7.3.3	11	▪ Describe the methods used to conserve air, manage the air line, and communicate in a confined space. (pp. 280–283) ▪ Determine the need for utilizing an SAR. (pp. 284–285) ▪ Discuss the proficiencies of an SAR. (pp. 280, 284) ▪ Discuss the identified constraints of an SAR. (pp. 280, 284) ▪ Manage air lines and communicate with outside sources while wearing an SAR. (pp. 281–283) ▪ Demonstrate air supply management while wearing an SAR. (p. 286)
7.3.4	10	▪ Discuss the proficiencies of short spine immobilization equipment. (pp. 264, 265–266) ▪ Discuss the identified constraints of short spine immobilization equipment. (pp. 264, 266) ▪ Describe the methods used to secure and move victims onto an immobilizer. (pp. 266–267) ▪ Demonstrate the process of immobilizing a victim in a confined-space incident. (pp. 265–266)
7.3.5	10, 11, 12	▪ Discuss the effects of hazardous atmospheres on victims and rescuers. (pp. 267–268) ▪ Discuss the protocols for medical and psychological evaluations of responders. (pp. 257–264, 268, 270–272) ▪ Demonstrate the process of monitoring a confined space. (pp. 268–269) ▪ Describe the methods of entry used in confined-space incidents. (pp. 278–279) ▪ List the types and operational requirements of supplying respirators. (pp. 279–280) ▪ Ensure victim and rescuer communications are established. (pp. 281–283) ▪ Discover entry and exit points. (pp. 308–309)
7.3.6	10, 11, 12	▪ Describe the methods for patient care. (pp. 257–264) ▪ Discuss the importance of having operational protocols related to communication and medical treatment. (pp. 259–264) ▪ Assess the scene for PPE and equipment requirements. (pp. 261–262) ▪ Ensure that operational protocols are instituted and reassessed as needed. (p. 260) ▪ Critique the scene and choose appropriate medical protocols to determine treatment priorities. (p. 262) ▪ Describe hazard-specific PPE. (pp. 279–280) ▪ Describe the importance of controlling the entry and egress routes of a confined space. (pp. 308–309)

Appendix B

A Sample Confined-Space Entry Permit

ENTRY PERMIT

Address: ______________________________

RP name: ______________________ Title: ______________________

RP or witness account of incident: ______________________

____

If no witness, clues available at the site: ______________________

____

Space type: Tank: ________ Pipe: ________ Silo: ________ Evacuation: ________

Confined-Space Permit obtained? ❑ Yes ❑ No

Product involved: ______________________________

Product hazards: LEL ______ % TLV ______ ppm IDLH ______ ppm

Explosive? ❑ Yes ❑ No

Establishment of zones? ❑ Yes ❑ No Isolation of area: ________ hr(s)

Lockout completed: ________ hr(s)

Number of victims: ________ Time victims trapped: ________ (24-hr clock)

Victim status: ______________________________

____

Victim #	Age	Name	Medical HX
1.			
2.			
3.			
4.			

Victim #	Location	Priority
1.		1 2 3 4
2.		1 2 3 4
3.		1 2 3 4
4.		1 2 3 4

Communications established with victims? ❑ Yes ❑ No

 NFPA 1006 (p. 1 of 2)

ENTRY PERMIT

Access points	Location	Entry accessible		Ventilation point	
1.	______________________	Yes	No	Yes	No
2.	______________________	Yes	No	Yes	No
3.	______________________	Yes	No	Yes	No
4.	______________________	Yes	No	Yes	No

Is there an adequate number of trained personnel on the scene to perform the tasks associated with the rescue? (minimum of eight required) ❑ Yes ❑ No

Is the proper equipment present at the scene to complete the operation?

	Yes	No
Atmosphere monitoring equipment	❑	❑
Explosionproof lighting	❑	❑
Explosionproof communications	❑	❑
Supplied-air breathing apparatus or remote aid	❑	❑
Cascade system	❑	❑
Victim removal systems/equipment	❑	❑
Ventilation equipment with CFM of 4000 – 5000 and necessary duct work	❑	❑

Request for hazmat and/or rescue units: ________ hr(s)

Diagram of confined space (including access and egress locations):

 NFPA 1006 (p. 2 of 2)

Appendix C

A Sample Confined-Space Entry Checklist

CONFINED-SPACE ENTRY TEAM CHECKLIST

Entry Team member's name: ______________________________

Filled out by: ______________________________

- ❑ Confined-space atmosphere evaluated
- ❑ Medical checkout by ALS unit
- ❑ Jump suit donned
- ❑ 2.2 cylinders on remote air topped off
- ❑ Escape bottle topped off
- ❑ Remote air tested and operational
- ❑ Communications check
- ❑ Life line attached
- ❑ Atmosphere monitors attached and on
- ❑ Helmet on
- ❑ Gloves on

ENTRY TEAM MEDICAL CHECKLIST

Entry time: ________ BP ____ / ____ Pulse ________ Resp. ________ Skin ________

Notes: ______________________________

Exit time: ________ BP ____ / ____ Pulse ________ Resp. ________ Skin ________

Notes: ______________________________

Appendix D

Sample Safety Data Sheet

Issuing Date January 5, 2015 **Revision Date** June 12, 2015 **Revision Number** 1

1. IDENTIFICATION OF THE SUBSTANCE/PREPARATION AND OF THE COMPANY/UNDERTAKING

Product identifier

Product name **XXXXX Regular-Bleach$_1$**

Other means of identification

EPA registration number 5813-100

Recommended use of the chemical and restrictions on use

Recommended use Household disinfecting, sanitizing, and laundry bleach

Uses advised against No information available

Details of the supplier of the Safety Data Sheet

Supplier address

The XXXXX Company
1221 Broadway
Oakland, CA 94612

Phone: 1-510-XXX-XXXX

Emergency telephone numbers

For medical emergencies, call: 1-800-446-1014

For transportation emergencies, call Chemtrec: 1-800-424-9300

2. HAZARDS IDENTIFICATION

Classification

This chemical is considered hazardous by the 2012 OSHA Hazard Communication Standard (29 CFR 1910.1200).

Skin corrosion/irritation	Category 1
Serious eye damage/eye irritation	Category 1

GHS Label elements, including precautionary statements

Emergency Overview

Signal Word Danger

Hazard Statements

Causes severe skin burns and eye damage

Causes serious eye damage

Appearance Clear, pale yellow **Physical State** Thin liquid **Odor** Bleach

Precautionary statements – Prevention

Wash face, hands and any exposed skin thoroughly after handling.
Wear protective gloves, protective clothing, face protection, and eye protection such as safety glasses.

Precautionary statements – Response

Immediately call a poison center or doctor.
If swallowed: Rinse mouth. Do NOT induce vomiting.
If on skin (or hair): Take off immediately all contaminated clothing. Rinse skin with water.
Wash contaminated clothing before reuse.
If inhaled: Remove person to fresh air and keep comfortable for breathing.
Specific treatment (see supplemental first aid instructions on this label).
If in eyes: Rinse cautiously with water for several minutes. Remove contact lenses, if present and easy to do. Continue rinsing.

Precautionary statements – Storage

Store locked up.

Precautionary statements – Disposal

Dispose of contents in accordance with all applicable federal, state, and local regulations.

Hazards not otherwise classified (HNOC)

Although not expected, heart conditions or chronic respiratory problems such as asthma, chronic bronchitis, or obstructive lung disease may be aggravated by exposure to high concentrations of vapor or mist.

Product contains a strong oxidizer. Always flush drains before and after use.

Unknown toxicity

Not applicable.

Other information

Very toxic to aquatic life with long lasting effects.

Interactions with other chemicals

Reacts with other household chemicals such as toilet bowl cleaners, rust removers, acids, or products containing ammonia to produce hazardous irritating gases, such as chlorine and other chlorinated compounds.

Chemical Name	CAS-No	Weight %	Trade Secret
Sodium hypochlorite	7681-52-9	5 -10	*

*The exact percentage (concentration) of composition has been withheld as a trade secret.

4. FIRST AID MEASURES

First aid measures

General advice — Call a poison control center or doctor immediately for treatment advice. Show this safety data sheet to the doctor in attendance.

Eye contact — Hold eye open and rinse slowly and gently with water for 15–20 minutes. Remove contact lenses, if present, after the first 5 minutes, then continue rinsing eye. Call a poison control center or doctor for treatment advice.

Skin contact — Take off contaminated clothing. Rinse skin immediately with plenty of water for 15–20 minutes. Call a poison control center or doctor for treatment advice.

Inhalation — Move to fresh air. If breathing is affected, call a doctor.

Ingestion — Have person sip a glassful of water if able to swallow. Do not induce vomiting unless told to do so by a poison control center or doctor. Do not give anything by mouth to an unconscious person. Call a poison control center or doctor immediately for treatment advice.

Protection of first-aiders — Avoid contact with skin, eyes, and clothing. Use personal protective equipment as required. Wear personal protective clothing (see section 8).

Most important symptoms and effects, both acute and delayed

Most important symptoms and effects — Burning of eyes and skin.

Indication of any immediate medical attention and special treatment needed

Notes to physician — Treat symptomatically. Probable mucosal damage may contraindicate the use of gastric lavage.

5. FIREFIGHTING MEASURES

Suitable extinguishing media
Use extinguishing measures that are appropriate to local circumstances and the surrounding environment.

Unsuitable extinguishing media
CAUTION: Use of water spray when fighting fire may be inefficient.

Specific hazards arising from the chemical
This product causes burns to eyes, skin, and mucous membranes. Thermal decomposition can release sodium chlorate and irritating gases and vapors.

Explosion data

Sensitivity to mechanical impact	None.
Sensitivity to static discharge	None.

Protective equipment and precautions for firefighters
As in any fire, wear self-contained breathing apparatus pressure-demand, MSHA/NIOSH (approved or equivalent) and full protective gear.

6. ACCIDENTAL RELEASE MEASURES

Personal precautions, protective equipment and emergency procedures

Personal precautions	Avoid contact with eyes, skin, and clothing. Ensure adequate ventilation. Use personal protective equipment as required. For spills of multiple products, responders should evaluate the MSDSs of the products for incompatibility with sodium hypochlorite. Breathing protection should be worn in enclosed and/or poorly-ventilated areas until hazard assessment is complete.
Other information	Refer to protective measures listed in Sections 7 and 8.

Environmental precautions

Environmental precautions	This product is toxic to fish, aquatic invertebrates, oysters, and shrimp. Do not allow product to enter storm drains, lakes, or streams. See Section 12 for ecological Information.

Methods and material for containment and cleaning up

Methods for containment	Prevent further leakage or spillage if safe to do so.
Methods for cleaning up	Absorb and containerize. Wash residual down to sanitary sewer. Contact the sanitary treatment facility in advance to assure ability to process washed-down material.

Precautions for safe handling

Handling Handle in accordance with good industrial hygiene and safety practice. Avoid contact with skin, eyes, and clothing. Do not eat, drink, or smoke when using this product.

Conditions for safe storage, including any incompatibilities

Storage Store away from children. Reclose cap tightly after each use. Store this product upright in a cool, dry area, away from direct sunlight and heat to avoid deterioration. Do not contaminate food or feed by storage of this product.

Incompatible products Toilet bowl cleaners, rust removers, acids, and products containing ammonia.

8. EXPOSURE CONTROLS/PERSONAL PROTECTION

Control parameters

Exposure guidelines

Chemical Name	ACGIH TLV	OSHA PEL	NIOSH IDLH
Sodium hypochlorite 7681-52-9	None	None	None

ACGIH TLV: American Conference of Governmental Industrial Hygienists - Threshold Limit Value. OSHA PEL: Occupational Safety and Health Administration - Permissible Exposure Limits. NIOSH IDLH: Immediately Dangerous to Life or Health.

Appropriate engineering controls

Engineering measures

Showers
Eyewash stations
Ventilation systems

Individual protection measures, such as personal protective equipment

Eye/face protection If splashes are likely to occur: Wear safety glasses with side shields (or goggles) or face shield.

Skin and body protection Wear rubber or neoprene gloves and protective clothing such as long-sleeved shirt.

Respiratory protection If irritation is experienced, NIOSH/MSHA approved respiratory protection should be worn. Positive-pressure supplied air respirators may be required for high airborne contaminant concentrations. Respiratory protection must be provided in accordance with current local regulations.

Hygiene measures Handle in accordance with good industrial hygiene and safety practice. Wash hands after direct contact. Do not wear product-contaminated clothing for prolonged periods. Remove and wash contaminated clothing before re-use. Do not eat, drink, or smoke when using this product.

9. PHYSICAL AND CHEMICAL PROPERTIES

Physical and chemical properties

Physical state	Thin liquid		
Appearance	Clear	**Odor**	Bleach
Color	Pale yellow	**Odor Threshold**	No information available

Property	**Values**	**Remarks/ Method**
pH	~12	None known
Melting/freezing point	No data available	None known
Boiling point / boiling range	No data available	None known
Flash point	Not flammable	None known
Evaporation rate	No data available	None known
Flammability (solid, gas)	No data available	None known
Flammability limits in air		
Upper flammability limit	No data available	None known
Lower flammability limit	No data available	None known
Vapor pressure	No data available	None known
Vapor density	No data available	None known
Specific gravity	~1.1	None known
Water solubility	Soluble	None known
Solubility in other solvents	No data available	None known
Partition coefficient: n-octanol/water	No data available	None known
Autoignition temperature	No data available	None known
Decomposition temperature	No data available	None known
Kinematic viscosity	No data available	None known
Dynamic viscosity	No data available	None known
Explosive properties	Not explosive	
Oxidizing properties	No data available	

Other information

Softening point	No data available
VOC content (%)	No data available
Particle size	No data available
Particle size distribution	No data available

10. STABILITY AND REACTIVITY

Reactivity

Reacts with other household chemicals such as toilet bowl cleaners, rust removers, acids, or products containing ammonia to produce hazardous irritating gases, such as chlorine and other chlorinated compounds.

Chemical stability

Stable under recommended storage conditions.

Possibility of hazardous reactions

None under normal processing.

Conditions to avoid

None known based on information supplied.

Incompatible materials

Toilet bowl cleaners, rust removers, acids, and products containing ammonia.

Hazardous decomposition products

None known based on information supplied.

11. TOXICOLOGICAL INFORMATION

Information on likely routes of exposure

Product information

Inhalation Exposure to vapor or mist may irritate respiratory tract and cause coughing. Inhalation of high concentrations may cause pulmonary edema.

Eye contact Corrosive. May cause severe damage to eyes.

Skin contact May cause severe irritation to skin. Prolonged contact may cause burns to skin.

Ingestion Ingestion may cause burns to gastrointestinal tract and respiratory tract, nausea, vomiting, and diarrhea.

Component information

Chemical Name	LD50 Oral	LD50 Dermal	LC50 Inhalation
Sodium hypochlorite 7681-52-9	8200 mg/kg (Rat)	>10000 mg/kg (Rabbit)	-

Information on toxicological effects

Symptoms May cause redness and tearing of the eyes. May cause burns to eyes. May cause redness or burns to skin. Inhalation may cause coughing.

Delayed and immediate effects as well as chronic effects from short and long-term exposure

Sensitization No information available.

Mutagenic effects No information available.

Carcinogenicity The table below indicates whether each agency has listed any ingredient as a carcinogen.

Chemical Name	ACGIH	IARC	NTP	OSHA
Sodium hypochlorite 7681-52-9	-	Group3	-	-

IARC (International Agency for Research on Cancer)
Group 3 - Not Classifiable as to Carcinogenicity in Humans

Reproductive toxicity No information available.

STOT – single exposure No information available.

STOT – repeated exposure No information available.
Chronic toxicity Carcinogenic potential is unknown.
Target organ effects Respiratory system, eyes, skin, gastrointestinal tract (GI).

Aspiration hazard No information available.

Numerical measures of toxicity – product information

The following values are calculated based on chapter 3.1 of the GHS document

ATEmix (oral)
54 g/kg
ATEmix (inhalation-dust/mist)
58 mg/L

12. ECOLOGICAL INFORMATION

Ecotoxicity
Very toxic to aquatic life with long lasting effects.

This product is toxic to fish, aquatic invertebrates, oysters, and shrimp. Do not allow product to enter storm drains, lakes, or streams.

Persistence and degradability
No information available.

Bioaccumulation
No information available.

Other adverse effects
No information available.

13. DISPOSAL CONSIDERATIONS

Disposal methods
Dispose of in accordance with all applicable federal, state, and local regulations. Do not contaminate food or feed by disposal of this product.

Contaminated packaging
Do not reuse empty containers. Dispose of in accordance with all applicable federal, state, and local regulations.

14. TRANSPORT INFORMATION

DOT	Not restricted.
TDG	Not restricted for road or rail.
ICAO	Not restricted, as per Special Provision A197, Environmentally Hazardous Substance exception.
IATA	Not restricted, as per Special Provision A197, Environmentally Hazardous Substance exception.
IMDG/IMO	Not restricted, as per IMDG Code 2.10.2.7, Marine Pollutant exception.

15. REGULATORY INFORMATION

Chemical inventories

TSCA	All components of this product are either on the TSCA 8(b) Inventory or otherwise exempt from listing.
DSL/NDSL	All components are on the DSL or NDSL.

TSCA - United States Toxic Substances Control Act Section 8(b) Inventory
DSL/NDSL - Canadian Domestic Substances List/Non-Domestic Substances List

U.S. federal regulations

SARA 313

Section 313 of Title III of the Superfund Amendments and Reauthorization Act of 1986 (SARA). This product does not contain any chemicals which are subject to the reporting requirements of the Act and Title 40 of the Code of Federal Regulations, Part 372

SARA 311/312 Hazard Categories

Acute Health Hazard	Yes
Chronic Health Hazard	No
Fire Hazard	No
Sudden Release of Pressure Hazard	No
Reactive Hazard	No

Clean Water Act

This product contains the following substances which are regulated pollutants pursuant to the Clean Water Act (40 CFR 122.21 and 40 CFR 122.42)

Chemical Name	CWA - Reportable Quantities	CWA - Toxic Pollutants	CWA - Priority Pollutants	CWA - Hazardous Substances
Sodium hypochlorite 7681-52-9	100 lb			X

CERCLA

This material, as supplied, contains one or more substances regulated as a hazardous substance under the Comprehensive Environmental Response Compensation and Liability Act (CERCLA) (40 CFR 302)

Chemical Name	Hazardous Substances RQs	Extremely Hazardous Substances RQs	RQ
Sodium hypochlorite 7681-52-9	100 lb	-	RQ 100 lb final RQ RQ 45.4 kg final RQ

EPA statement

This chemical is a pesticide product registered by the Environmental Protection Agency and is subject to certain labeling requirements under federal pesticide law. These requirements differ from the classification criteria and hazard information required for safety data sheets and for workplace labels of non-pesticide chemicals. Following is the hazard information as required on the pesticide label:

> **DANGER: CORROSIVE.** Causes irreversible eye damage and skin burns. Harmful if swallowed. Do not get in eyes, on skin, or on clothing. Wear protective eyewear and rubber gloves when handling this product. Wash thoroughly with soap and water after handling and before eating, drinking, chewing gum, using tobacco, or using the restroom. Avoid breathing vapors and use only in a well-ventilated area.

U.S. state regulations

California Proposition 65
This product does not contain any Proposition 65 chemicals.

U.S. state Right-to-Know Regulations

Chemical Name	New Jersey	Massachusetts	Pennsylvania	Rhode Island	Illinois
Sodium hypochlorite 7681-52-9	X	X	X	X	
Sodium chlorate 7775-09-9	X	X	X		

International regulations

Canada
WHMIS Hazard Class
E - Corrosive material

16. OTHER INFORMATION

NFPA	**Health Hazard** 3	**Flammability** 0	**Instability** 0	**Physical and Chemical Hazards** -
HMIS	**Health Hazard** 3	**Flammability** 0	**Physical Hazard** 0	**Personal Protection** B

Prepared by Product Stewardship
23 British American Blvd.
Latham, NY 12110
1-800-572-6501

Revision date June 12, 2015

Revision note Revision Section 14.

Reference 1096036/164964.159

General disclaimer

The information provided in this Safety Data Sheet is correct to the best of our knowledge, information and belief at the date of its publication. The information given is designed only as a guidance for safe handling, use, processing, storage, transportation, disposal, and release and is not to be considered a warranty or quality specification. The information relates only to the specific material designated and may not be valid for such material used in combination with any other materials or in any process, unless specified in the text.

End of Safety Data Sheet

This material was produced under grant 46F6-HT30 from the Occupational Safety and Health Administration, U.S. Department of Labor. It does not necessarily reflect the views or policies of the U.S. Department of Labor, nor does mention of trade names, commercial products, or organizations imply endorsement by the U.S. Government. Revisions were made to this material under grant number SH-31240-SH7 from the Occupational Safety and Health Administration, U.S. Department of Labor.

Courtesy of OSHA/U.S. Department of Labor.

Appendix E

Calculating the Turnover Rate of Ventilation

The turnover rate of ventilation, particularly in confined-space ventilation operations, can be calculated using the following mathematical formula:

$$\frac{\text{Cubic feet of space}}{\text{Fan output in cubic feet per minute (CFM)}}$$

Explanation of Terms

- **Cubic feet of space (ft^3):** This is the volume of the confined space that needs to be ventilated. It is calculated by multiplying the length, width, and height of the space.
- **Fan output in cubic feet per minute (CFM):** This is the volume of air that the ventilation fan can move in one minute. It is provided by the manufacturer of the ventilation equipment.

Example Turnover Rate Calculation

Let's say you have a confined space that is 10 feet (3 m) long, 8 feet (2.4 m) wide, and 6 feet (1.8 m) high. The fan being used has an output of 1200 CFM.

- Calculate the cubic feet of space (ft^3):

$$\text{Cubic feet of space} = 10 \text{ ft} \times 8 \text{ ft} \times 6 \text{ ft} = 480 \text{ ft}^3 \ (13.5921 \text{ m}^3)$$

- Use the formula to find the turnover rate:

$$\frac{480 \text{ ft}^3}{1200 \text{ CFM}} = 0.4 \text{ minutes (24 seconds)}$$

This result means that it takes 0.4 minutes, or 24 seconds, to completely exchange the air in the confined space once using the specified fan.

Recommended Air Changes per Hour (ACH)

To ensure safety and proper ventilation in confined spaces, the recommended air changes per hour (ACH) typically range from 6 to 12 ACH, depending on the specific situation and the type of contaminants present.

Calculating ACH

you want to calculate how many air changes per hour your ventilation setup achieves, you can use the following formula:

$$\frac{\text{CFM} \times 60}{\text{Cubic feet of space}}$$

Example ACH Calculation

Using the previous example:

$$\frac{1200 \text{ CFM} \times 60}{480 \text{ ft}^3} = 150 \text{ ACH}$$

This calculation shows that the setup achieves 150 air changes per hour, which is more than sufficient for most confined-space ventilation needs.

Key Takeaways

- The turnover rate helps you understand how quickly the air in a space is completely exchanged.
- The ACH gives you a measure of how many times the air in the space is replaced within an hour.
- Ensure your ventilation meets or exceeds the recommended ACH for safety.

Appendix F

Understanding and Calculating Mechanical Advantage in Rope Systems

The mechanical advantage (MA) in rope systems, such as those used with tripods for confined-space rescue operations, is a crucial concept in understanding how force is distributed. MA is essentially the factor by which a system multiplies the input force to achieve the output force. The basic formula for MA is as follows:

$$\frac{\text{Output force}}{\text{Input force}}$$

Calculating Mechanical Advantage in Pulley Systems

In the context of rope systems, particularly those using pulleys, the MA can be determined by counting the number of rope segments supporting the load:

- **Simple pulley (1:1):** If there is one pulley and one segment of rope supporting the load, the **MA = 1**.
- **2:1 system:** If the system has one pulley attached to the load and the rope is anchored above, with two segments of the rope supporting the load, the **MA = 2**.
- **3:1 system:** In a system where the load is supported by three segments of rope, the **MA = 3**.

Example Calculation of Output Force for a Pulley System with a Tripod

The formula for output force is as follows:

$$\text{MA} \times \text{Input force}$$

Suppose you are using a pulley system with a tripod for confined-space rescue, and the system is configured such that three segments of rope are supporting the load (MA = 3). If the input force applied is 100 pounds (45 kg), the output force that the system can lift is as follows:

$$3 \times 100 \text{ lb} = 300 \text{ lb (136 kg)}$$

Appendix G

An Example of Organizational Policy on Mental Health

[Organization Name] is committed to promoting and protecting the mental health and well-being of all employees. We recognize that mental health is an essential component of overall health and productivity. This policy outlines the organization's approach to supporting employees' mental health and ensuring a supportive work environment.

Objectives

- To create a workplace culture that supports mental health and reduces stigma associated with mental health issues.
- To provide employees with access to mental health resources and support services.
- To promote early identification and intervention for employees experiencing mental health issues.
- To ensure compliance with relevant legislation and best practices related to mental health in the workplace.

Scope

This policy applies to all employees, including full-time, part-time, temporary, and contract employees.

Definitions

- **Mental health:** A state of well-being in which an individual realizes their own abilities and can cope with normal stresses of life, work productively, and contribute to their community.
- **Mental health issue:** A broad range of mental health conditions, including but not limited to anxiety, depression, post-traumatic stress disorder (PTSD), and other mental illnesses.

Responsibilities

- **Management:** Responsible for fostering a supportive environment, providing necessary resources, and ensuring that the policy is implemented effectively.
- **Human resources (HR):** Responsible for providing access to mental health resources, coordinating training programs, and maintaining confidentiality.
- **Employees:** Encouraged to seek support when needed, participate in mental health programs, and support colleagues in creating a stigma-free workplace.

Policy Guidelines

Creating a Supportive Environment

- Promote a culture of openness and support regarding mental health.
- Encourage employees to discuss mental health concerns without fear of discrimination or stigma.
- Provide training to all employees on mental health awareness and sensitivity.

Providing Access to Mental Health Resources

- Offer an employee assistance program (EAP) that provides confidential counseling and support services.
- Provide access to mental health resources, including information on mental health conditions and coping strategies.
- Ensure employees know how to access mental health services and support within the organization and in the community.

Committing to Early Identification and Intervention

- Train managers and supervisors to recognize signs of mental health issues and respond appropriately.
- Encourage employees to seek help early if they are experiencing mental health challenges.
- Implement procedures for referring employees to appropriate mental health services.

Implementing Supportive Work Practices

- Offer flexible work arrangements, where possible, to accommodate employees' mental health needs.
- Provide reasonable accommodations for employees experiencing mental health issues, such as modified duties or adjusted schedules.
- Encourage work–life balance and discourage excessive overtime.

Establishing Crisis Intervention

- Establish protocols for responding to mental health crises, including immediate support and referral to professional services.
- Ensure that employees know how to access emergency mental health services.

Providing Confidentiality

- Maintain strict confidentiality regarding employees' mental health information.
- Ensure that any disclosure of mental health information is conducted with the employee's consent and in compliance with legal requirements.

Providing Ongoing Support and Education

- Provide ongoing mental health education and training programs for all employees.
- Regularly review and update mental health policies and programs to ensure they meet employees' needs and reflect best practices.

Implementation and Review

- **Implementation:** The HR department is responsible for implementing this policy, with support from management and input from employees.
- **Review:** This policy will be reviewed annually to ensure its effectiveness and relevance. Feedback from employees and changes in legislation or best practices will inform any necessary updates.

Conclusion

[Organization Name] is dedicated to fostering a mentally healthy workplace where all employees feel supported and valued. By implementing this policy, we aim to create a positive work environment that promotes mental well-being and enhances overall organizational performance.

Approved by: *[Name of chief executive officer/fire chief]*

Date: *[Effective date]*

Review Date: *[Next review date]*

Glossary

1:1 mechanical advantage system Also known as a *direct pull system*, a system that provides a mechanical advantage of 1:1, meaning that the force applied to the system equals the force exerted on the load. This system involves a single rope or cable directly attached to the load, with no additional pulleys or mechanical devices to amplify force.

2:1 mechanical advantage system A system that provides a mechanical advantage of 2:1, meaning that for every unit of force applied to the system, two units of force are exerted on the load; this system typically involves a pulley arrangement where the load is attached to one end of the rope, and the rope runs through a pulley that redirects the force. The other end of the rope is then pulled, requiring half the effort needed to lift the load directly.

3:1 mechanical advantage system A system that provides a mechanical advantage of 3:1, meaning that for every unit of force applied to the system, three units of force are exerted on the load; this system typically involves multiple pulleys arranged in a block-and-tackle configuration, with the rope threaded through the pulleys to create a mechanical advantage.

4:1 mechanical advantage system A system that provides a mechanical advantage of 4:1, meaning that for every unit of force applied to the system, four units of force are exerted on the load; this system typically involves a combination of pulleys in a block-and-tackle arrangement.

5:1 mechanical advantage system A system that provides a mechanical advantage of 5:1, offering even greater amplification of force compared to a 3:1 system; this system utilizes additional pulleys and ropes arranged in a compound pulley system, with multiple stages of mechanical advantage to achieve a 5:1 ratio.

A

access control The processes of making an area safe for qualified people to enter or keeping unqualified people out.

advanced life support (ALS) interventions Medical interventions such as advanced airway management, intravenous access, administration of medications, and cardiac monitoring.

airway patency The ability of a victim's airway to remain open and unobstructed for sufficient airflow.

American National Standards Institute (ANSI) A nonprofit organization that administers and coordinates the U.S. voluntary standards and conformity assessment system.

angular force The force exerted on an anchor point when the load is pulled at an angle relative to it.

arc flash The light and heat produced during an arc fault, which is an electrical explosion or discharge.

ascender A rope grab device used to ascend a fixed rope.

ascending techniques Techniques used in confined-space operations to access elevated or vertical areas within the confined space, such as vertical shafts, pits, or risers. They allow rescuers to reach victims, conduct assessments, or perform rescue operations in areas that are inaccessible by other means.

atmospheric hazard A hazard present in the air of a confined space, including toxic gases, flammable gases, or oxygen deficiency.

atmospheric monitoring A method of evaluating the ambient atmosphere of a space, including but not limited to its oxygen content, flammability, and toxicity.

attendant An individual who is stationed outside one or more permit-required confined spaces and monitors the people within and the conditions in the area around the space.

attendant-based rescue A rescue attempt in which the attendant participates as part of the rescue service in a nonentry capacity once entry rescue procedures have been initiated, provided that the attendant performs no duties that might interfere with the attendant's primary duty to effectively monitor and protect all of the authorized entrants for all of the permit spaces they are attending and for whom they are responsible.

authority having jurisdiction (AHJ) An organization, office, or individual responsible for enforcement of the requirements of a code or standard or for approving equipment, materials, installations, or procedures.

authorized entrant A term used to describe a U.S. federally regulated industrial worker designated to enter confined spaces who meets specified training requirements for each specific space they enter.

AVPU scale A scale used to measure victim condition, with A = Alert, V = responds to Voice, P = responds to Pain, and U = Unresponsive.

awareness level responder A first responder who understands the basics of confined-space safety and knows how to safeguard personnel from fire, explosion, unqualified rescue attempts, and other health hazards that are uniquely associated with confined spaces.

B

basket stretcher A type of stretcher designed for transporting patients, particularly in rugged or outdoor environments; its design features a sturdy, elongated, and open frame, typically made of metal or composite materials, that is covered with a durable, often waterproof, fabric mesh or canvas. Also known as a *Stokes basket* or *rescue stretcher.*

beam clamp A portable anchor system device that is designed to securely attach to horizontal beams or structural members, providing a stable anchor point for various rescue operations.

belay A rope fixed around an object intended to secure it.

belay device A braking mechanism through which a secondary line is rigged.

belay line A secondary (backup) line to protect against a fall or system failure. Also called a *safety line.*

beneficial energy source An energy source, the presence of which does not pose a hazard or the hazards are controlled, that is of assistance during an emergency (electricity for lights and ventilation equipment, natural gas for generators, etc.).

blinding The capping of product lines that enter a confined space.

C

calibrated Carefully assessed, set, or adjusted.

carabiner A load-bearing metal connector with self-closing gates used to link the elements of a high-angle system.

cervical collar A rigid or semirigid device placed around the victim's neck to restrict movement and maintain alignment of the cervical spine.

change-of-direction pulley Also known as a redirection pulley, a system used to redirect the direction of force in a rope system.

chest ascender attachment point A specific location on a climbing or rescue harness designed for attaching a chest ascender, which is a device used to assist in ascending a rope; the chest ascender helps provide upward mobility while climbing—particularly in vertical environments—and plays a crucial role in ensuring safety and efficiency during ascents.

***Code of Federal Regulations* (C.F.R.)** The codification of the general and permanent rules published in the *Federal Register* by the departments and agencies of the U.S. government.

colorimetric tubes Monitoring equipment that contains chemical reagents that undergo a color change when exposed to specific gases or vapors; the length of the color stain correlates with the gas concentration. Also known as *colorimetric chips.*

combustible gas detector A monitoring device specifically designed to detect the presence of flammable gases or vapors in the atmosphere.

communications management The systematic planning, implementation, monitoring, and control of all channels of communication within a project, organization, or team to ensure that information is efficiently and effectively exchanged.

communications plan A plan for communications to flow between rescuers, attendants, and all other parties involved in a confined-space rescue.

confined space A space with all three of the following characteristics: It must be large enough for a person to enter to perform a specific job, have limited means of access and a restricted entrance or exit, and not be designed for people to spend significant time within or for continuous occupancy.

confined-space entry program An employer's overall program for controlling and protecting employees from permit space hazards and for regulating employee entry into a permit space.

confined-space rescue technician A skilled individual trained to operate at both the operations and technician levels within confined spaces.

confined-space supervisor An individual who supervises an operation in a confined space; is responsible for knowing the hazards that may be faced during entry, including the modes, signs (or symptoms), and consequences of the exposure; and is responsible for informing the attendant and entrant(s) of these hazards.

critical incident stress A strong emotional or physical reaction a person may have in response to a traumatic event or incident that can affect the ability to function at work, at home, or in other areas of life.

D

davit arm A type of anchor system device that consists of a vertical post or mast with an extendable boom arm that can rotate horizontally and is typically made of durable materials, such as steel or aluminum, that can withstand the loads encountered during rescue operations. They may be fixed or portable.

descending techniques Techniques used in confined-space operations to exit elevated or vertical areas within the confined space or to lower victims to safety. They allow rescuers to retreat from elevated positions or evacuate victims from hazardous environments with controlled descent.

descent control device (DCD) A rappel device that creates friction by means of a rope running through it.

dorsal attachment point A primary connection point for a rigging system located at the back of a harness.

dynamic analysis for force application Calculation of the forces exerted on a structure or anchor point when there is movement or acceleration involved.

dynamic force The additional forces exerted on an anchor point resulting from changes in velocity or direction of motion.

E

egress route The means of exit from a confined space for rescuers and victims.

energized hazard Exposed wires, energized equipment, and electrical machinery that may be present in a confined space.

engulfment The surrounding and effective capture of a person by a liquid or finely divided (flowable) solid substance that can be aspirated to cause death by filling or plugging the respiratory system or that can exert enough force on the body to cause death by strangulation, constriction, or crushing.

entry permit system A formal procedure used to control and manage access to confined spaces, hazardous areas, or specific work sites; this system ensures that only authorized personnel are allowed entry and that all necessary safety measures are in place to protect workers from potential hazards.

environmental air monitoring The systematic assessment of pollutant levels in air by measuring the quantity and types of pollutants present.

explosive range The range between the lower exposure limit and the upper exposure limit for a particular gas or vapor.

extremities The arms, hands, legs, and feet.

F

Federal Emergency Management Agency (FEMA) An agency of the U.S. federal government with a primary purpose of coordinating responses to disasters that overwhelm the resources of local and state authorities.

flame ionization detector (FID) A monitoring device that uses a hydrogen flame to ionize carbon-containing compounds in a gas sample; the resulting ions produce a measurable electrical current.

flammable range The range of concentrations of a mixture of flammable gas or vapor in air over which a flash will occur or a flame will travel upon ignition.

flash point The lowest liquid temperature at which, under certain conditions, a liquid emits vapor that is flammable.

flexible stretcher A type of medical transport device designed to safely and comfortably carry patients in various emergency situations, particularly when navigating through narrow or confined spaces; unlike traditional rigid stretchers, flexible stretchers are made from lightweight, flexible materials that allow them to conform to the contours of the patient's body and the surrounding environment.

force application The distribution of forces exerted on a portable anchor system during rescue operations.

frostbite A medical condition in which body tissues are injured as a result of exposure to extreme cold, typically affecting the nose, fingers, or toes.

G

guiding line A rope line designed to help prevent a load from swinging or spinning during the operation.

H

haul line A type of tag line that a haul team uses for pulling on a load.

haul system An arrangement of ropes, pulleys, and connectors that is arranged in a manner to provide mechanical advantage to raise a load under controlled circumstances.

hazard and risk assessment An assessment that determines what could possibly cause harm and the degree of likelihood that harm will be caused.

hazard identification The process of recognizing and documenting potential sources of harm or adverse effects in the workplace, environment, or any operational setting. It is a critical step in risk management because it allows organizations to anticipate and mitigate risks before they result in accidents, injuries, or damage.

head immobilization device A device, such as padded blocks or head blocks, used to stabilize the victim's head and prevent lateral movement.

Health Insurance Portability and Accountability Act of 1996 (HIPAA) A federal law that sets standards for the protection of individuals' health information, known as *protected health information* (PHI).

heat cramps A medical condition in which painful muscle spasms occur during activity in hot environments.

heat exhaustion A medical condition in which the body responds to excessive losses of water and salt, usually through excessive sweating.

heat stroke A medical condition in which the body's temperature rises rapidly, the sweating mechanism fails, and the body is unable to cool down.

Hierarchy of Controls A system developed by the Occupational Safety and Health Administration (OSHA) that provides a means of determining ways to implement systems or controls, from most effective to least effective, that protect workers from injuries, illnesses, and fatalities.

high-angle rescue A rescue that occurs in situations where the slope angle exceeds 50 degrees.

high-point anchor A secure anchoring system used in rope rescue and confined-space operations, positioned above the working area or entry point. This anchor provides a high point of attachment for ropes, allowing for efficient and controlled vertical movement of personnel and equipment.

hitch A method of tying a rope in such a way that when the object or rope is removed, the knot falls apart.

horizontal force The force exerted on an anchor point when the load is pulled or pushed horizontally away from it.

hot work Work involving electric or gas welding, cutting, brazing, or similar flame- or spark-producing operations.

hypothermia Lowering of the body's core temperature.

I

immediately dangerous to life or health (IDLH) The maximum level to which one could be exposed to a toxic substance or situation and still escape without experiencing any effects that may impair escape or cause irreversible health effects.

incident action plan (IAP) A plan developed by the incident commander (IC) that establishes goals and objectives for the emergency, identifies the resources to be used, and provides the means to accomplish the goals and objectives.

incident commander (IC) The individual responsible for overseeing the entire incident response, making critical decisions, developing objectives, and ensuring the safety and effectiveness of the operation.

Incident Command System (ICS) A recognized system for providing management of personnel, resources, and activities during emergency operations by defining roles and responsibilities and standard operating procedures.

incident command The overall system or framework used to manage emergency response operations, ensuring effective coordination, communication, and resource management.

Incident Management System (IMS) A recognized system for providing management of personnel, resources, and activities during emergency operations by defining roles and responsibilities and standard operating procedures.

ingress route The access to and means of entry into a confined space for rescuers and equipment.

intrinsically safe A term used to describe battery-powered or electrical equipment that is safe to operate in hazardous areas because it emits levels of electrical or thermal energy that are too low to cause ignition.

K

Kendrick extrication device (KED) A multipart immobilization device that provides full-body support and stabilization. It consists of a vest, padded straps, and a head immobilization device, allowing rescuers to secure the victim's entire body while maintaining spinal alignment.

kernmantle rescue rope A rope design consisting of an interior core (the kern), which usually supports the major portion of the load on the rope, and an outer sheath (the mantle) that serves to protect the core but also may support a minor portion of the load.

L

lifting bridle A safety device used in rescue operations and lifting scenarios to securely attach a load or patient to a lifting mechanism, such as a helicopter, hoist, or crane; it is designed to evenly distribute the weight of the load, ensuring stability and safety during the lifting process.

litter basket A stretcher or basket designed for horizontal retrieval that provides support and stability for the victim during transportation.

lockout/tagout The process of eliminating and securing sources of potential physical or chemical hazards by isolating them from a confined space and/or controlling flowable materials in the space.

logistics The process of planning, implementing, and controlling the efficient and effective flow of goods, services, and information from the point of origin to the point of consumption; it encompasses all the activities required to move and store products, raw materials, or information, ensuring that they are delivered to the right place, at the right time, in the right condition, and at the right cost.

lower flammable limit (LFL) or lower explosive limit (LEL) The lower end of the concentration range over which a flammable mixture of gas or vapor in air can be ignited at a given temperature and pressure.

LSP Half Back A vertical extrication device designed for rugged use, combining the functionality of a full-body harness with the ability to immobilize a patient in cases of suspected cervical or spinal trauma; it consists of two main components: a durable, padded full-body harness and a removable aluminum back stay with a head immobilizer.

M

mechanical advantage A force created through mechanical means, including but not limited to a system of levers, gearing, or ropes and pulleys, usually creating an output force greater than the input force.

mechanical Prusik A mechanical type of friction hitch used to attach a loop of cord around a rope.

medical monitoring Basic medical evaluation of emergency response personnel by determining and recording basic vital signs such as blood pressure, respirations per minute, and pulse.

monopod A single-leg support structure designed to provide vertical stability and load-bearing capacity as a portable anchor system device.

multi-gas monitor A monitoring device that is capable of simultaneously detecting and measuring multiple gases (typically, oxygen, combustible gases, hydrogen sulfide, and carbon monoxide). Also known as a *multi-gas detector*.

N

National Institute for Occupational Safety and Health (NIOSH) The agency within the U.S. Department of Health and Human Services that identifies work-related diseases and injuries and the potential hazards of new work-related technologies and practices.

negative-pressure ventilation (NPV) A method of ventilation that creates a lower pressure inside the space, drawing fresh air in through designated entry points and exhausting contaminated air through a controlled outlet.

non-permit-required confined space A confined space that does not contain—or in the case of atmospheric hazards, does not have the potential to contain—any hazard capable of causing death or serious physical harm.

O

OATH A method of silent communication involving tugs on a rope (O for okay = one tug; A for advance = two tugs; T for take up slack = three tugs; H for help = four tugs).

Occupational Safety and Health Administration (OSHA) A U.S. government regulatory agency with a mission of ensuring that workers have safe and healthful working conditions free from unlawful retaliation.

operations level responder A first responder who understands rescuer safety, team evaluation, preplanning, atmospheric monitoring, and the proper use of personal protective equipment and is also capable of assembling a portable anchor system for application of a high-point attachment, following lockout/tagout procedures, preparing for entry into a vertically oriented confined space, and entering horizontally and vertically oriented confined spaces. The operations level also covers techniques for patient packaging and recovery.

oxygen monitor A monitoring device, also known as an *oxygen-deficiency monitor* or *oxygen-deficiency indicator*, that measures the concentration of oxygen in the atmosphere.

P

particulate matter (PM) A mixture of solid particles and liquid droplets found in air.

permissible exposure limit (PEL) A regulatory limit on the amount or concentration of a substance in the air.

permit-required confined space A confined space that is designated as being known to contain—or in the case of atmospheric hazards, has the potential to contain—any hazard capable of causing death or serious physical harm.

personal gas detector A wearable device designed to provide individual personnel with real-time alerts and alarms in the event of hazardous atmospheric conditions.

personal protective equipment (PPE) The equipment provided to shield or isolate a person from the chemical, physical, or thermal hazards that can be encountered at a specific rescue incident.

personnel accountability systems Systems used to track the location and status of all personnel involved in the operation, often by the use of personnel tags, check-in/checkout procedures, and accountability boards.

personnel rehabilitation The process of restoring and supporting the physical, mental, and emotional well-being of individuals who have experienced stress, injury, trauma, or other adverse conditions, particularly in demanding environments such as military operations, emergency services, or high-intensity work environments; this process aims to ensure that individuals can return to their duties or normal life in a healthy, effective, and sustainable manner.

photoionization detector (PID) A monitoring device that is used to detect volatile organic compounds (VOCs) and other hazardous substances based on their ability to ionize when exposed to ultraviolet (UV) light.

pH paper Also known as *litmus paper* or *pH indicator paper*, a simple and inexpensive tool used to measure the acidity or alkalinity (pH) of a solution.

physical hazard A hazard present in a confined space, including electrical equipment, agitators or paddles, fire-suppression equipment, sharp edges, areas that are difficult to access because of height or other limitations, and the physical state of the material present in the space, as well as other considerations.

portable winch system A portable anchor system device that typically consists of a winch unit, rigging attachment, and anchor point.

positive-pressure respirator An air-saving device that maintains a slight positive pressure in the mask, reducing the likelihood of inward leaks but using air more efficiently.

positive-pressure ventilation (PPV) A ventilation technique that uses a fan to create a higher pressure inside the confined space than outside, pushing contaminants out through designated exit points.

post-traumatic stress disorder (PTSD) A mental health condition triggered by experiencing or witnessing a traumatic event that is characterized by symptoms such as flashbacks, severe anxiety, and uncontrollable thoughts about the event.

progress capture device (PCD) A device that holds a rope in place to prevent a load from sliding back down when the pulling force is released.

Prusik hitch A type of friction hitch used in ascending and belaying.

Prusik loop A continuous loop of rope in which a Prusik hitch is tied.

pulley A device with free-turning, grooved metal wheels used to reduce rope friction.

purging The removal of gases or particulate matter from air within an enclosed space by means of air displacement.

R

rapid victim harness A piece of rescue equipment designed to quickly and securely immobilize and extract victims from confined spaces. Also known as a *webbing harness.*

recovery The process of retrieving the body of a deceased person from a confined space.

Reeves™ Sleeve A medical device used in emergency situations, particularly in rescue and transport scenarios. It is a type of extrication device designed to immobilize and protect a patient's spine during transport.

rehabilitation area A place where responders can rest, rejuvenate, and receive emotional support during a rescue.

rescue The process of safely extricating a living person from a confined space.

rescue attendant An individual who is stationed outside a confined space during a rescue attempt who monitors the people within and the conditions in the area around the space.

rescue entrant An individual who is qualified to enter a confined space and assist in a rescue attempt.

rescue sled A specialized piece of equipment designed for transporting patients in emergency situations, particularly in environments where traditional stretchers may be impractical.

resource management The process of determining the proper allocation of resources based on operational needs and priorities, taking into account the nature and severity of identified hazards.

retrieval winch A mechanical device designed to allow the operator to wind up or wind out a rope or cable to support a person being raised or lowered into or out of a confined space.

risk assessment The systematic process of identifying, evaluating, and analyzing potential risks that could negatively affect individuals, assets, operations, or the environment; the goal is to determine the likelihood of these risks occurring and the potential consequences, enabling organizations to implement appropriate measures to mitigate or eliminate them.

risk–benefit analysis A process involving identifying and evaluating the potential risks associated with various rescue strategies against the benefits of successful victim extraction.

risk management The process of identifying, assessing, prioritizing, and mitigating risks to minimize their impact on an organization, project, or individual; it involves a series of steps designed to manage uncertainty by taking proactive measures to avoid, reduce, or transfer risks and prepare for their potential consequences.

S

safety briefing A briefing in which the safety procedures for entry into a confined space are discussed.

safety/communication rope A specialized rope used in various rescue, climbing, and industrial applications to provide both a safety line and a means of communication between team members during operations; this type of rope is designed to ensure the safety of individuals working at heights or in hazardous environments while allowing them to communicate effectively.

safety management The systematic process of developing, implementing, and maintaining policies, procedures, and practices that ensure the safety and well-being of individuals within an organization or environment; it involves identifying potential hazards, assessing risks, and taking proactive measures to prevent accidents, injuries, and illnesses.

scene perimeter The area surrounding a rescue scene that must be cordoned off to protect onlookers, rescuers, and victims.

scoop stretcher A medical device specifically designed to safely lift and transport patients with minimal movement, making them ideal for victims with suspected spinal injuries.

self-contained breathing apparatus (SCBA) A device worn to provide breathable air in environments with toxic atmospheric hazards.

self-retracting lanyard A vertical fall-protection lifeline that allows the user to move around easily and freely while the line stays consistently taut, effectively preventing falls.

self-retracting lifeline (SRL) A fall arrest device that contains a drum-wound line that pays out from and automatically retracts back onto an enclosed drum during use and that automatically locks at the onset of a fall.

sheave A free-turning, grooved metal wheel used to reduce rope friction on a pulley.

shock A life-threatening condition that occurs when the body and organs do not receive sufficient blood flow.

single-gas monitor A monitoring device that detects and measures the concentration of a single gas, such as oxygen, combustible gases (lower explosive limit [LEL]), hydrogen sulfide, or carbon monoxide. Also known as a *single-gas detector*.

size-up The ongoing evaluation of problems confronted in a confined-space rescue operation.

snap hook A hook with a spring that allows the entrance but prevents the escape of something.

spine board A rigid, flat surface made of plastic or wood to which a victim is secured using straps or a harness.

splint A device used to shore up and protect injured limbs during a rescue.

standard operating procedures (SOPs) Written guidelines for handling a specific situation.

standby assistance Ensuring the safety of the people within a confined space and maintaining communication with the workers in the space.

static analysis for force application Calculation of the total weight or force exerted on a structure or anchor point when all forces are in equilibrium.

sternal attachment point A primary connection point for a rigging system located at the front of a harness.

strengths, weaknesses, opportunities, and threats (SWOT) analysis A strategic planning tool used to identify and evaluate the strengths, weaknesses, opportunities, and threats related to a project, business venture, or organization.

SUDRH A method of communication involving signals using air horns (S for stop = one short blast; U for up = two short blasts; D for down = three short blasts; R for rope free = four short blasts; and H for help = one long blast).

supplied-air breathing apparatus (SABA) A combination of a supplied-air respirator and an escape self-contained breathing apparatus.

supplied-air respirator (SAR) A low-profile respirator worn to provide breathable air in environments with toxic atmospheric hazards.

survivability profile An attempt to gain a clear understanding of the threats to health and safety within a confined space, followed by an analysis of what procedures and actions can be made to mitigate the threats and, in effect, make the confined-space entry survivable.

T

tag line A non-load-bearing line that connects to a litter to help it move from side to side in order to avoid obstacles, prevent swing-fall, or attain a more advantageous position.

technical rescue The application of special knowledge, skills, and equipment to resolve unique and/or complex rescue situations.

technician level responder A first responder who is capable of preplanning a confined-space incident; the responder must apply and use a supplied-air respirator or supplied-air breathing apparatus as a rescue entrant and must be well versed in using tripods and other artificial high anchor points, as well as lockout/tagout procedures. These responders are skilled in using rope systems specifically designed for confined-space entry and rescue scenarios and using communications systems.

tie-back An additional safety measure for tripods in which the anchor point or anchoring system used in the rescue setup is secured with an additional line or anchor.

triage A systematic process used to prioritize and categorize patients based on the severity of their condition and the urgency of their medical needs, especially in emergency situations or mass-casualty incidents; the goal of triage is to ensure that limited medical resources are allocated effectively, allowing for timely care to those who need it most.

tripod A portable anchor system device typically meant to support people as they are raised or lowered into and out of a space.

two-tensioned rope system (TTRS) A system used in confined-space rescues to provide enhanced safety and redundancy by providing two ropes that are both tensioned and actively managed, providing a backup in case one fails.

U

upper flammable limit (UFL) or upper explosive limit (UEL) The upper end of the concentration range over which a flammable mixture of gas or vapor in air can be ignited at a given temperature and pressure.

V

vacuum mattress A mattress-like structure made of a durable, flexible material, typically reinforced with plastic or metal inserts to maintain rigidity, that conforms to the shape of the victim's body when air is extracted, providing customized support and immobilization.

vertical force The force exerted on an anchor point when a load is suspended directly below it.

victim profile Detailed summaries of individuals affected by a confined-space incident, including their identity, role or function, health status, and other special considerations.

volatile organic compounds (VOCs) Organic compounds that have a high vapor pressure and low water solubility at room temperature, which allows them to evaporate into air.

W

work/rest cycle A rotation schedule for confined-space technical rescue teams and their backups.

Y

Yates Spec Pak A victim immobilization and evacuation system that also functions as a lifting harness, eliminating the need for a separate victim harness; its self-adjusting (sliding) cervical restraint addresses the challenges associated with lifting patients vertically while they are secured to a rigid backboard.

Z

Z-rig system A 3:1 mechanical advantage rope system that allows rescuers to lift or lower a load with one-third of the force that would otherwise be needed.

Index

Note: Page numbers followed by *f* or *t* indicate material in figures, or tables, respectively.

M

N

O

P

Q

R

S

T

U

V

W

Y

Z